THE TURNING OF THE TIDE
RELIGION IN CHINA TODAY

THE TURNING
OF
THE TIDE

RELIGION IN CHINA TODAY

Edited by
JULIAN F. PAS

HONG KONG BRANCH
ROYAL ASIATIC SOCIETY
in association with
HONG KONG
OXFORD UNIVERSITY PRESS
OXFORD   NEW YORK
1989

Oxford University Press

Oxford   New York   Toronto
Petaling Jaya   Singapore   Hong Kong   Tokyo
Delhi   Bombay   Calcutta   Madras   Karachi
Nairobi   Dar es Salaam   Cape Town
Melbourne   Auckland

and associated companies in
Berlin   Ibadan

First published 1989
Published in the United States
by Oxford University Press, Inc., New York

British Library Cataloguing in Publication Data

The Turning of the tide: religion in China today.
1. China. Religions
I. Pas, Julian F. II. Royal Asiatic Society of Great
Britain and Ireland. Hong Kong Branch
291'.0951
ISBN 0-19-584101-8
ISBN 0-19-585117-X pbk

Library of Congress Cataloging-in-Publication Data

The Turning of the tide: religion in China today/edited by Julian
F. Pas.
p. cm.
Includes bibliographical references.
ISBN 0-19-584101-8: $28.00 (U.S.). — ISBN 0-19-585117-X (pbk.):
$18.00 (U.S.)
1. China — Religion — 20th century. 2. Hong Kong — Religion —
20th century. I. Pas, Julian F.
BL 1802. T87 1989
200'.95125 — dc20   89-39603 CIP

Printed in Hong Kong by Yee Tin Tong Printing Press, Ltd.,
Tong Chong St, Quarry Bay, Hong Kong
Published by the Hong Kong Branch, Royal Asiatic Society,
in association with Oxford University Press, Warwick House,
Hong Kong

PREFACE

Like many organisations in present day Hong Kong, the Royal Asiatic Society, Hong Kong Branch, is pondering the future. Hopes and fears attend the impending return of the territory to Chinese sovereignty in 1997 and the establishment of the promised highly autonomous Hong Kong Special Administrative Region in that year. We believe that reliable information about what is happening in China today is vital to shaping the policies that will assist in producing a satisfactory outcome in 1997, for individuals and associations as well as for our future government.

This book is about one aspect of the national life: religion in China today. It has been produced by scholars who are acquainted with the contemporary scene, some of them Chinese by residence as well as race. Between them, they have provided a first-hand and as far as possible accurate, in depth account. They are mindful of the incomplete nature of their work, its limited geographical cover, and of the fast changing, unfolding nature of current developments in this field. However, they feel, as we do, that an authoritative statement at this time will be of value to many people, especially those belonging to religious groups in Hong Kong.

Their account enables us to see how religion has been affected by, and is responding to, the modernisation policies and more outward looking stance of the present leadership of the People's Republic. Its handling by the authorities is a reflection, in part, of the human rights situation in the country. In this wider context, it provides a yardstick by which to measure China's progress towards modernisation, and even more, its intentions towards the individual and non-government agencies in a Communist, turning socialist, country.

The fact that the text takes us only up to 1985 or thereabouts is, to my mind, of considerably less importance than the light shed upon these opening and in some ways formative years of the resurgence of overt religious activities in mainland China, and the less tangible spiritual movements underlying them. In this respect, it is hoped that this book of essays represents a benchmark that will assist others to assess the events and developments of the intervening period in this important field, and to interpret those occurring in the years immediately to come. Indeed, our editor and some of the contributors are already talking of another volume! Surely, this is all for the good.

It was for these reasons that the Royal Asiatic Society, Hong Kong Branch, decided to finance the publishing of this work. However, publication would not have been possible without a grant covering half the cost from the Chinese General Charities Fund. This was made available by courtesy of the Chinese Temples Com-

v

mittee, a statutory public body serviced by the City and New Territories Administration of the Hong Kong Government. The application to the Committee was made in the belief that the work will be of benefit to Hong Kong, as well as to the much wider group of persons outside Hong Kong with an interest in the subject. The RAS Council wishes to record its great appreciation of this help, so timely given.

The Council also wishes to express its appreciation to the President's publication fund (University of Saskatchewan) for a grant in aid to assist the editor in the editorial work .

Finally, a special word of appreciation is due to the editor. Dr Julian Pas is a member of the Hong Kong Branch, Royal Asiatic Society and a contributor to our annual Journal. He conceived the idea of this volume a few years ago following one of his visits to China, and has been tireless in making all the editorial arrangements leading to completion of the finished typescript. Without his enthusiastic and infectious leadership I doubt whether the project could have been realized.

Hong Kong, March 1989   *JAMES HAYES*
President, Hong Kong Branch,
Royal Asiatic Society

POSTSCRIPT

Since these words were written, momentous events have taken place in China. Their impact has made itself felt world-wide, but it is still impossible to forecast the outcome in China itself. If anything, they have enhanced the value of the contributions to this book, with their part record of the unfolding religious and human rights scene up to mid 1989.

I have also to record the agreement of Oxford University Press (Hong Kong) to co-publish this work. The assistance of the Press was sought to facilitate publicity and marketing, which clearly would be on a wider basis than anything the RAS Hong Kong Branch might contrive unaided. This is the second time that we have entered into co-publication with Oxford (Hong Kong), and I wish to thank them for their courtesy, consideration and assistance in agreeing to the joint venture at a late stage in our publishing arrangements.

I have also to record our appreciation of our printers, Messrs Yee Tin Tong, who have also handled the last few issues of the Journal of the Branch, and especially their Mr. Henry Law.

Last but not least, I have to thank Timothy Woo, our Member and Senior Assistant Librarian at Lingnan College Library, for preparing the Index. JH.

CONTENTS

CONTRIBUTORS

ALVIN P. COHEN is a professor of Chinese at the department of Asian Languages, University of Massachusetts, in Amherst, Mass. He received his Ph.D. in Oriental Languages (Chinese) from the University of California, at Berkeley (1971). Since that year he has been a professor of Chinese in Amherst. He was a visiting professor of Chinese at Tunghai University (Taichung, Taiwan) in 1980-81. His research interests are in Chinese popular religion, Chinese bibliography, and history and culture of the North-South Dynasties period.

Among his publications are *Grammar Notes for Introductory Classical Chinese* (1975, 1980); *Tales of Vengeful Souls* (Paris and Taipei, 1982); "Coercing the Rain Deities in Ancient China", *HR* 17 (1978), 244-265; "Solar Eclipses recorded in China during the Tang Dynasty", *Monumenta Serica* 35 (1981-83), 347-430.

KENNETH DEAN holds a doctorate from Stanford University and is now Assistant Professor at McGill University, Montreal, Canada. His interest is in the area of Taoism. One of his articles on the revival of *jiao* rituals in the People's Republic of China has appeared in *Cahiers d'Extrême-Asie, 2 (1986)*.

JOE DUNN (or Deng Zhaoming), the present editor of *Bridge*, Hong Kong bi-monthly journal issued by the Christian Study Centre on Chinese Religion and Culture, reporting on the situation of Christianity in China. Studied at the Ch'eng-kung University in Taiwan, the Lutheran Theological Seminary in Hong Kong, and earned his M.Th. at Hamburg University, Germany (1969). He was the editor for the theological books program for the Association of theological schools in South East Asia (1969-73), and became a research fellow at the Hong Kong Christian Study Centre on Chinese Religion and Culture (1973).

DAVID FAURE is a lecturer in history at the Chinese University of Hong Kong. He is the author of *The Structure of Chinese Rural Society, Lineage and Village in the Eastern New Territories, Hong Kong* (Hong Kong: Oxford University Press, 1986). He is currently working on the social history of the Pearl River Delta.

THOMAS HAHN was born in West Germany, where he studied German literature, cultural anthropology, philosophy and sinology. He wrote his master's thesis on the Laozi commentaries by Lu Dongbin. From 1984 till 1987 he resided in China on a German scholarship, doing research in the area of the religious geography of Taoism (especially the sacred mountains) and on the history of Taoism in the 20th century. After more than one year's study of Taoism at the centre of religious studies of Sichuan University, Chengdu, followed by a stay at the Beijing University, he plans to complete his research at the University of Heidelberg.

JAN YÜN-HUA 冉雲華 was born and educated in China. Following his master's and doctoral dissertations, he became a lecturer in Chinese Studies at Visra-bharati University in Santiniketan, India. Invited by McMaster University of Canada in 1967, he settled down in Hamilton, Ontario. Teaching in the department of Religious Studies at McMaster, he has been very active both in research and in various international academic organizations. Besides three books and chapters in books, he has published more than sixty research articles in various academic journals.

His current interests are in Huang-Lao Taoism, in Ch'an Buddhism and the early history of Taoist Religion. He is presently working on a book on Tsung-mi 宗密 (780-841) and has recently completed an English translation of *Yuan-jen lun* 原人論 .

MAN KAM LEUNG 梁文金 is a graduate of the University of Hong Kong and the University of Hawaii. He is currently an Associate Professor of Chinese History, History Department, at the University of Saskatchewan, Saskatoon, Canada. He specializes in modern Chinese political and intellectual history.

JULIAN F. PAS was born in Belgium and studied theology at the University of Louvain. He also studied Chinese language and culture in Taiwan and obtained his Ph.D. in Asian religions at McMaster University, Canada. Since 1969 he has been teaching at the University of Saskatchewan, where he is a professor of religious studies. He has published a dozen articles on Chinese religion and is now preparing several book manuscripts on Taoism and Chinese popular religion. He is the current editor of the *Journal of Chinese Religions.*

JOSEPH J. SPAE, c.i.c.m. (Scheut missionaries), is the current editor of *China Update*, a quarterly newsletter about religious (and other cultural) events in China. He studied at the Universities of Louvain (Belgium) and Tokyo, and spent many years in China and Japan. He is the author of several works on Japanese Religiosity and other books and articles.

PAULA SWART received her B.A. in Chinese Studies from the University of Leiden, Holland and an M.A. degree in Oriental Art from the University of Amsterdam. She studied one year in the History Department of Nanjing University. At present she works as an Asian Art consultant and has written many articles on Chinese Art and Culture.

TANAKA ISSEI 田 仲 一 成 is a professor of literature at the Institute of Oriental Culture, University of Tokyo. He holds a B.A., an M.A. and a Litt.D. in Chinese Literature (1983) from the University of Tokyo. His scholarly expertise is in the popular aspects of Chinese literature, as his publications indicate: *Ritual Theatres in China* (in Japanese) 中 國 祭 祀 演 劇 研 究 , 1981; *Lineage and Theatre in China* (in Japanese) 中 國 の 宗 教 て 演 劇 , 1985. Both books were published by the University of Tokyo Press.

BARRY TILL received his M.A. degree in Far Eastern Studies in 1975 from the University of Saskatchewan, Canada. He did research work in Chinese Archaeology at the University of Nanking. At present he is the Curator of Asian Art at the Art Gallery of Greater Victoria, British Columbia, and has published extensively in the field of Chinese culture.

BARTHOLOMEW PUI MING TSUI studied Chinese Buddhist religious history at McMaster University, Canada and obtained the Ph.D. degree with a dissertation entitled "Li Ch'un-fu and His Discussions of the Collected Plaints on Tao". He now lecturers on Chinese religions at the Chinese University of Hong Kong. His current interest includes Buddhism, Taoism, and popular religions in the Hong Kong area.

INTRODUCTION:
CHINESE RELIGION IN TRANSITION

Julian F. Pas

Preliminary Statements

Omnis comparatio claudicat (each analogy or comparison limps). The analogy of "Turning of the Tide" reflects the recent withdrawal of anti-religious sentiment, and even of religious persecution which had for a long time engulfed China. At times the repression was violent when during the *Ten Years' Chaos* (a better expression than the totally inaccurate and meaningless "Cultural Revolution") mobs of red guards, driven to iconoclastic insanity by political leaders, stormed temples and monasteries as well as private homes to carry off and destroy cultural treasures accumulated over the centuries: sculpture, paintings, books, ritual garments, devotional articles, etc. It starts to gradually emerge now how many temples have been totally destroyed, badly damaged, or expropriated, and how many religious personnel have been chased away or tortured. Now that the Chinese people realize the mistakes committed by these ill-advised and misguided "young heroes", perhaps more or less accurate statistics will eventually be compiled. But then, perhaps nobody really wants to know? Now a period of calm has returned. The waves of anti-religious feeling have withdrawn and religious practices are re-emerging. It is a *symphony in D-minor* : D stands for the note *Re* , since so many of the religious phenomena can be indicated with *re* words: revival, recover, rebuild, restore, return, etc. *Minor* indicates moderate optimism, a wait-and-see attitude, since it would be presumptuous to prophesy.

It is also presumptuous and, perhaps utterly hopeless, to attempt a descriptive analysis of the religious situation in the People's Republic of China today. Admittedly, it is presumptuous, yet in this volume we are trying to do it just as Confucius was said to be ". . . one who knows a thing cannot be done and still wants to do it." (*Analects* 14:41)

For centuries, China appeared to the West as an unchanging, giant society. Yet, it is far from true. China has not stopped changing for the last 150 years, the changes were fast, drastic, and continuous, caused by factors from within and from without.

Religion, as one aspect of the Chinese reality, has also changed considerably and it is still changing. What, then, is the use of catching this one floating moment of historical change? Tomorrow, everything we have said may already belong to the past.

1

Yet, to catch this one glimpse of today has its use. Just as a snapshot of a seven-year old youngster perpetuates a particular and never returning stage of development and growth, so will our treatment of Chinese religion as it appears to us today, remain forever valid and useful, even if it is transitory. This momentary glimpse, even immature and superficial, will help to gain insight in one stage of change, as long as we remember its transitoriness.

For the first time in thirty-five years, the religious scene in China appears to present something more than a passing mood. There have been major changes in economic policy, in political ideology, and in social structure, with a daring move towards modernization. Marx is, after all, not infallible. If he could be wrong in his political-economic views, who says he did not make a mistake in his interpretation of religion?

There is no decisive proof to say that a religious interpretation of life is the only one possible; a materialist-atheist one, whether one names it Marxist or socialist or humanist, has possibly equal chances of being true. Therefore, tolerance of the viewpoint of an opponent is the only way to achieve harmony in a society where everyone's mind thinks for itself and resents the imposition of so-called infallible truths. Has the Chinese government once and for all realized that freedom of thinking, including freedom of religious belief, is a basic human right? Possibly, yet the reversal of China's policy with regard to religion may have come as an incredible shock to many people. Can it be trusted? Will it last? Too many earlier experiences have taught the people to be cautious. Today, after six or seven years of more tolerant leadership, it appears that the change is sincere and lasting, and that the government will abide by its principle of tolerance.

The last few years have been a turning point in China's modern history, including her religious history. The hexagram No. 24 of the *I Ching: FU* , "Return" comes to mind:

The idea of a turning point arises from the fact that after the dark lines have pushed all of the light lines upward and out of the hexagram, another light line enters the hexagram from below. The time of darkness is past. The winter solstice brings the victory of light.[1]

Another analogy, equally rich in meaning, is the "Turning of the Tide" : after a period of low tide, when fishes were dying on the beach (when religious life dried up), the tide returns and with it, new life. But the analogy can be reversed, as was said above. The religious persecution was like a period of fierce floods, when storm waves covered the land and destroyed buildings, personnel, and religious treasures. Whichever analogy is being used, the questions remain. Is the future going to be safe and secure? Does the new tide not come too late?

2

The concept of collecting a volume of articles on the situation of religion in the People's Republic of China today, arose in December 1985 during a symposium on Taoist Ritual and Music of Today held at the Chinese University of Hong Kong. Several presentations dealt with aspects of contemporary religious life in China and several participants had visited China in the recent past. When executive members of the Hong Kong Branch of the Royal Asiatic Society showed great interest in assuming responsibility for the publication of such a volume, tentative plans were made to go ahead and do it. This volume is, therefore, an exceptional case, in that it does not contain the text of papers presented at some symposium. It is far from complete and adequate. It is indeed, a presumptuous effort. Yet in its inadequacy and modesty, we hope to do a service to many readers who have no first hand knowledge of the Chinese situation. A snapshot indeed! Yet fixing for the future, a moment of truth.

Historical Background

"In China, religion belongs to the domain of politics" (Joseph Spae in his article on the Catholic Church in China below). Neither the political situation, nor the condition of religion in China, can be fully understood without a recourse to history. Over the centuries, starting with the Shang (or even earlier), government and religion have had intimate ties and occasionally lived in a symbiotic relationship with each other. At other times they were at each other's throats, or in more diplomatic terms, they used each other to further their own goals. In other words, government used religion for its own reasons, while religion also used government for its objectives. Often religion posed a threat to the rulers and, consequently, the rulers tried to subdue religion, diffuse its potential dangers of rebellion or interference, and control its manifestations. Scholars like J. de Groot interpreted this as persecution, but that may be too strong an expression. There have been brief periods in Chinese history when some religious groups were persecuted, especially Buddhism, and sectarian movements like the White Lotus Society which did foster uprisings, but in general, the imperial government was very careful in checking the possible dangers of religious activities. The overall attitude through the centuries was that of cautious control and careful tolerance.

It does not seem exaggerated to state that most kings and emperors of China used religion or certain aspects of it as a tool to further their own goals. Four, or possibly more, different types of using religion can be easily spotted.

Firstly, religion was frequently used to establish royal or imperial authority, divine sanction, or divine legitimation of power.

The "Mandate of Heaven" invoked by the Duke of Chou is the most eloquent example, but the adoption of Confucianism by Han Emperor Wu in 136 B.C. may have had the same purpose. Confucianism believed in the heavenly appointment of rulers, and thus played into the hands of Emperor Wu and his later successors to the throne. The performance of the Feng and Shan sacrifices by Ch'in Shih-huang-ti and Han Wu Ti served a similar purpose. Even the claims of the T'ang emperors to be descendants of Lao-tzu can be seen as a variation of the same theme. Their claimed religious and biological connection with the Taoist sage boosted the imperial family's prestige and political authority.

Secondly, religion occasionally provided an expedient escape from political disaster, as when the Sung Emperor Chen-tsung claimed to have received divine revelations from the Jade Emperor. These revelations countered the loss of prestige incurred through his defeat in war against the Northern barbarians.

Thirdly, religious organizations, especially Taoism and Buddhism, were frequently supported by rulers in order to ensure the sympathy and the co-operation of the masses. During the period of disunity, most Northern dynasties supported Buddhism (K'ou Ch'ien-chih's Taoist experience during the Northern Wei was rather a short interlude). The Sui and T'ang emperors favoured Buddhism and/or Taoism, depending upon each emperor's predilection. Yet most of them did not totally favour just one religion for fear of antagonizing the people. The Yüan emperors favoured Lamaism, although Genghis Khan had been impressed by Taoism, and gave it extreme support; he hoped that in Taoism, he would find a method to reach longevity or immortality.

The political, or rather diplomatic, use of religion has also had its modern applications. The Japanese occupying Manchuria in the 1930's, appealed to the people to rally around Buddhism as a common religious faith to ensure their collaboration. Chiang Kai-shek's New Life Movement can also be seen as a quasi-political manoeuver to rally public support for his regime. And in modern times, it is common practice in Taiwan for candidates in an election campaign to visit temples and offer incense as one way to gain votes.

Fourthly, and lastly, many Chinese emperors from T'ang through Ming times have summoned prominent Buddhist or Taoist masters to their palaces to serve as counselors or to lecture to the imperial household. As a result, there have also been many cases of religious interference in matters of state, sometimes, as in the case of Taoist priest Lin Ling-su during the Northern Sung, with disastrous results for the dynasty.

All these are different types of interaction between the imperial government and the religious establishment. Also, it is obvious that in most of these relationships, the rulers used or exploited religion for their own advantage. Cases of persecution or proscription have not been uncommon and show that if religion was perceived to be a danger to the state, the government did not have the least hesitation to suppress it.

This relationship between government in traditional China and established religion has not significantly changed in the present times: only the actors are different. Looking at the socialist period (since 1949), this attitude has not basically changed, except that the need for control and manipulation has become more urgent due to the Marxist view on religion. The imperial government was not anti-religious (it was sometimes *anti* some religions, or *anti* some manifestations of religion), nor clearly atheist. Emperors and mandarins took an active part in the state religion, performing sacrificial rites on behalf of the whole nation or for the wellbeing of a more restricted area. This seems to be incompatible with atheism. With the coming of Marxism and the "Liberation" of China by the communist party and the People's Liberation Army, a new period started in which religion at first was seen as an enemy, as a reactionary group which might boycott the goals and objectives of socialist reconstruction.

The Marxist view does not favour religious tolerance, nor does it necessarily lead to religious persecution. Yet in times of social-economic collapse, military upheaval, warlords, and civil wars (the period of Japanese aggression, civil war, post-war confusion), one cannot expect an emotional balance and fairness among the competing rivals. As the communist armies gradually swept over large areas of China, and eventually totally defeated the Nationalists, the victorious party men were not too restrained in dealing with religion. A flood of literature was produced by exiled missionaries, accusing the communist rulers of atrocities. Not to be condoned, can they not be seen as side effects of the woes of transition to a new society? What would one expect to happen in such a time of take-over? It was a time of enthusiasm and of great pride in the hard won victory. Finally, the yoke of the hated regime had been thrown off. The communists had a sense of mission. They liberated the people from oppression and exploitation. And who was the hated enemy? Imperialism in its various aspects: a political regime of oppression, identified with the nationalist party; but also the foreign nations, who had tried to cut up China like a melon and had certainly drained away a vast amount of China's wealth. The religions, per se, free from political activities, yet had some obvious ties with all that the old regime stood for. Confucianism was a remnant of the feudal society during which the people were oppressed. It was also used by Chiang Kai-shek as an instrument of moral reform, ultimately with political goals. Taoism and Buddhism seemed grossly superstitious and parasitic; the majority of their clergy were poorly educated and were not examples of moral greatness. Christianity seemed like a tool in the hands of imperialist countries. If one remembers how Christianity in the 19th century entered China after the forced concessions of the opium wars, one can understand the deep resentment and bitter feelings against this foreign religion. As soon as China became free to act, the foreign missionaries had to go. It was something that most governments would have done in similar circumstances.

In view of China's ancient and more recent past, the fate of religion since liberation becomes more understandable, although violations of human rights have been made (especially during the the Years of Chaos). Today the situation appears to be more relaxed. Yet, what exactly is the present government's and party's attitude toward the religious question?

People's Republic of China Government's and Chinese Communist Party's Present Religious Policy: New Constitution of 1979 and 1982 Policy Paper

The official view concerning religion in today's China can be explored in two ways (later, we will discuss the practical implementation of this policy): the declaration of freedom of religious belief in the new constitution; and an official party policy paper issued in 1982 and made generally known in 1985. The second document is a further commentary on what the party understands to be the correct way of interpreting the constitution.

The New Constitution
The Constitution adopted on December 4th, 1982 by the Fifth National People's Congress is the third revision of the 1st Constitution issued in 1954. Former revisions had been made in 1975 and 1978. The third revision was carefully prepared but greeted with popular apathy, despite the important changes it contains:
In a nutshell, China's New Constitution sharply alters the commune system, formally ends the right to strike, sets a 10 year limit on the tenure of top officials, bans any attempt at "sabotage" of China's socialist system, and formally declares the independence of Chinese religious believers from the Vatican and other foreign religious bodies.[2]
The new policy toward religion is outlined in article 36, one of the articles which define the "Fundamental Rights and Duties of Citizens":
Article 36
Citizens of the People's Republic of China enjoy freedom of religious belief.

6

No state organization, public organization, or individual may compel citizens to believe in, or not to believe in, any religion; nor may they discriminate against citizens who believe in, or, do not believe in, any religion.
The state protects normal religious activities. No one may make use of religion to engage in activities that disrupt public order, impair the health of citizens or interfere with the educational system of the state.
Religious bodies and religious affairs are not subject to any foreign domination.[3]

Although the article is extremely short (the whole constitution is very short: only 12 pages in English translation!), the questions arising from it are endless.[4] It is a document that apparently can be interpreted in many ways, depending on what suits the current party line. One expression which appears to be disturbing is "normal religious activities" (*zhengchangdi zhongjiao huodong* 正 常 的 宗 教 活 動). Does it mean, as it was previously interpreted, that pure religious belief is acceptable, but "superstition" is outlawed?[5]

Many of the possible questions arising from the new constitution have been rediscussed and officially answered in a later document, a policy paper concerning religion in China's socialist era.

Policy Paper Concerning Religion

In 1982, the Central Committee of the Chinese Communist Party published an official communique "Concerning the Basic Viewpoint and Policy toward the Question of Religion during China's Socialist Period".[6] It contains a fairly systematic summary of the experiences and lessons (i.e. successes and failures) of the Chinese Communist Party in handling the question of religion since 1949. As it determines the official view and policy toward religion at the present moment, affecting the future as well, it is an important source from which to gain a better understanding of the religious situation in China today. It is worthwhile to quote and/or summarize some of the more relevant passages.

Religion is seen as a result of social frustrations and exploitation and as a tool to manipulate the masses. In China's socialist society, this base has been destroyed so that with the advances in education, culture, science, and technology, religion will gradually disappear of its own accord. It would therefore be against Marxist principles to directly destroy religious thinking and activities:

The opinion of those who believe that religion can be wiped out in one stroke through executive orders or by other heavy-handed action, is in direct contradiction with the basic views of Marxist thought concerning the problem of religion; this opinion is completely wrong and extremely harmful. (p.1221)

China is a country of religious pluralism: Buddhism in China has a history of circa 2,000 years; Taoism (*Daojiao*), circa 1,700 years; Islam, circa 1,300 years; Catholicism and Protestantism had their greatest spread after the opium wars.

The numbers of believers are estimated as follows: at liberation: Islam — 8 million; today — 10 million (mostly through natural population increase of 10 minority nationalities). Catholicism at liberation 2.7 million vs 3 million today, with Protestantism with 700,000 believers vs 3 million today

Buddhism (including Lamaism) is the common religion of the total population of Tibet, Mongolia, the Tai nationality and other minorities; Buddhism and Taoism still have a definite influence on the Han Chinese. Although those who believe in ghosts and spirits are numerous, yet the number of true religious believers is not very high.

Since liberation, the party has scored several successes. It has broken the power of imperialism within the churches, and by establishing the Three Self Movement, has effectively made the churches independent and free from the interference of imperialist aggression.

Whereas the period of the "Cultural Revolution" was a reversal of the true Marxist view and policy toward religion, restoration has set in after the smashing of the gang of Jiang Qing. Freedom of religious belief has again been implemented.

The party's basic policy toward the problem of religion is to respect and to protect the freedom of religious belief: this is the party's long-term policy until the time arrives that religion will expire of its own accord. Although all citizens have the freedom of religious belief, "we party-members are atheists and should energetically attempt to spread atheism" (p. 1225) but without using violence or administrative action.

Also, no pressure may be exercised in schools to force youngsters under the age of 18 to enter religion, to enter the priesthood, nor to reinstate feudalism and colonialism, nor may religion be used to oppose the party leadership, the socialist system or the unity of our country. On the contrary, religion may be used to rally all believers into realizing the modernizations and building up a strong socialist country.

One of the most important tasks of the party toward religious education is the recruitment and education of religious personnel. As of today (i.e. 1982), the total number of religious personnel for all religions is over 59,000. This breaks down as follows:

Buddhist monks, nuns and lamas	27,000
Taoist monks and nuns	2,600
Islam (imams)	20,000

| Catholicism | 3,400 |
| Protestantism | 5,900 |

This number is far below the number of personnel at the time of liberation. Most of these persons are truly patriotic and law-abiding citizens and promote the socialist system. The number of opponents and of those who cooperate with foreign revolutionaries is negligible. Many members of the religious personnel exert a strong influence upon the "spiritual life" (*jingshen shenghuo* 精 神 生 活) of the masses which cannot be ignored.

An important condition for the normalization of religious activities consists in the party's efforts to provide suitable places for religious activities. At the time of liberation, there were about 100,000 such places in the whole of China; today, the number of temples, churches, places for general activities, and centres established by the believers themselves number just over 30,000.

Some very influential and culturally very valuable temples and churches should be restored if conditions permit. But it would not be reasonable to build temples and shrines without consideration of economic feasability and appropriate the country's resources without good reason.

Nobody should interfere with "normal" religious activities taking place according to religious customs in places of worship or at the homes of the believers themselves, like worship of the Buddha(s), chanting sutras, burning incense, worship, prayer, Holy Mass, baptism, ordination, fasting, celebration of holy days, etc.

As for the Protestants, although home meetings are in principle not allowed, yet they should not be inflexibly forbidden, but alternative methods should be worked out.

All the places of religious activities are under the administrative leadership of the government's Religious Affairs Bureau. Nobody is allowed to propagate atheism in those places; on the other hand outside the places of worship, no one is allowed to spread or propagate theistic beliefs; also the publication of religious books or periodicals has to be approved by the government.

All the donations presented to temples and churches should, under the guidance of the government and the religious organizations, be used for the restoration and upkeep of the buildings and partially for the training of better qualified religious personnel.

The function of all patriotic religious organizations is to promote the religious policy and to normalize religious activities. There are eight national patriotic religious bodies:

Buddhist Association of China (BAC)　中國佛教協會
Taoist Association of China (TAC)　中國道教協會
Islamic Association of China (IAC)　中國伊斯蘭教協會

Patriotic Catholic Association of China (PCAC) 中國天主教
愛國會

Administrative Commission of the Catholic Church of
China (ACCC) 中國天主教教務委員會

Catholic Bishops' Conference of China (CBCC) 中國天主教
主教團

Patriotic Protestant Three Self Movement of China
(PPTSC) 中國基督教三自愛國運動委員會

Protestant Association of China (PAC)[7] 中國基督教協會

There are also regional and local religious organizations. The basic duties of all these bodies are to assist the party and the government to implement the policy of the freedom of religious belief, to help the masses of religious believers and the religious personnel to raise their awareness of patriotism and socialism.

The training of young patriotic religious personnel is of great importance, to which we have to give our strong support. Candidates have to make their own choice and should have a sufficient educational basis to enter religion. The young religious personnel ought to raise their spirit of patriotism and be alert to socialism and be prepared to implement the party's policy toward religion.

Although the Chinese Communist Party promotes a policy of freedom of religious belief, still it does not mean that party members have such a freedom. That freedom applies only to common citizens. A party member is in this respect, different from other citizens; as a follower of Marxism, he should be an atheist. The party has repeatedly stated very clearly that party members should not be religious believers and should not participate in religious activities. Those who persist in doing otherwise should be persuaded to give up party membership (p. 1233).

This is the general rule for party members. However, the rule should not always be inflexibly applied. Among minority populations it is not possible to make such demands since religion is often an essential part of social life and popular customs; for instance wedding and funeral rituals and traditional festivals. If party members would not participate in those, they would totally isolate themselves from their social group. In other words, one has to take into account the concrete circumstances of those party members. Moreover, very often these customs have become merely secular events.

All party members should be deeply aware of the fact that China is a socialist country consisting of many nationalities. The relationship between nationality and religion is different for each minority religion, for instance in Islam or Lamaism. For them, nationality and religion are one and the same thing. In the case of the Han nationality however, it is different. There is no fundamental relationship between Han people and Buddhism, Taoism, Catholicism, or Protestantism. This difference of circumstance has to be kept clearly in mind. At any rate, religion should never be used as a means to divide the people. If our party leadership cannot firmly handle these problems, it will not be able to unify all the various national minorities in its common march forward.

Whereas all "normal" religious activities are strongly protected, at the same time, those activities that are against the law or counter-revolutionary or that are outside the realm of religion, or supersitious activities that harm the well-being of the country or the life of the people, have to be firmly smashed.

All reactionary practitioners like (unorthodox) Taoists, spiritual charlatans or shamans, who have already been suppressed, should not be allowed to revive their activities. All those who, with empty talk, delude the masses and harm the people by cheating them out of their money, should be vigorously suppressed. All those who make a living through physiognomy, fate calculation, or geomancy, should be educated and assisted to give up such a life, to earn a living through their work and not to take advantage of superstition to cheat people. If they do not comply, they should also be suppressed in accordance with the law.

All levels of party members and government officials should exert great vigilance with regard to the presence of counter-revolutionary elements or law-breaking elements within religious bodies and should reveal their activities to protect the normal religious activities. Only then can religious activities be normalized.

In China, there are four religions which occupy an eminent position: Buddhism, Islam, Catholicism, and Protestantism. These four are also prominent amongst the world religions internationally. Catholicism and Protestantism are the strongest in Europe, North America, Latin America, and other places; Buddhism is strongest in Japan and South-East Asia; Islam flourishes in the countries of Asia and Africa: all of these exert considerable social influence (p. 1236).

Since China's external relations are becoming more important every day, so also do the external relations of the religious world every day grow in importance. This, of course, implies the danger of the power of imperialist religions such as the Roman Vatican and the Protestant 'House Churches'.

We have to develop friendly international relations with those religions, but at the same time, strongly resist the interference of reactionary foreign religions so as not again to be controlled by them. We cannot tolerate foreign religious personnel propagating their faith in China nor allow them to sneak in religious propaganda material.

Religious organizations or personnel in China should not ac-

11

cept monetary donations offered by foreign religious organizations. However, donations made by foreign religious believers or overseas Chinese may be accepted by temples and churches in China, but only after being approved by the respective bodies of the people's government.

We should seriously watch the adverse powers of foreign religions to establish underground churches or other illegal organizations in our country under the cloak of religion, nor let them undertake spying and destructive activities. The government's institutions which deal with religious affairs have to be strengthened; moreover they should make all cadres who work in this area systematically study Marx's essays concerning religion to gain a deep understanding of the party's basic view of religion and its basic policy, have close contacts with the masses of believers, and cooperate with religious personnel.

It is an important task of the party to advance in the scientific study of the religious problem from the viewpoint of Marzism; to criticize, from a Marxist philosophical viewpoint, idealism (which includes theism); to provide to the masses of the people and especially to our youth, an education in the scientific world view of materialism and historical materialism (which includes atheism).

An important aspect is publication. It is the party's duty to speed materialism and atheism without however, attacking the believers' religious sentiment; the party should respect their theism, but likewise the believers have to respect the party's atheism.

This goal will take a long time to achieve and is not a matter of only one, two, or three generations. But that time will certainly arrive. China will be prosperous and free from the need to rely on religion to achieve the true goals of society. It is the party's tasks to strive for this great achievement (pp. 1239-40).

Implementation of Religious Policy

If the above official document outlines in detail the party's and government's theoretical viewpoint on how to deal with religion in principle, what is their practical attitude toward the various aspects of religion in China today? How is the policy of religious freedom being implemented? This is an even more difficult question to answer, since factual information is scarce.

Both implicit and explicit statements in the constitution and in the 1982 policy paper indicate that the party and government of the People's Republic of China are extremely serious about the issue of religion and that their control of all expressions of religious life will continue to be very strict. While the Religious Affairs Bureau is the main government agency in overall control of religious matters throughout the country, under the direct leadership of the State Council, each particular religious body, like the eight associations listed above, has a certain autonomy in decision making. Yet, among the membership of each of these associations, there are party members, ensuring that policies and directives are following the officially approved party views. If one considers that government and party control (or supervise?) such crucial matters as recruitment of new candidates, temple economy and finances, religious activities in general, the propagation of religion through printing, etc., one has the strong impression that true freedom of religion is just empty talk.

Religion is tolerated, yet not encouraged. As J. Spae says, Beijing's good will for Islam is "bald pragmatism".[8] The same principle seems to apply to the official attitude toward all religions. In earlier years, Buddhism was supported since it had diplomatic usefulness; now, the same applies to Christianity: ever since the Four Modernizations were launched, China hoped to gain assistance from Western countries, especially the United States, and since Christianity is the predominant religion of the West, it became politically expedient to adopt a tolerant attitude towards Christianity in China.

If the official way of handling the problem of religion is pragmatism, or in other words, expediency, it follows that government's attitude is different towards each religion. It is worthwhile to look at this in more detail.

First, how does the government deal with the native traditions, **Confucianism** and **Taoism**? The latter is trying its best to revive its ancient roots, but since many folk practices (or "superstitions") have been often incorrectly identified with Taoism, it has suffered a great deal. It appears that at the present, the Taoist priestly traditions come back to life: monastic Taoism (as in the Quanzhen order) and Heavenly Master Taoism (Zhengyi order). Since Taoism does not appear to be a threat to political rule today, it is not the object of extreme government vigilance, beyond the presence of party members in the TAC. Confucianism has been reassessed: Confucius had been under severe attack during the seventies "as the most reactionary representative of feudal thinking"[9] but is presently considered a great educator, thinker, and politician. This is again of course not from a religious point of view, i.e., involving worship, but as a good example of a leader/statesman promoting unity and harmony, benevolence and ethics, all of which are a beneficial influence on society. In other words, the moderates presently at the fore in government are reaching back to and promoting China's roots, Confucianism, as a stabilizing influence on society. Also, the spirit of learning so important in Confucianism is once again being encouraged and fostered as China strives to modernize rapidly. Although the hierarchical societal system advocated by Confucius is still officially rejected, filial piety is strongly endorsed and children are urged to care for their elderly parents, thus easing the financial burden on the state as regards social security costs. These ethics are said to be in tune with socialism.[10]

Buddhism has had a different fate but the same principle of expediency applies: the BAC has been used as an instrument to promote political goals. Through its contacts with Buddhists abroad, the Association carries on public relations work for the PRC government and engages in international diplomacy activities: its delegations act as government spokesmen abroad. It emphasizes the Chinese Buddhists' apparently strong support for the government, thus providing legitimization, and provides a concrete example of the government's benevolent religious policy which in turn acts as an affirmation of Communism.[11]

Buddhism and Taoism are also seen as sources of cultural pride: since the "Ten Years of Chaos" many historic sites, temples, and cave grottoes with their sculpture and murals have been repaired or reconstructed: they are monuments of the working people's creative talent, and are also important relics of China's cultural heritage. These repairs are still continuing today but in the eyes of the government, they are more significant to attract foreign tourists and enhance China's international prestige.

The situation of Tibetan Buddism is very different. Here religion and politics (nationalism) are intricately connected, and although the government's interest has been mainly political, yet their harsh treatment of the Tibetan people may have been interpreted as religious persecution. The problem of the Dalai Lama in exile is still a difficult dilemma for both sides.

The two other originally non-Chinese religions, **Islam** and **Christianity**, pose different problems. Islam provides the greatest challenge to the CCP. Although from Central Asian origin, a large group of Moslems, the Hui nationality, are being considered as indigenous Chinese (together with the Han, the Man, the Meng, and the Tsàng: the five major ethnic groups of China). Another group of Moslems is composed of various minority nationalities. As with the Tibetans, the Moslem religion is intricately linked with their nationalism and territoriality which is seen as a permanent threat to the central government's position of authority. Joined together, the Moslems form a strong power group in China and have a keen sense of their own culture. This makes the government's position very delicate; moreover Chinese Moslems have important links with the global Pan-Islamic movement. This relationship can be exploited by the government, as in the case of Buddhism, and Chinese Moslems could be manipulated as instruments of foreign policy.

14

The case of Christianity[12] is different again. On the one hand, it reminds China of the painful 19th century with all the humiliation of imposed treaties and concessions; on the other hand, Christianity, if isolated from foreign interference, can also serve as a welcome diplomatic tool in view of China's foreign policy and economy. Since two articles deal with Christianity in detail, no more has to be said here.

In different ways, all major religions are used by the government as foreign policy tools, of course also depending on the internal power struggles between radicals and moderates. Once again, the difference between the present and China's imperial past is one of the variations of the same theme: control and manipulation; the degree of control depends on who is in power at the moment.

Some reports from visitors to China are rather sobering: an Australian reporter whose article was published in *The Australian* [13] wonders what the reality of human rights is in China:

Two months of stay in China convinced me that it remains one of the world's more repressive and totalitarian regimes ... [but] In order to sell itself to the West (to attract Western technology and capital) it is useful to project a more Liberal image. (pp.56-57)

...China lacks all kinds of other freedoms (besides political freedom) as well. Religion suffers through repression in China. The Chinese constitution guarantees citizens freedom of religion, but the entire constitution is an elegant travesty, guaranteeing, in lofty prose, a number of rights which have never existed in reality. (p.58)

This sort of statement is perhaps too harsh, especially since most visitors to China praise the relaxed political atmosphere today and are impressed by the moderate freedom that religious groups enjoy.[14] The government is probably aware that the number of true believers in religion is very limited, that their influence on political life is rather restricted, and that they therefore pose no real threat to the regime, as long as their efforts are focussed on the spiritual affairs of the people.

People's Attitudes Toward Religion: Commoners, Scholars

How does the new government policy toward religion affect the people? (By "people" should be understood various segments of the population, the uneducated as well as the educated). Several of the articles included in this volume deal with this topic. Here, I only want to add some additional information and comments.

Among the masses of the common people, there seems to be a lack, perhaps total lack, of interest in religion. This is based on my own (limited) observations of the Chinese people, especially in the cities. The majority of the people apparently do not have a clear world view; I would dare to say that even the Marxist view of society and human destiny leaves them indifferent. Daily life has apparently nothing to do with religion. One person told me that what families at home and colleagues at work usually talk about is clothes, food, refrigerators, television sets, and other luxuries of life, now more abundantly available than even two or three years ago. In a special report on China in *Asiaweek*, the author refers to a quotation of Friedrich Engels to the effect that "the socialist man wants first of all the bare essentials of living, secondly the luxuries of life, and thirdly self-fulfillment."[15]

The Chinese people have had for too long to do without some of the bare essentials; now that more and more consumer goods are available, they put all their ingenuity and energy into getting their share. There seems to be little need for other self-fulfillment yet, at least among the masses of people. Perhaps, once people's wishes have been fulfilled, they will become more aware of a spiritual gap, although self-fulfillment does not only mean religious fulfillment.

The 35 years of promotion of atheism, and especially the scare of the Ten Years' Chaos has contributed a great deal to weaken religious sentiment. This view is shared by another China visitor, Alvin Cohen, who writes:

> My general impression, based upon my very limited obser-
> vations, is that as the government has become more tolerant
> of religious activities, perhaps some people are revealing
> their interest in this aspect of life. However, after 35 years of
> government intolerance and anti-religious education and
> propaganda, as well as ten years of outright destruction dur-
> ing the Cultural Revolution, the entire generation of middle-
> age and young people simply do not know what it is all about
> and do not even know how to go through the motions. There
> is curiosity, interest, and even efforts at ritual practice
> among some people, but this does not necessarily indicate a
> revival of religion. It seems to me that the Buddhist and
> Taoist clergy will have to actively engage in some form of
> public instruction (or even proselytism, if they will be al-
> lowed to do so) in order to re-introduce their religious beliefs
> and practices to an entire generation that at best knows little
> or nothing about religion, and at worst has a large reservoir
> of negative feelings about it.[16]

This situation does not necessarily apply to the countryside (see my article in this volume).

Among the intellectuals, there is a revived interest in the study of religion, sponsored, amazingly, by the government itself. With in the Chinese Academy of Social Sciences

中 國 社 會 科 學 院 which is a national body located in Beijing, there are about 30 institutes, among which is the Institute for Research on World Religions 世 界 宗 教 研 究 所 . This institute was first established in 1964 in order to study the principles (doctrines) of the three major world religions: Christianity, Buddhism, and Islam, as well as their history and present situation. Today (1986), it is composed of seven departments: Buddhist Studies, Christian Studies, Islamic Studies, Taoist Studies, Confucian Studies, Studies of Marxism-Leninism — Mao Zedong Thought, and Ancient Chinese Philosophy. As of 1986, the institute had a staff of over one hundred, of which about two thirds were research personnel.

Besides, each province has its own Academy of Social Sciences, but only Shanghai and Xinjiang had (in 1986) a deparment for the study of religion.

The Shanghai unit is very dynamic and has been engaged in the recording of Taoist rituals. These provincial academies are independent from the national academy in Beijing, but are under the jurisdiction of their respective governments.

All this concentration of full-time research scholars of course, produces results. Many publications, books, and journal articles, have been published and other serious works are under way (see Jan Yun-hua's article). One wonders however, how seriously scholars are obliged to interpret religion and religious history from a Marxist viewpoint. This certainly was the case a few years ago and is clearly visible in works like Qin Xitai's first volume on Taoist History, published in 1981. His second volume, of 1985, is relatively free from such a bias.

This lack of academic freedom is probably felt by many researchers. An opinion article appeared in *China Daily* of February 26, 1985: "Being bold in social sciences":

In China, research in the social sciences used to be a risky business. Social scientists who aired views and opinions differing from current propaganda were relentlessly criticized during the "cultural revolution." As a result, some people engaged in the social sciences still fear making "political" mistakes.

Greater freedom needs to be granted, and social scientists should be given encouragement and support in advancing new theories. The social sciences should be regarded more as a science than a tool that can be used for any purpose.[17]

As social scientists in the past were often obliged to let official Marxist viewpoints dictate their studies, so we also find artists and authors subject to the same kind of presssure. In a recent article by Chen Yaoting of the Shanghai Academy of Social Sciences, this anti-religious bias in literature is clearly pointed out.[18] The anti-religious attitude which had prevailed in government and party circles since 1949, has certainly inhibited many kinds of expressions of religious sentiment. The author analyzes thirteen pieces of recent literature, published between 1979 and 1981 dealing with religion as their central theme. His criticism is that they all share a prejudiced view of religious faith and do not portray a truly realistic picture of the people's religious life. The projected image is not credible: if Buddhist monks, Taoist or Catholic priests appear, they are portrayed as spies or bullies; to recite the Buddha's name is good for old ladies of the landowners class; those who believe in religion are mostly depicted as stupid men and uneducated women. Chen sees this as a reflection of political life during recent years. After the Ten Years of Chaos (1966-1977), the theme of religion has re-entered the area of literature, yet religion is still treated in a negative way.

A concrete example, a story of "The Party Secretary and the Chief Engineer" (1981), deals with the dilemma of how to reconcile being a good patriot, a model worker, and a competent scientist with being, at the same time, a devoted religious believer. The engineer is such a person, but has to go through a long series of problems to prove his point. In his own way, he succeeds in harmonizing his religious conviction with patriotic and professional responsibility. Religion cannot be seen to be an obstacle toward socialism. However, this view is not shared by all literary artists.

Chen also encourages literary artists to study religion in a more objective way in order to understand the true nature of religion and also to use religious symbols and rituals more correctly in their writings. This will add a new and rich dimension to contemporary literature.

Contributions to this Volume

The present volume, admittedly, is far from exhaustive or complete. Yet it covers some important aspects of the changing religious situation in today's China. In several articles, the tension between the official Marxist world view and the persistent religious beliefs of some segments of the Chinese people are clearly attested.

The contributions include two essays on the academic study of religion (Jan Yun-hua and Man Kam Leung); four articles on aspects of religious revival: the return of the Taoist Jiao festival (Kenneth Dean); new developments in Buddhist and Taoist monasteries (T. Hahn); the revival of temple worship as well as some folk religious traditions in rural China (J. Pas); and a short report on the recently reinstated sacrifice to Confucius at his birthplace temple in Qufu (P. Swart and B. Till). The article on religious amulets (A. Cohen) may be a surprise to some readers but its rationale is that it shows the old silver amulets confiscated by the red guards, which are now being replaced with cheap models (see article by J. Pas). What happens within the Christian churches is discussed in two essays: one on the Catholic Church (J. Spae) and one on Protestantism (J. Dunn).

There are some obvious gaps. There is no article about Islam, nor one about Tibetan Buddhism. I had hoped to have these included, but my efforts were not successful and I apologize for it. There also is no article about Judaism in China. But then, Judaism has no longer a significant presence there. Jews may have immigrated to China as early as the 8th century. They settled in several cities, but Gaifeng (Kaifeng), the capital of the Northern Song (960-1127) became their major center. For a long time, they kept their traditions alive; they had a synagogue with a rabbi, but with the passing of time and various circumstances the Jewish community disintegrated. They intermarried with the Chinese and have almost completely been absorbed in the Chinese mainstream.

Today, it appears that there are at least 200 residents of Gaifeng who claim to be of Jewish descent, although externally they completely look like Chinese and they no longer engage in Jewish practices. With the new religious climate, Jews from outside the country are trying to establish new contacts.[19]

There are, on the other hand and surprisingly enough, three articles about religion in Hong Kong. This is not only as a courtesy to the Hong Kong branch of the Royal Asiatic Society which has undertaken to publish this volume, but it also serves as a concrete example of Chinese religion in a different political, social, and economic climate. Religion in Taiwan could also have been included, but there are already quite a number of recent studies dealing with it: there is no need for repetition. The article by Tanaka Issei provides an excellent contrast to the contribution of Kenneth Dean: both describe the Taoist-popular rituals of the Jiao festival. David Faure's report on popular religion in Hong Kong shows how different circumstances affect expressions of the popular faith, while the same applies to Bartholomew Tsui's article on Buddhism. Religion is like a tree: depending on the circumstances of weather, soil, and human care, it grows strong or weak, or it may even be cut down. The three Hong Kong contributions moreover, bring out the contrast between the fate of religion in a climate of freedom and religion in a situation of rather strong government control.

19

Conclusions: Religion in China's Socialist Society, Potential and Dangers

It is time to wind up this introduction with a number of general considerations. The usual question popping up at the end of this kind of discussion is about the future of religion in China. This is an issue without a definite answer. Too many predictions have turned out to be wrong. K. Schipper, an authority on Taoism, wrote that Taoism (and Chinese religion in general) had been virtually wiped out with the advent of Communism.[20] Similarly, H. Welch predicated that Buddhism would be pretty well dead in the 80's.[21] One forgets, however, that the Chinese people are very tenacious and resilient; like the loyalists of early times, when the Tao prevails in the country, they come forward; when the Tao is in darkness, they go into hiding. Whether the Tao prevails or not in matters of religion is the same as to say whether the government is radically leftist or moderate. If the present climate of moderation lasts, religion will become stronger even if the material conditions of the people improve. Marx's own prediction that religion will eventually disappear of its own accord, may be a mistake: if he was wrong in some economic matters, he may have been wrong elsewhere. But certainly the Taiwan example does not support Marx's prophecy. Taiwan has never been as prosperous as it is now, yet religion flourishes.

China has not reached a high standard of living yet; some segments of the population still live in poverty. That may be one reason why religion continues to attract followers. This whole matter of the continued existence of religion in socialist China is of great concern to the government. Also academics have studied the issue. One example is the address by Ruan Renze of the Shanghai Academy of Social Sciences to the delegates attending a symposium of Taoism and Music in Hong Kong, December, 1985.

Mr. Ruan stated that there are three kinds of reasons why religion continues to exist during the socialist phase of social development: social, ideological, and psychological reasons.

As social reasons, he points out that since 1949, despite the overwhelming transformations taking place in China, the country remains largely underdeveloped on the economic, educational, and scientific level. Due to China's size, there are marked regional differences and inequalities. "At the same time, miscalculations in government policy as well as the faulty implementation of government policy have also had a deleterious effect," occasionally causing manmade disasters. All these circumstances have driven some people to find consolation in religion and "in it to search for escape".

As ideological reasons, Mr. Ruan points out that even in a

socialist society, there are a myriad of influences at work which affect the lives and thought patterns of individuals. "Finding ourselves in such a complex world of individual thought, it is hardly surprising that some people still believe in religion", just as others are materialist.

As psychological reasons, he says that "Daily life gives rise to no end of personal frustrations and disappointments." Among those are family and love problems, or problems related to education or employment, from which some people try to escape through religious belief.

Although it appears that Marxist orthodoxy does not approve of all these escape lines, yet patience is required. Religion will eventually die out, when the causes of its emergence are eliminated. In the meantime, the party and government adopt a policy of tolerance as long as religious believers are truly patriotic and support the government's policy of realizing modernization and build up a strong country. Religion in fact, is seen as a positive force in the nation: it encourages patriotism in the believers, it sets high moral standards which benefit society and continues a tradition of public service. Also, through international contacts, it contributes to the furthering of world peace. Of course there are individuals who use religion as an excuse to boycott the nation's progress. These "evil elements" must be suppressed by law so that they cannot pervert the positive activities of religious organizations.

Taking the view of Mr. Ruan as a model besides several others,[22] one is impressed by the earnestness of the Chinese leadership to justify their policy toward religion. Because it is a moderate leadership, their attitude is one of tolerance. If a more leftist government would come to power, intolerance would prevail again. Based on the same Marxist doctrines, one can justify either policy, depending on one's preconceived opinion. This being so, perhaps we will see another turning of the tide some day.

However, while religion is enjoying a new freedom, political campaigns have not fully come to an end. In 1982-84, a campaign was launched which seemed to have religious overtones: the campaign against moral pollution. In its negative expression, it was intended to fight moral depravity, said to have infiltrated from the West through films, magazines, etc. At the same time, China cracked down on criminality and corruption, and a wave of executions took place all through 1983.

In its positive expression, a campaign for "a spiritual civilization" was launched as well by the 12th Party Congress in September, 1982. In Western news reports about China, one often came across "spiritual culture" or "spiritual civilization" (*jingshen wenming* 精神文明), which caused confusion and amazement in my mind; as if the Chinese government promoted a revival of religious values. Indeed, "spiritual" is normally thought of as the opposite of "material", and has in most cases, a religious connotation. The Chinese term *jingshen* means "spirit" in one of its more restricted English meanings such as "disposition, temper", or "a pervading animating principle", etc. so that "spiritual culture" is an inaccurate and misleading translation since the party's intention had nothing to do with "spiritual" values but had in mind to promote a material culture:

By so doing we will try to avoid the chronic maladies of capitalist modernization such as spiritual voidness, moral dejection, increased criminal activities, and emphasizing profit-making above all else.[23]

It seems that the Chinese Communist Party wanted to launch a positive campaign, perhaps of providing a more ethical way of life, yet we see it expressed in a negative way: against "spiritual" pollution. This makes one wonder on what a truly "spiritual" culture could be based. Does Marxism provide a sufficiently attractive foundation for ethics? Has the absence of religion in the lives of most Chinese people created a spiritual vacuum that can now readily be filled with the restoration of religion? Or is it that the ancient Confucian way of life, or Buddhist and Taoist ethical presciptions still continue to influence society, even unawares? These are some problems that are as yet difficult to answer.

To conclude, although religion in China is on the ascent, it still faces several dangers or uncertainties: can the long interruption of monastic training (Buddhism, Taoism) or theological training (Christianity) be overcome? Can religion avoid being mummified and reduced to a mere presence in the museum? Many restored or redecorated temples are just showcases for tourists bringing in a very welcome revenue. Many are still empty, without monastic life. In others, monks and novices are busily engaged in receiving tourists, and do not have sufficient time for their own training or spiritual cultivation. Is it not possible that such a situation will hollow out true religious life and do more harm in the end than persecution or harassment by government agencies?

Religion in China still faces many uncertainties; the greatest one is the fundamentally irreconcilable contradiction between Marxism and a religious world view. Both sides are aware of this, but it testifies to the practical common sense of the Chinese people that theoretical contradictions can be somehow accommodated if that is the most practical solution toward co-existence and social harmony.

22

NOTES

1 *The I Ching* or *Book of Changes*. Wilhelm/Baynes Translation (Princeton: Princeton University Press Bollinger Series, No. 1-9, 1950) (new printing: 1985).

2 *China Update*, No. 3 (Spring 1983), p. 31.

3 *Beijing Review*, 25 (No. 52, December 27, 1982).

4 *China Update*, No. 3 (Spring 1983), pp. 31-42 devotes these 12 pages to a discussion of reactions from various circles to Article 36.

5 See Donald E. MacInnis. *Religious Policy and Practice in Communist China*. (New York: Macmillan Co., 1972), pp. 37-89 and p. 48 in particular.

6 中共中央印發《關於我國社會主義時期宗教問題的基本觀點和基本政策》的通知。
Published by the Chinese Communist Party Central Committee, March 1982 and made public by its inclusion in Beijing: Central Party School *Selection of Documents on the United Front in Recent Times*, November 1985, pp. 1220-1240. 新時期統一戰線文獻選編

7 BAC: 中國佛教協會
TAC: 中國道教協會
IAC:中國伊斯蘭教協會
PCAC:中國天主教愛國會
ACCC:中國天主教教務委員會
CBCC:中國天主教主教團
PPTSC:中國基督教三自愛國運動委員會
PAC:中國基督教協會

8 *CUP*, 11 (1985), p. 115.

9 Suzanne Garmsen. "The Religious Policy of the Government of the People's Republic of China" (unpublished research paper, 1984; 29 pages in typescript), p. 7.

10 S. Garmsen, p. 7.

11 S. Garmsen, p. 9.

12 In the eyes of the Chinese government, Christianity constitutes two different religions: Catholicism, *Tianzhujiao* 天 主 教 : The Teaching of "God"; and Protestantism, *Jidujiao* 基 督 教 : The Teaching of "Jesus Christ".

13 It was reprinted in *Free China Review*, April 1986, pp. 56-59.

14 Other recent reports are more moderate: S. Wesley Ariarajah. "Religion in China: Some Impressions". *Inter-Religio* (Nagoya, Japan), No. 10 (Fall 1986), 50-65. Seri Phongphit, "A New Dawn for Religion in China", *Inter Religio*, No. 10 (Fall 1986), 66-78.

15 *Asiaweek* (Hong Kong), "New New China", April 19, 1985, p. 30.

16 A. Cohen, unpublished manuscript, 1986.

17 The editorial in fact comments on an article by Jiang Binghai of the Shanghai Academy of Social Sciences, published in *WenHui Bao*.

18 Chen Yaoting 陳 耀 庭 "Shilun jinnian wenyi chuangzuo zhong de zongjiao wenti" 試 論 近 年 文 藝 創 作 中 的 宗 教 問 題 , in *Dangdai wenyi sichao*, 當 代 文 藝 思 潮 6 (1983). 65-71. An enlarged version appeared in other journals in 1984 and 1985.

19 See two recent articles about the Jews in China: "Chinese Jews" in *The Jerusalem Post*, International Edition, May 6-13, 1984; "Toronto

Professor Visits the Fabled Jews of Kaifeng", in *The Canadian Jewish News*, January 22, 1987.

[20] K. Schipper, *Le Corps Taoiste* (Paris: A. Fayard, 1982), pp. 31-32.

[21] H. Welch, "Buddhism Under the Communists", *China Quarterly*, April-June 1961, No. 6, p. 13

[22] See below Jan Yun-hüa's article which discusses an essay (in Chinese) by Xiao Zhitian, "A Tentative Study of the Problem of Coordination Between Religion and the Socialist Society in China". *Quarterly Journal of the Shanghai Academy of Social Sciences*, 1 (1985), 137-146.

[23] *Beijing Review* (Oct. 1, 1983). For a discussion of "spiritual civilization" see *China Update*, 7 (Spring 1984), 125-134.

RECENT CHINESE RESEARCH PUBLICATIONS ON RELIGIOUS STUDIES

JAN Yün-hua

Since the publication of a paper on a similar topic by this author three years ago,[1] a number of new books and articles, as well as new trends, have emerged in the Chinese studies of religions. Some of these new developments not only seem significant to the academic advancement of the discipline, but they also shed new light on the freedom of faith, as well as Chinese intellectual activities at large. All these new trends seem quite worthwhile for a new survey. It might be recalled that this author used the phase, "An Incomplete and Imbalanced Picture", as a subtitle for the previous paper, as the two adjectives clearly indicate the limited nature of the paper. To a certain extent those two words are still applicable to the present survey though the meaning differs somewhat between the two publications. The main difference is that the present work concentrates more on academic publications and does not touch any religious life, as the previous paper did. Although it is impossible to have a complete list or review of all periodical publications on the subject, as far as the main trends of the research and academic debates are concerned, this paper is more complete than the previous one.

For example, the scope of this paper finds that books and representative articles of various religious traditions, as well as general studies of the discipline, are covered; whereas only Buddhism and religious Taoism have been the subjects of the previous review. A survey of the general studies of religions seems an important new addition as this may reflect on the future scholarship in China. The paper will be divided into two sections. First it will review Buddhist and Taoist studies, updating the previous survey. And second, it will discuss the recent researches of other traditions, as well as the general trends of the field, apart from Buddhism and Taoism. This would, however, exclude Confucianism and Taoist philosophy, as Chinese scholars have classified those two under the umbrella of 'philosophy'. Apart from this theoretical problem, the large numbers of publications and the complicated problems involved in these two subjects require more space for an accurate survey. Moreover, some surveys on the subject, such as Taoism, have been published.[2]

1

Judging from the number of publications, it seems that Buddhist Studies has still retained its leading role in religious studies. Major projects continue and are published more or less on schedule although some delays have taken place. Two major projects under the editorial leadership of Ren Jiyu are good examples. The *Zhonghua Dazang-jing (Hanwen bufen)* [3] has begun publication. At the time of the writing of this survey, ten volumes have so far been published. According to the announced schedule, four volumes should have been published in 1984, and 10 volumes per year thereafter. It seems that the completion date of this monumental work of an estimated 220 volumes will be behind schedule unless the printing capacity of the press is increased. Apart from the delay of publication, the editorial work is also narrower in scope than scholars would like to see.

The published volumes presented to readers are not punctuated, nor is this a modernized edition, but photographic reproductions of early Chinese editions of the Buddhist Tripitakas. According to the pre-publication prospectus of this project, this new edition of Chinese Buddhist scriptures will use the *Jinzang* as the basic copy for reproduction. The *Jinzang* is a reproduction of the *Kaibaozang*, made during the Northern Song period. The *Jinzang* itself was printed during 1148-1173 A.D. Apart from the *Kaibao* edition, this collection also includes a number of Buddhist scriptures that were printed in the later ages of Chinese history. Centuries after its publication, the *Jinzang* was discovered in Zhaocheng in Shanxi province in 1933. Out of nearly 7,000 *juan* of original copies, 5,380 are still extant. The *Zhonghua dazang-jing: Hanwen bufen* will be divided into two series: the Regular Collection *(zhengbian)* and the supplementary *(xubian)*. The Regular Collection uses *Jinzang* as the basic texts for reproduction, and some works which have been omitted in *Jinzang* will be added from other collections, such as the Korean Edition. The reason for choosing the Korean Edition as the principal source for supplement is that both the Jin and the Korean editions were based on the same source: *Kaibao zang*. Apart from these two editions, some other additional works from Chinese editions will be used, if there are special works comparable to, but excluded from, the aforementioned two editions. The Supplementary Collection would select the remaining scriptures from other editions of the Tripitaka, in Chinese, but exclude the works that have already been included in the Regular Collection. These include the Stone Scriptures from Fangshan (engraved between A.D. 628-ca. 1640), to the *Wanzi Xuzangjing* (originally printed in Japan during the 19th century and reprinted in China in 1912). According to the editor's estimate, this new collection will include 4,200 works in 23,000 *juan*. Although this is a photographic reproduction, without modernized punctuation and footnotes, there are collated notes at the end of each chapter, except for the text of which no other edition is available. The editors hope that the whole project will be concluded in 1993.

The delay in printing has also occurred in the publication of Ren's other multiple volume project: *Zhongguo Fojiao shi* ("A History of Buddhism in China"). Since the appearance of volume 1 in 1981, only the second volume has been recently published.[4] It covers the period of the Southern and Northern Dynasties (265-589 A.D.) At the time of this survey, the author had not yet seen the latest volume and was therefore not in a position to assess the achievement. Since Tang Yongtong's monumental work covers the period, it will be interesting to see how the new history differs from Tang's work.[5] In this connection, one may compare Ren's work with the history of Buddhism in China, written by Shigeo Kamata, a well-known Japanese expert on the subject, of which three volumes have been published to date.[6] A preliminary comparison of the Chinese and Japanese projects clearly indicates the different styles of scholarship in the two countries. Kamata gives more attention to the historical problems and sources of the subject; while Ren pays more attention to Buddhist ideas and their relation to the Chinese religions. It seems that Kamata concentrates on history, while Ren and his colleagues give more attention to the history of ideas and their relation to, and comparison with, the Chinese native religions. One should not forget that Kamata has already made it clear that volumes 4 and 6 of this forthcoming history will concentrate on doctrinal problems. This indicates that he considers the periods of the South-North Dynasties and Sui-Tang to be important to the doctrinal development of Chinese Buddhism. Both works are essential to students of Chinese religions. As far as the first volume of Ren's work is concerned, it is more detailed than Tang's history of Buddhism during the Han to the Three Kingdoms period. However, Tang's main achievement is on the history of Buddhism during the Southern and Northern Dynasties, and this is the reason why the second volume of Ren's book, covering the same period, will be of great interest to scholars.

The serial publication edited by Lou Yulie et. al., *Zhongguo Fojiao sixiang ziliao xuanbian* is also in progress.[7] So far the four parts of the second volume have been published in their entirety. Although no notes are given, the selections on Buddhist documents from Sui-Tang periods, and the introductory notes to the material are very valuable to students of Chinese Buddhism. The division of works in volume 2 is as follows: part 1 is concerned mainly with *Tiantai* Buddhism with the selection of Sanlun master Jizang (549-623) as a secondary item. Part 2 is reserved for the selected writings of the five *Huayan* patriarchs. Part 3 consists of writings by the leaders of *Weishi*, the *Pure Land*, and writings by the Buddhist historians such as Daoxuan(596-667) and others. Part 4 deals with the documents of the *Chan* school of Buddhism. Appended to this section is a collection of imperial documents on Buddhism, including government orders, etc. The works attributed to the Tang emperors include Taizong, Gaozong, Wu Zetian and Xuanzong.

Among original research works, Fang Litian has published his recent work, *Huiyuan jiqi Foxue*.[8] This work exclusively concentrates on the life and doctrines propounded by the famous Chinese monk of the 4th-5th century A.D. The book comprises ten chapters, which cover all aspects of Huiyuan's life as well as his beliefs and teachings. A chronological table of the monk's life and age is appended at the end. The book has another welcome feature: the Index.

Zhongguo Foxue lunwenji is another publication,[9] which contains twenty articles on various aspects of religion, by Ren Jiyu, Ji Xianlin, Huang Xinchuan, Shi Jun, Fang Litian, Guo Peng and others. These articles were originally presented at a Buddhist Symposium held in 1980. Some views expressed therein still reflect the state of ideology prevailing at that time.

Chinese studies of Buddhism have opened up some new ground. *Tangdai wenxue yu Fojiao* by Sun Changwu,[10] and *Fojiao yu Zhongguo wenxue*, by Zhang Zhongxing, are two representatives of this tendency.[11] Both works concentrate on the relations between Buddhism and Chinese literature, the former being more substantial. Shi Cun's *Yinming shuyao* is an introduction to Buddhist logic.[12] Although this is a small book, it is quite useful, even for advanced students. Similarly, another small book, *Fayuan tancong*, written by an elder statesman of Buddhism, Zhou Shujia, is also a delightful work.[13] It is aimed at popular readers. It touches upon various topics of the Chinese Buddhist establishment, such as monastic architecture, management, rites, pilgrimage centres, arts and monuments, scriptures, popular *bianwen*, arhats and Chinese Buddhist priestly dress, etc. The great value of this book is its reliability and accuracy. A number of articles and books on Buddhism in Tibet and Inner Mongolia have been published. Wang Furen's *Xizang Fojiao shilue* is one such book.[14]

28

Apart from the books on Buddhism, there are other books on Chinese philosophy, which include some substantive contributions to Buddhist studies. Of these, the two series of *Zhongguo gudai chuming zhexuejia pingzhuan*, volume 2, and its supplementary (*xubian*) volumes 2-3, make good contributions to the studies of some eminent Buddhist thinkers.[15] Feng Qi's book on the development of logic in ancient Chinese philosophy, *Zhongguo gudai zhexue de luoji fazhan*[16] may be regarded as another good example in this direction. Most of the second volume of this work concentrates largely on Buddhist and Neo-Taoist contributions. Like most of the historical surveys, this work covers a long period of history, from the beginning of Chinese philosophy until the thought of Dai Zhen (1723-1777). As a consequence of this broad coverage, there is no discussion of the characteristics of Buddhist logic and its special contribution to Chinese philosophy. Nevertheless, since there are only a few works published on this subject, the publication of Feng's work is still very valuable.

There are a few modernized Chinese Buddhist scriptures that have come out during the last few years. Of these, the publication of the new edition of *The Buddhist Record of the Western World*, edited by Ji Xianlin and his colleagues, is the best example.[17] This new edition includes a 138 page introduction by Ji. The work is a joint project by a number of eminent scholars such as Fan Xiangyong (for textual collations), Yang Tingfu (for technical terms), Zhang Guangda and Geng Shiming (for Central Asian history), Zhu Jieqin and Zhang Yi (on Indian history), Jiang Zhongxin (on Sanskrit and Pali terms): and Ji himself directed and shared the work throughout its progress. This is the first scholarly and modernised Chinese publication of this important historical record on Buddhism, India and Central Asia which is carefully and expertly collated, annotated, punctuated, indexed with references from Chinese, Indian, Japanese, and European research works. One would be happy if more classical texts were published in this scholarly manner. This work has a sister volume, *Datang xiyuji jinyi*, which is a modern Chinese translation of the Buddhist record.[18] The work is carried out under the direction of Ji and his colleagues from Beijing University and the Institute for South Asian Studies in the Chinese Academy of Social Sciences. This is another high quality and scholarly production. Other works of a similar nature are Daoxuan's (596-667) *Shijia fang zhi*, ed. by Fan Xiangyong,[19] *Tanjing jiaoshi* by Guo Peng,[20] *Hua yan jinshizhang jiaoshi* by Fang Litian,[21] and *Wudeng huiyuan* in 3 volumes edited by Su Yuanlei.[22] The works by Fan and Su are punctuated, edited with annotations, whereas the other two have some additional works. More works of this kind are expected in the near future as the aforementioned works belong to a series of publications. Of these serial publications, the series of *Zhongguo Fojiao dianji xuankan* [23] directly focuses on the subject, and *Zhongwai jiaotong shiji congkan* [24] includes the works related to Buddhism, Islam and Christianity.

Shijie zongjiao yanjiu still stands out as the leading periodical in the academic study of religions; and Buddhism also remains the leading subject as far as the number of articles is concerned. Some studies published in this periodical are quite solid research, some with new material. Other studies may not be so new to international scholarship but quite new to the scholars in China. The *Fayin* [25] published by the Buddhist Association of China has been expanded from a quarterly into a bi-monthly periodical, but the space for scholarly research is still very limited. It seems that the scope will remain the same as this is the official organ of the Association. News and policy directions along with short studies on Buddhism run side by side.

Compared with Buddhist works, the recent publications on religious Taoism still look weaker in number. Recent work, significant to the subject, is led by Wang Ming and Qing Xitai. Wang's *Daojia he Daojiao sixiang yanjiu* , is a collection of his published papers, past and recent, along with a few unpublished papers.[26] The study of *Taiping jing, Yinfu jing, Huangting jing* , as well as papers on Ge Hong (283-363 A.D.), Dao Hongjing (456-536 A.D.) and legendary Li Hong, are essential to scholars of Taoist religion. It is, therefore, a very useful and convenient volume. After a delay of some time, Wang Ming's *Baopuzi neipian jiaoshi* , revised edition has been published.[27] A number of new editions have been added to the list of collations, and this work seems to be one of the standard editions. As this work is part of a large project, *Xinbian zhuzi jicheng* (New Editions of Collected Works of the Philosophers),[28] more works on Taoist texts can be expected in this series.

The second volume of *Zhongguo Daojiao sixiang shigang* , by Qing Xitai[29] was published after a period of delay. Because of his preoccupation with other research projects, this volume is a collective accomplishment. Apart for Qing, his colleagues Ding Yizhuang, Zhao Zongcheng, Zeng Zhaonan and others, also took part in the second volume. This volume covers the period from Sui to the Northern Song period, and involves many new topics.

The volume comprises three chapters (5th-7th) on the relationship between religious Taoists and the politics of these dynasties, the doctrinal development and inter-religious relations with Confucianism and Buddhism. Of these chapters the one on doctrinal development is more substantive and has, rightly so, the largest share of the space: about one third of the volume (pp.514-744). The Taoist thinkers discussed here include Sun Simiao (ca. 581-682 A.D.), Cheng Xuanying (7th century), Wang Xuanlan (626-697), Sima Chengzhen (646-735), and another seven Taoists who lived during that period. As far as the philosophical development is concerned, the book contributes new knowledge to the understanding of the tradition. The non-philosophical aspects, many of which are the main achievements of the religion, are outside the scope of this new book. One would like to see Qing's new project, the history of Taoism in general, materialise in the near future.

The two major compilations reported in the previous survey, namely *Daozang tiyao* and *Daojia jinshi lu*, are not yet on the market. Like some other academic publishing programmes, their appearance seems delayed by the limited printing capacity.

A recent book by Cheng Yishan entitled *Zhongguo gudai yuanqi xueshuo* [30] focusses on the concept of *yuanqi* or "primordial breath". Since the concept is one of the principal ideals in Taoist cultivation for immortality, the Taoist contribution should be an essential part of the book. However, at the time this survey was taken the author had not received the book and therefore, was not in a position to make a judgment on it.

Among the periodicals which made major contributions to Taoist studies, *Daoxue huikan*, the official organ of the Taoist Association of China and *Zongjiao xue yanjiu*, published by the Institute for Religious Studies of Sichuan University, are still the main publications. *Shijie zongjiao yanjiu* mentioned above has also published a substantial number of articles on various topics of Taoism. Apart from these specialized periodicals, some university and provincial publications often turn out good articles in religious studies in general, and Taoism in particular. Those periodicals which contained research articles on Taoism during the last two years,[31] include the Journals from Normal University of Hebei, Normal University of Tianjin *Sixiang zhanxian, Zhongguo zhexueshi yanjiu, qiusu.*

2

It is a pleasure to see that the Chinese study of religions has expanded its scope and deepened its research. Apart from the

31

publications on Buddhism and Taoism, some works in other directions have been revived. Of the established religions, books on Islam and Christianity have come to notice, although the number is still marginal. Studies on the religious faiths of minority nationalities of China, such as Shamanistic studies, have made remarkable progress. Other topics, such as atheism, have also received scholarly attention. The debates on religion in a socialist society are a new feature among scholarly discussions. In comparison with the past, recent discussions on the current situation of religion in a socialist society are comparatively free and scholarly. Although some of the writers still use Marxist terminology and lean heavily towards classical Marxist views on religion, they are freer and more spontaneous than the debates in the past. Ancient Chinese religion is another subject that has attracted some scholarly attention. All these new developments are welcome features for the study of Chinese religions.

Among the books in Islamic Studies in China, selections from the articles presented at three conferences are representative of current Chinese scholarship in the field. The books of *Yisilanjiao zai Zhongguo* or "Islam in China", cover the spread and development of the religion in China.[32] The other two conferences focussed on more limited topics, namely, Islam in China during the Qing Dynasty; and the conference on Islam in Quanzhou , an important seaport in the East-West trade during the medieval period.[33] Works published by individual scholars or by the research institutes have also revived. Some of these works are general discussions and some on specific topics. Li Ximi's historical survey on Islam,[34] the book on the Shi'ah sect compiled by the Institute for Research on World Religions,[35] as well as the translation of a Syrian work on the story of the Kuran,[36] are representative works in this category. Scholarly publications by individual scholars include Ma Tong's work on the history of relations between Islam and Chinese aristocratic systems:[37] and *Zhongguo Yisilan shi cungao* by Bai Shouyi,[38] the well-known veteran expert in the field. The former touches an important aspect of Chinese religious history in general, and Islam in particular. Unless one knows the aspects of medieval Chinese society, many problems related to the ups and downs of certain religions would not be clear. The publication of Ma's work, therefore, is a welcome addition, even though it is only an outline history. Bai's book is largely a reprint of his early books and articles on Islamic history, originally published during the forties and fifties. Some of these early books or papers will stand as the standard Chinese reference on the subject. It is very convenient to have them in one volume. Apart from Bai's own writings, the book also contains a collection of six articles or statements by other scholars or religionists. They also are important material for historians of the Islamic tradition in China. Pang Shiqian's article on Islamic education, Zhao Zhenwu's account of the 30 years of Chinese Muslim culture, and the autobiographical account of an eminent Chinese Muslim named Wang Jingzhai are included in the appendix. Fewer books are published on Christianity and most of them are translations. A history of Christianity in China before 1550 A.D., translated by He Zhenhua,[39] and another book on Martin Luther's Reformation by Li Pingye[40] are samples of this effort. Only one original Chinese work on the subject has been published in the last two years, the book by Jiang Wenhan. This is a study of Christianity in pre-modern China, and of the Jews in Kaifeng. A periodical entitled *Zongjiaoxue* was published by the Nanjing Theological Seminary, but had not been seen by the author at the time of this survey.

Compared with other subjects, Chinese studies on the religious life of the minority nationalities are very significant. A number of field works on the tribes and nationalities in southwest China, especially Yunnan and Sichuan provinces, has been published, and most of these publications include religious life as an important component. Besides these general surveys, there are a few books now available to general readers. Two are good illustrations of this achievement. One of them is a study of Shamanism, edited by Qiu Pu.[41] It is a collective work by a group of scholars from the Institute for Nationality Research (Minzu Yanjiusuo) in the Chinese Academy of Social Sciences. The book comprises eight chapters, plus an introduction and a conclusion. The subjects discussed in the book include the origin of religion, the cults of nature, totems and ancestral worship; the system and institutions of shamans, the imprint of social class as well as comparisons between shamanism and ancient Chinese religion. A larger and richer book has been published in Yunnan province, called *Zhongguo shaoshuminzu zongjiao chubian* , edited by Song Enchang.[42] The book is a collection of 46 articles. It covers surveys and studies of the religious lives of all major nationalities except the Chinese (Hanzu). Geographically, it covers the subjects from Northwest China to Tibet in the southwest; from Inner Mongolia to China's southernmost border with Burma and Thailand; and from Xinjiang to Guangxi Zhuang Autonomous Region. This, as well as the other book mentioned above, will enrich our knowledge of shamanism in China and its relation to Chinese religions, as well as of tribal beliefs. A careful study of the old and new materials presented in the two works would also help readers to understand ancient and folk religions among the Han people, as there are parallels and historical connections between these traditions. Some puzzling problems in ancient Chinese beliefs still seem to remain as living factors in the primitive and tribal society, and some anthropological problems in the history of religions cannot be fully understood through rational views as recorded in the book of history and philosophy. As the book itself indicates that the collection is the first in this series (*chubian*), more valuable information can be expected in future publications. Among the books on comparative studies of tribal religion, Zhao Lu's study on the myths of the Bai tribe and its relation with Tantrism is an interesting study.[43] Unfortunately, the book was not available to this author at the time this paper was being written.

Compared with studies of primitive religions, the books on ancient Chinese religion are much fewer. Only one book by Zhu Tianshun called "A Preliminary Research of Ancient Chinese Religion" has been published.[44] Although the book deals with various aspects of Chinese religious beliefs from archaic to the Qin-Han periods, it is suitable neither for general readers nor for scholars. It is rather technical for general reading. A lack of references to scholarship in the field and footnotes on documentation, makes the book less useful to the experts in the field. It is a pity that such rich sources as various archaeological reports published during recent years are almost ignored by the authors. Nor is there a comparison with tribal religions, as mentioned in the preceding paragraphs. It seems that Chinese scholars are, at least at the moment, pre-occupied with the 'Great Traditions', whereas the study of early Chinese religion receives much less attention.

The studies of atheistic beliefs in China are mainly contributed by Wang Yousan. He has published an outline history of atheist tradition in China, has edited a conference volume and compiled a collection of ancient Chinese material on the subject.[45] One would expect that a topic, such as atheism, would receive more favourable attention among scholars in a socialist country. These findings prove that it is not the case.

Among the general histories of Chinese philosophy, which include religious thought as part of the coverage, the works by Feng Youlan and a group of historians lead by Ren Jiyu are outstanding examples. The first volume of Feng's revised version of his monumental "New History of Chinese Philosophy",[46] covers religious thought during Shang and Zhou as well as the transformation from religion to philosophy during Chunqiu and Zhanguo times. The second volume concentrates on the Classic period; and the third on the thought of the Han dynasty. Some chapters are relevant to religious philosophy, especially the chapter on the peasant rebellion and *Tai ping jing*. Ren and his colleagues have published two of their projected seven-volume book, *Zhongguo zhexue fazhan she*.[47] The first volume covers ancient Chinese thought during the pre-Qin times, the second focusses on the Qin and Han periods. As far as the periods and scope of this book and Feng's history are concerned, they are identical in coverage. At the same time the two books are different. Feng's book stresses the principal trends and his own reflections on the philosophy, but rarely refers to other scholars' opinion on the subject; whereas Ren's book is more focussed on the development of Chinese philosophy as well as scholarly opinions and researches on the subject. The chapters on the transformation of religious thought from Shang to Zhou in the first volume, and the religious philosophy in the *liji* the theology of Dong Zhongshu and Baihu Tong, the development in medical philosophy as well as the transformation from the late Taoist philosophy to religious Taoism, are good chapters for students of Chinese religions. The reference to other researches, and indexes of names and titles found in the book, are a welcome new feature of Ren's book.

A number of books on religion at large have been published, but most of them are for the general public rather than for academic research. The books by Huaizhu, Yan Zhenchang and Zhang Wenjian are good examples.[48] There is a book of more serious nature under the title of *Zongjiao, Kexue, Zhexue* ("Religion, Science and Philosophy").[49] The book was compiled by the theoretical research team of the Institute for Research in World Religions and published from Henan in 1982. However, this author has not yet seen a copy, nor has any book review of it been available. Under normal situations, the book should stand as an authoritative guide to most religionists on the mainland, but the mood among Chinese intellectuals has changed fast during the last few years. Therefore, no one knows where it stands now. If one wishes to understand the mood among scholars in religious studies, one has to look to periodical articles for some indications.

The indications of a new direction in religious studies in China may be seen from a Symposium held in Beijing, during December, 1984. The Symposium commemorated the 20th anniversary of the founding of the Institute for Research of World Religions. Ren Jiyu, the Director of the Institute, addressed the Symposium and called for new research in religious studies in accordance with the new historical period. He admitted that in the past, Chinese studies in religion had been affected by 'Leftism' and had thus failed to discuss some very important problems. He also stated that simplistic copying of Marxist theory does not work. One has to study new problems, make new conclusions, produce substantial research works. These things are possible only if scholars dare to search for the new, and advance with courage. Among the problems that need scholarly attention, Ren has listed a number of questions. In the past most people said that religions supported the ruling class and thus became the ideology of society. Can one now say that religion is the superstructure of the socialist society? Again, it was often said that religions possess a high degree of class characteristics. Is this true when the old ruling class has now been eliminated in the socialist society? What is the place and function of religion in a socialist society? Is there a socialist religion or not, etc?[50]

Substantive discussion on some problems raised by Ren were produced in the next year. Some discussants have presented their views and some authors have impressed readers by their works, namely, Jiang Wenxuan from Guangxi, Xiao Zhitian and Yan Boming from Shanghai, and Zhang Ji'an from Beijing.

Jiang discussed two basic problems: 1) the root of the existence of religions in a socialist society, 2) conditions and reasons for the existence of religions.[51] Jiang argues that the existence of religions in a socialist society is a fact deeply rooted in the long history of the country. Some of the reasons include an inadequate supply of the material needs of the people, the complexity of the class struggle within and without society, and man's dream for salvation. He branded the last item as an epistemological problem. From this historio-social analysis, Jiang concludes that religions inevitably exist in China. He further pointed out that religion will not merely hold its position, but even development is possible. This possibility is based on the following conditions: a) religions continually reform themselves and are thus useful to Chinese society; b) there are various weaknesses in the application of Chinese government policies which provide conditions for religious influence. Jiang has provided some new materials to support his argument. In one place he used the statistics of a certain town which lists 39 new recruits to Christianity during August, 1984. Of these 39 new recruits, 12 are male and 27 are female. Their occupational classification is as follows: 24 workers (including 5 retired), two teachers (one retired), 5 domestic workers and 8 unemployed youths. The number from the working class is predominant and the time of recruitment is quite recent. This fact seems more convincing than any of the ideological cliches.

Xiao's article on "The Problem of Coordination between Religion and the Socialist Society in China"[52] comprises four parts: the definition of terms, the bases and conditions for coordination, the process and demonstration of coordination, and continuous efforts for coordination to promote the 'four modernizations.' The article points out that the cooperation between religious organizations and the Chinese government during the last four decades has already proven that coordination between the two is not only possible, but a reality. The unity of religious believers and the people who do not believe in religion is another fact. Based on socialism and patriotism, some of the religious teachings of the doctrine — faith, morality and practice — are concurring with socialist life. The article considers that the correction of leftist policy on religion is the most urgent problem of today.

Another veteran scholar of Chinese philosophy, Yan Beiming, also discusses the problem of religious morality in the socialist period.[53] Yan considers that moral restraint, taught by most religions, has a great control over its believers. Any study of the motives and effects of religious morality must be based on the analysis of whether actual benefits are received; empty words alone cannot explain any problem. In Yan's view, the principal contradiction in today's China no longer is the conflict between religion and society. In fact, the teaching to do good and to eliminate evil, to observe the law and obey the discipline, etc., by various religious traditions, not only is a good motivation, but also has a good effect on society at large. He regards these compatibilities to be bases for coordination between religion and socialism. He condemns the artificial and simplistic classification of theism versus atheism, or science versus superstition, etc., and considers them to be extreme views of the leftists. He concedes that religious morality is lower than communist morality. But this will pass out in the future. Yet at the same time, the end of religion passes through the process of development and selective elimination during the stage of socialism. He considers that Communist morality does not include religious morality nor is the latter a supplement to the former. Under the circumstances, religion is still capable of playing a positive role in today's China.

Zhang Ji'an has made a preliminary enquiry as to the place of religion in the socialist period.[54] The article starts with the classic Communist view of religion as stated by Marx, Engels and Lenin; and then discusses the transformation of religion in China. He points out that political support of socialism, religious reformation after the establishment of the People's Republic and the socialist reformation in the economic system of the country are the main and important points of religious transformation. In the subsequent sections of the article, the author continues his enquiry into the social roots of the existence of religion in a socialist society and asks whether religious thought could become a component of the superstructure of socialism. The final question concerns the basic religious policy of the Chinese government in the socialist period. As far as the social roots for the existence of religion are concerned, this paper concurs with others. Can religious thought become a component of and merge with socialism? The answer is 'no'. Because the author of the paper considers all religious ideas to be based on the "old tradition", although with minor modifications here and there, the basic stand has never changed. Socialist ideas are aimed at establishing a new social structure through the destruction of the old establishment. The basic positions of religion and socialism are, therefore, incompatible. He argues that freedom of religious faith is a basic policy in a socialist society. This is a fact one should respect and the situation cannot be ignored. He agrees that there will be a contradiction between the religious believer and those who do not believe in religion; yet he considers this contradiction to be secondary because both sides concur in political and economic interests. What should the basic policy on religion be?

The author of this paper proposes the following points: a) freedom to choose or not to choose a religion; b) the government should support all normal religious practices, but be watchful and suppress unlawful activities performed under the cover of religious practice; c) encourage religious organizations and individuals to play positive roles in public works; d) analyse religious phenomena from the Marxist viewpoint, and thus educate the youth; e) criticize religious ideas that are incompatible with Marxist views through scientific analysis and research, but at the same time, never handle the treatment of criticism in a stormy manner as if it were a political campaign, nor unlawfully harm other people, as has occurred during the Cultural Revolution.

3

When the study of religion in China during the past two or three years is reviewed as a whole, some impressions become clear. In the first place, the Chinese contribution to the study of Buddhism still overshadows other areas. But the scope of coverage is much enlarged. A number of works of Islamic research are a welcome feature, while the study of Christianity is much less prominent. The most valuable contributions are the publications on minority religious beliefs. These new works provide worthwhile reading for scholars in the area of religious studies. In the judgement of this author, this kind of publication is significant not only for the understanding of tribal religion, but also to give more insight into ancient Chinese religion as well as other related subjects such as anthropological, sociological and comparative enquiries. The discussion of the place and the future of religion in a socialist society is a new development in Chinese studies of religion. Whereas there is a general consensus on the existence of religious faiths in a socialist society, scholars are rather cautious about the future of religion. However, all participants are agreed about the value and contribution of religion to the socialist construction at the present. In other words as far as the present stage of Chinese society is concerned, the evaluations are positive. This clearly indicates that the extreme leftist views on religion can no longer hold their ground in academic circles. However, even among the scholars who have engaged in the current discussions, one still finds the distinction between "soft" and "hard" expressions toward religion. Nevertheless, the study of religion is no longer regarded as a 'black cat' in China.

NOTES

1 Jan, "The Religious Situation and the Studies of Buddhism and Tao-ism in China: An Incomplete and Imbalanced Picture", *Journal of Chinese Religions*, No. 12 (1984), 37-64.

2 For an example, Barbara Hendrischke, "Chinese Research into Daoism after the Cultural Revolution". *Asiatische Studien Etudes Asiatiques*, 38 (1984), 25-42,

3 See "*Zhonghua dazangjing (Hanwen bufen) gailun* 中 華 大 藏 經 (漢 文 部 分) 概 論 ", *Shijie zongjiao yanjiu* 世 界 宗 教 研 究, 1985/4, 1-4. Hereafter, this journal will be abbreviated as *SZY*.

4 *Zhongguo Fojiaoshi* 中 國 佛 教 史, vol. 2 (Beijing: *Zhongguo shehui kexue chubanshe*, 1985).

5 *Hanwei liangjin nanbeichao Fojiaoshi* 漢 魏 兩 晉 南 北 朝 佛 教 史 : *Tang yongtong lunzhuji zhiji* 湯 用 彤 論 著 集 之 一 2 vols, Beijing: Zhonghua shuju, 1983.

6 Shigeo Kamata 鎌 田 茂 雄, *Chugoku bukkyo shi* 中 國 佛 教 史, 3 vols. Tokyo: Iwanami shoten, 1982, 1983, 1984.

7 Lou Yulie 樓 宇 烈, ed., *Zhongguo Fojiao sixiang ziliao xuanbian* 中 國 佛 教 思 想 資 料 選 編 Beijing: Zhonghua shuju, 1983.

8 Fang Litian 方 立 天, *Huiyuan jiqi Foxue* 慧 遠 及 其 佛 學 Bei-jing: Zhongguo renmin daxue chubanshe, 1984.

9 *Zhongguo Foxue lunwenji* 中 國 佛 學 論 文 集 Xi'an: Shaanxi renmin chubanshe, 1984.

10 Sun Changwu 孫 昌 武, *Tangdai wenxue yu Fojiao* 唐 代 文 學 與 佛 教. Xi'an: Shaanxi renmin chubanshe, 1985.

11 Zhang Zhongxing 張 中 行, *Fojiao yu Zhongguo wenxue* 佛 教 與 中 國 文 學. Hefei: Anhui jiaoyu chubanshe, 1984.

12 Shi Cun 石 村 *Yinming shuyao* 因 明 述 要. Beijing: Zhonghua shuju, 1982.

13 Zhou Shujia 周 叔 迦, *Fayuan tancong* 法 苑 談 叢. Beijing: Zhongguo Fojiao xiehui, 1985.

14 Wang Furen 王 輔 仁, *Xizang Fojiaoshi lue* 西 藏 佛 教 史 略. Xining: Qinghai renmin chubanshe, 1983.

15 *Zhongguo gudai chuming zhexuejia pingzhuan* 中 國 古 代 著 名 哲 學 家 評 傳 Jinan: Qilu shushe, 1980; and 續 編 Xubian, 1982.

16 Feng Qi 馮 契, *Zhongguo gudai zhexue de luoji fazhan* 中 國 古 代 的 邏 輯 發 展 vol. 2, Shanghai: Renmin chubanshe, 1984.

17 *Datang xiyuji jiaozhu* 大 唐 西 域 記 校 註 by Ji Xianlin 季 羨 林 *et al.* Beijing: Zhonghua shuju, 1985.

18 *Datang xiyuji jinyi* 大 唐 西 域 記 今 譯 by Ji Xianlin 季 羨 林 *et al.* Xi'an: Shaanxi renmin chubanshe, 1985.

19 *Shijia fangzhi* 釋 迦 方 志 edited by Fan Xiangyong. Beijing: Zhonghua shuju, 1983.

20 Guo Peng 郭 朋, *Tanjing jiaoshi* 壇 經 校 釋. Beijing: Zhonghua shuju, 1983.

21 Fang Litian 方 立 天, *Huayan jinshi zi zhang jiaoshi* 華 嚴 金 獅 子 章 校 釋. Beijing: Zhonghua shuju, 1983.

22 *Wudeng huiyuan* 五 燈 會 元 ed. by Su Yuanlei 蘇 淵 雷. 3 vols.

Beijing: Zhonghua shuju, 1984.

23 *Zhongguo Fojiao dianji xuankan* 中國佛教典籍選刊 published by the Zhonghua shuju.

24 *Zhongwai jiaotong shiji congkan* 中外交通史籍叢刊 published by the same publisher as mentioned above.

25 *Fayin* 法音 .

26 Wang Ming 王明 , *Daojia he Daojiao sixiang yanjiu* 道家和道教思想研究 . Beijing: Zhongguo shehui kexue chubanshe, 1984.

27 *Baopuzi neipian* 抱朴子內篇 ed. by Wang Ming. Beijing: Zhonghua shuju, 1985.

28 Xinbian zhuzi jicheng 新編諸子集成 .

29 Qing Xitai 卿希泰 , *Zhongguo Taojiao sixiang shigang* 中國道教思想史綱 . Chengdu: Sichuan renmin chubanshe, 1985.

30 Cheng Yishan 程宜山 , *Zhongguo gudai yuanqi xueshuo* 中國古代元氣學說 , 1986.

31 For discussions on these periodical papers, see *Zhongguo zhexue nianjian* 中國哲學年鑑 , 1983, 563ff.; 1984, 155, 166ff., 584; and 1985, 144-155, 476f., 483f.

32 *Yisilanjiao zai Zhongguo* 伊斯蘭教在中國 . Yinchuan: Ningxia renmin chubanshe, 1983.

33 *Quanzhou Yisilanjiao yanjiu lunwen xuan* 泉州伊斯蘭教研究論文選 . Fuzhou: Fujian renmin chubanshe 1983.

34 Li Ximi 李希泌 *Yisilanjiao shihua* 伊斯蘭教史話 . Beijing: Shangwu Yinshuguan, 1983.

35 *Shiyepai* 十葉派 Beijing: Zhongguo shehui kexue chubanshe, 1983.

36 *Gulanjing de gushi* 古蘭經的故事 . Beijing: Xinhua shudian, 1983.

37 Ma Tong 馬通 , *Zhongguo Yisilan jiaopai yu menhuan zhidu shilue* 中國伊斯蘭教派與門宦制度史略 . Yinchuan: Ningxia renmin chubanshe, 1983.

38 Bai Shouyi 白壽彝 , *Zhongguo Yisilan shi cungao* 中國伊斯蘭史存稿 . Yinchuan: Ningxia renmin chubanshe, 1983.

39 *Yiwuwuling nian qian de Zhongguo Jidujiao shi* 一五五〇年前中國基督教史 translated from the English by He Zhenhua. Beijing: Zhonghua shuju, 1984.

40 Li Pingye 李平曄 , *Ren de faxian: Mading Lude yu zongjiao gaige* 人的發現：馬丁路德與宗教改革 Chengdu: Sichuan renmin chubanshe, 1983.

41 Qiu Pu 秋浦 , ed. *Shamanjiao yanjiu* 薩滿教研究 . Shanghai: Renmin chubanshe, 1985.

42 Song Enchang 宋恩常 *et al., Zhongguo shaoshu minzu zongjiao: chubian* 中國少數民族宗教：初編 . Kunming: Yunnan renmin chubanshe, 1985.

43 Zhao Lu 櫓 , *Lun baizu shenhua yu mijiao* 論白族神話與密教 . Beijing: Zhongguo minjian wenyi, 1985.

44 Zhu Tianshun 朱天順 , *Zhongguo gudai zongjiao chutan* 中國古

代 宗 教 初 探 . Shanghai: Renmin chubanshe, 1982.

45 Wang Yousan 王 友 三 , *Zhongguo wushenlun shigang* 中 國 無
神 論 史 綱 Shanghai: Renmin chubanshe, 1982. Among books
edited by him and others are *Zhongguo wushenlun wenji* 中 國 無
神 論 文 集 . Wuhan: Hubei renmin chubanshe, 1982; and *Zhongguo
wushenlun ziliao xuanbian* 中 國 無 神 論 資 料 選 編 . Beijing:
Zhonghua shuju, 1983.

46 *Zhongguo zhexuesi xinbian* 中 國 哲 學 史 新 編 (revised edition)
by Feng Youlan 馮 友 蘭 , 3 vols., Beijing: Renmin chubanshe, 1982,
1984, 1985.

47 Ren Jiyu 任 繼 愈 , *et al., Zhongguo zhexue fazhan shi*
中 國 哲 學 發 展 史 . vol. 1: *Xianqin* 先 秦 ; vol. 2: *Qinhan*
秦 漢 , Beijing: Renmin chubanshe, 1983, 1985.

48 Huaizhu 懷 竹 , *Zongjiao qianshuo* 宗 教 淺 說 . Hefei: Anhui
renming chubanshe, 1983. Yan Zhenchang 閻 鎮 閭 , *Zhongjiao
gailun* 宗 教 概 論 . Zhengzhou: Henan renmin chubanshe, 1984.
Zhang Weijian 張 文 建 , *Zongjiao shihua* 宗 教 史 話 . Changchun: Jilin renmin chubanshe, 1983.

49 *Zongjiao, kexue, zhexue* 宗 教 , 科 學 , 哲 學 . Zhengzhou: Honan
renmin chubanshe, 1982.

50 This is reported in *SZY* 1985/1, p. 155.

51 Jiang Wenxuan's 蔣 文 宣 article is in *SZY* 1985/3, 112-120.

52 Xiao Zhitian's 肖 志 恬 article is published in *Xueshu qikan*
學 術 季 刊 , the official organ of the Shanghai Academy of Social
Sciences, 1985/1, 137-146.

53 For the article by Yan Beiming 嚴 北 溟 , see *ibidem,* 1985/3, 137-
146.

54 For Zhang Ji'an's 張 繼 安 paper, see *SZY* 1985/4, 58-69.

THE RELATIONS BETWEEN RELIGION AND PEASANT REBELLIONS IN CHINA: A REVIEW OF THE INTERPRETATIONS BY CHINESE HISTORIANS

Man Kam Leung

The study of Chinese peasant rebellions and peasant wars in the People's Republic of China can be divided into three periods: 1949-1958; 1959-1966; and 1976 to the present. From 1949 to 1958 was the beginning period when the study of peasant rebellions was a new field for exploration by Chinese historians. From 1959 to 1966 was the period of searching for a theoretical understanding of the nature of the peasant rebellions in Chinese history. From 1966 to 1976 not much work was done because of the Cultural Revolution (1966-1976). After 1976, there was a revived interest in the study of peasant rebellions. In their search for a theoretical understanding of the peasant rebellions, a difficult and thorny problem for Chinese historians to solve has been the role of religion in those rebellions.

It is a well known fact that Chinese religion had played important roles in Chinese peasant rebellions. This, however, has led to a difficult problem for modern Chinese historians, namely: how to explain away Karl Marx's dictum that religion is the opiate of the people. In the 1950s religion was assigned a positive role in Chinese peasant rebellions by Chinese historians. Sun Zuomin 孫祚民, in his classic study of Chinese peasant wars, emphasized that religion had played three important roles in Chinese peasant rebellions.[1] First, religion helped the peasants to organize. In Sun's opinion, in pre-modern Chinese society, the peasantry was engaged in individual labour production and scattered in rural areas. They lacked any mass organization. Since modern political parties could not appear in a feudal society, religious organizations and secret societies became the major mass organizations which could rally the peasantry together. Religious belief could also be the spiritual force to forge the peasantry into one common belief. Sun Zuomin mentioned the Yellow Turban Rebellion (184 A.D.) as a typical example in which religion had played an important role. Sun further compared the rebellions which were not influenced by religion with those which were, and came to the conclusion that peasant rebellions without religious influence were not able to flourish quickly into nation-wide movements. A rebellion would need a prolonged period to nourish its momentum into a mass movement.

Second, Sun argued that Chinese peasants were basically passive in nature but religious mysticism could serve as a stimulant to instigate action. Religious mysticism gave the peasants a sense of invincibility and the confidence of victory. Religious mysticism, mingled with political prophecy and political rumours, sometimes managed to destabilize the political and intellectual order of an existing regime. Religious mysticism, once accepted by the peasantry, could destroy the myth and political justification of a dynasty. It further gave the peasantry a sense of spiritual freedom in their struggle for power.

Third, when a peasant rebellion failed, religion had performed the function of preserving the power and resources of the defeated by hiding its remnant members and shattered ideology in their underground network of secret societies and religious orders. Sun specifically pointed out the White Lotus Sect and those rebellions associated with it. In Sun's opinion, many peasant rebellions were instigated by the White Lotus Sect since the Yuan Dynasty (1271-1368). This sect built up a massive following and had deeply entrenched itself in the junction territories of modern day Shandong, Hebei, and Henan. From the beginning of the Yuan Dynasty to the end of the Qing Dynasty (1644-1911), the White Lotus Sect and its splinter groups repeatedly organized peasant uprisings.

Sun Zuomin then comes to the conclusion that in spite of severe suppression by the authorities, the White Lotus Sect continued to survive and continued to influence peasant rebellions. Sun's book was published in 1956 and during the period from 1959 to 1966 his views were repeatedly discussed in China. In March and April, 1960, two conferences were held in Shanghai on Chinese peasant rebellions. During the conferences, the relation between peasant rebellions and religion was heatedly debated. The debate continued to attract the attention of Chinese historians and many articles appeared in the newspapers and journals. Conferences were organized in various places to discuss this topic.[2] The enthusiasm only stopped with the onset of the Cultural Revolution in 1966.

After 1976, the debate was revived. In short, the central themes of the debate are three:[3]

1. What was the nature of the religions which were followed by the peasant rebels?

2. Did peasant rebellions have to be associated with religion in order to be successful?

3. What were the functions of religion in the peasant rebellions?

1. What was the nature of the religions used by the peasant rebels? This has been the first basic question that the Chinese historians have asked themselves in their discussion of the relations between religion and peasant rebellions. Of course, the important point here is the ways in which Chinese historians have tried to explain the Marxist dictum that religion is the opiate of the people in the context of the peasant rebellions. In addition to this theoretical requirement, historians have also to abide by the analysis of Mao Zedong that a man in traditional Chinese society has been dominated by three systems of authority: (1) the state system (political authority), (2) the clan system (clan authority), (3) the supernatural system (religious authority), and for the females, besides being long dominated by these three systems of authority, they were also dominated by the men (authority of the husband). Mao concludes:

These four authorities—political, clan, religious and male—are the embodiment of the whole feudal-patriarchal system and ideology, and are the four thick ropes binding the Chinese people, particularly the peasants.[4]

Before 1966, any explanation of the relation between religion and peasant rebellions had to be fitted into the conceptual framework of Marx and Mao. Sun Zuomin was probably one of the first to put forth the theory that there were two kinds of religions—one for the upper class and one for the lower class. An upper class religion was the religion which the ruling class used to delude the masses. The lower class religion was the heretic sect splintered from the upper class religion, which had rooted itself in the peasant masses and protected their interests, and also expressed their desires.[5]

Yang Kuan 楊寬 , in a series of articles written between 1959 and 1961 said that in traditional China, the peasantry, in order to overthrow a feudal regime, must also overthrow the religious authority of that regime. Heretic religious sects would be the intellectual weapons capable of doing so. Therefore, in Chinese history, armed peasant rebellions were often forged with heretic religions in order to be effective. Like Sun Zuomin, Yang Kuan also emphasized the importance of religion as organizational and propaganda weapons in a peasant rebellion. Yang also pointed out that in many of the peasant rebellions, the rebels put forth demands for social equality and equal shares of property. In Yang's opinions, the two concepts were the basic key doctrines in any primitive religions. The Chinese peasantry, longing for their liberation from the yoke of poverty and exploitation, had transferred these two ideals into realities by staging rebellions. Consequently peasant uprisings in China often combined an attack on political feudal power and feudal religious authority.[6]

In his article *Lun Taiping jing* 論太平經 (On the *Taiping Classic*), Yuan Kuan said that *Taiping* jing was the first revolutionary theoretical work of the Chinese peasant rebellion, because this book contains two key concepts of social equality and equalization of property.[7] However, Chen Shou-shi 陳守實 in his article *Lun Cao Wei tuntian* 論曹魏屯田 (On the land colonization of the Wei Dynasty) rejected Yang's argument that the *Taiping jing* contained any revolutionary ideas for the peasant rebellion. Instead, the ideals of social equality and equalization of property were the reflection of the legacy of the early land tenure system in human history—the primitive commune system—when everything was shared equally by each member of the commune. Chen agreed that religion had been used as an organizational weapon by the peasantry at the early stage of a rebellion. Once the rebellion started, religion ceased to play an important part when the struggle turned political. In Chen's view, peasant wars and religious superstitions are basically antagonistic.[8]

In 1980, Zhang Zuoyao 張作耀 took a view compromising between Chen and Yang, saying that during the Yellow Turban Rebellion, the leader Zhang Jiao did take some ideas from the *Taiping jing* to form his Taiping Religion. Zhang further agreed with Yang that the ideas of social equality and equalization of property reflected the demands and desires of the peasants in an agricultural society. Zhang also agreed that there were two kinds of religions in China—religions of the ruling class and the heretic religions which became the spiritual pillars of strength for the peasant rebellions.[9]

2. Did peasant rebellions in China have to be associated with religion in order to be effective or successful? Shao Xunzheng 邵循正 pointed out as early as 1961 that those historians who argued that there were two kinds of religion, one upper and one lower class, had misunderstood Friedrich Engels' words on the subject.[10] Engels says:

> This supremacy of theology in the realm of intellectual activities was at the same time a logical consequence of the situation of the church as the most general force coordinating and sanctioning existing feudal domination.
>
> It is obvious that under such conditions, all general and overt attacks on feudalism, in the first place attacks on the church, all revolutionary, social and political doctrines, necessarily became theological heresies. In order to be attacked,

46

existing social conditions had to be stripped of their aureole of sanctity.[11]

According to Shao, in the above passage, Engels never meant that there was a religion for the peasants, and a religion for the landlords. Shao also pointed out that what Engels discussed were the conditions in Germany. Chinese historians should not use his argument to apply to Chinese situations in their explanation of peasant rebellions. Shao asked the following questions:

> Chinese culture, the Chinese educational system, even the Chinese political and legal doctrines were not dominated and monopolized by any religious church. None of the Chinese peasants were Catholics. Then why did the revolutionary doctrines of the peasantry have to be heretic religions? Why did the revolutionary passions of the peasant masses have to be nourished by religions?[12]

Shao concluded that the target of the Chinese peasant rebellions was not the religious authority of the regime, but its political power. For this reason, many major peasant rebellions in China did not have connections with any religion. Shao's view was shared by many others. For example Xiong Deji 熊 德 基 commented that there had never been any religious texts written specifically for peasant revolutionary theories in China, nor were there any religious doctrines devoted to the revolutionary ideals for Chinese peasant rebellions. It was only after a peasant rebellion had started that the rebels used a religious text and its ideals to justify their cause. There had never been any intrinsic relations between peasant rebellion and religion.[13]

3. What were the functions of these religions used by the peasant rebels? As for the functions of religion in the peasant rebellion, Chinese historians differed. Some said that religion had played a positive role in Chinese peasant rebellion, but some historians disagreed. Rong Sheng 戎 笙 , Long Shengyuan 龍 盛 遠 , and He Lingxiu 何 齡 修 said that religion did play a significant part in peasant rebellion by helping them to organize into a fighting force and to instill into them a fighting ideology. However, religion, by its very nature, is still a superstition, and a passive ideology. Religion could not give the peasants a correct interpretation of the political realities. In the end, all political prophecies and religious pretexts would be exposed in the face of the harsh realities of political power struggle. These negative aspects of religion would become more evident when a peasant rebellion was successful and victorious. The leaders, drunken by victory and hypnotized by their own brand of religion, deluded themselves about being in-

47

vincible and infallible.[14] Yang Kuan said that:

Revolutionary belief and revolutionary bravery based on religious superstition were not solid and therefore unreliable. Under the cruel trial of struggle and battle, people would gradually lose their belief in religion. If there were no further realistic and specific political goals to put forth in a rebellion the masses would lose their confidence in the rebellion, its own strength. As a result, people would be downhearted and the rebellion would fail.

A revolutionary leader used religion to organize and to lead his people. In the process of doing so, he would create a personality cult. His followers would then worship his personal power and authority. If some peasant leader oversold his religious power and used this power to aggrandize himself, the religion as a weapon of instigation and of organization for revolution would fail. Religion would eventually become the poison to destroy the revolution.[15]

Yang Kuan had the events of 1856 in the Taiping Rebellion (1850-1865) in mind. In that year, there was a power struggle among the Taiping leaders. Both Hong Xiuquan 洪秀全 (1813-1864) and Yang Xiuqing 楊秀清 (d.1856), claimed that they could communicate with God and the result was a blood bath between these two spokesmen for God and their individual followers. Although Hong was the winner, the purge of 1856 had fatally wounded the strength of the Taiping Rebellion.

Perhaps because of the conflicting impact of religion on the Chinese peasant rebellion, Zhang Zuoyao came to the conclusion:

As for the function of religion in Chinese peasant wars, if we want to have a general summary, we can say that its negative and "opiate" effect has far outweighed its positive result. Although religion has helped promote the outbreak of peasant uprisings, it has also obstructed the development of peasant wars. Sometimes religion even pushed peasant revolutionary wars to a blind alley and death.[16]

In summary, it is very evident that modern Chinese historians, in their study of the relation between religion and peasant rebellion, have been required to conform to Marxist and Maoist ideological guidelines in their analysis. However, changes are coming. Zhao Fusan 趙復三 , vice-president of the Chinese Academy of Social Sciences, published an article in the May 1986 issue of *Social Sciences in China* (Chinese edition) entitled *How to Understand the Essence of Religion* .[17] In this article, for the first time since 1949, a scholar and also an official of the government, openly questions the common interpretation of the Marxist dictum that "religion is the opiate of the people." In Zhao's view, what Marx really meant was that religion, like opium, was a "painkiller" for those who sought solace in times of pain, suffering and despair. There has never been such a connotation that the ruling class had used religion to delude and to pacify the people. Zhao also points out that opium was used as a pain-killer in Marx's time. Opium as a recreational and addictive drug came in later days. Since the Opium War (1839-42), Zhao says, the Chinese people have understood opium only as an addictive drug and therefore interpreted Marx's words accordingly.

Zhao's view represents more or less the official view of the Chinese government towards religion at the present time. We can assume that new interpretations of the role of religion in Chinese peasant rebellions will be forthcoming in the People's Republic of China in the near future.

NOTES

1 Sun Zuomin 孫 祚 民 , *Zhongguo nongmin zhanzheng wenti tansuo* 中 國 農 民 戰 爭 問 題 探 索 (Shanghai, 1956), 79-87.

2 Wenhui bao bianji bu 文 滙 報 編 輯 部 , Shanghai lishijie guanyu Zhongguo nongmin zhanzheng yu zongjiao guanxi wenti de taolun 上 海 歷 史 界 關 於 中 國 農 民 戰 爭 與 宗 教 關 係 問 題 的 討 論 in Shi Shaobin 史 紹 賓 ed., *Zhongguo fengjian shehui nongmin zhanzheng wenti taolunji* 中 國 封 建 社 會 農 民 戰 爭 問 題 討 論 集 (Beijing, 1962), 480-485; see also a list of these articles on pp. 515-527. Shi's book is hereafter cited as Shi's *Wenti taolunji*.

3 *"Lishi yanjiu"* bianji bu 歷 史 研 究 編 輯 部 , *Jianguo yilai shixue lilun wenti taolun juyao* 建 國 以 來 史 學 理 論 問 題 討 論 舉 要 (Shandong: Jinan, 1983).

4 Mao Zedong, "Report on an Investigation of the Peasant Movement in Hunan," *Selected works of Mao Tse-tung*, I (Beijing, 1967), 44.

5 Sung Zuomin, 75-79.

6 Yang Kuan 楊 寬 , "Lun Zhongguo nongmin zhanzheng zhong geming sixiang de zuoyong ji qi yu zongjiao de guanxi 論中國農民 戰爭中革命思想的作用及其與宗教的關係" in Shi's *Wenti taolunji*, 321-339.

7 *Xueshu yuekan* 學 術 月 刊 , No. 9 (Sept. 1959), 26-34.

8 *Xueshu yuekan* 學 術 月 刊 , No. 2 (Feb. 1960), 56-62; 67, especially p. 67.

9 *Jindai shi yanjiu* 近 代 史 研 究 No. 2 (1980), 98-115.

10 Shao Xunzheng 邵 循 正 "Mimi shehui zongjiao he nongmin zhanzheng 秘 密 社 會 ，宗 教 和 農 民 戰 爭 " in Shi's *Wenti taolunji,* 369-384.

11 Friedrich Engels, *Peasant War In Germany* (New York, 1926), 52.

12 Shao's article in Shi's *Wenti taolunji*, 371.

13 Xiong Deji 熊 德 基 "Zhongguo nongmin zhanzheng yu zongjiao ji qi xiangguan zhu wenti 中 國 農 民 戰 爭 及 其 相 關 諸 問 題 " in *Lishi luncong* 歷 史 論 叢 (Beijing, 1964), 79-102; see especially p. 101.

14 "Shilun Zhongguo nongmin zhangzheng yu zongjiao de guanxi 試 論 中 國 農 民 戰 爭 與 宗 教 的 關 係 " in Shi's *Wenti taolunji,* 340-352.

15 "Lun Zhongguo nongmin zhanzheng zhong geming sixiang de zuoyong ji qi yu zongjiao de guanxi 論 中 國 農 民 戰 爭 中 革 命 思 想 的 作 用 及 其 與 宗 教 的 關 係 " in Shi's *Wenti taolunji,* 338-339.

16 *Jindai shi yanjiu* 近 代 史 研 究 No. 2 (1980), 108.

17 *Zhongguo shehui kexue* 中國社會科學 (No. 5, May 1986).

REVIVAL OF RELIGIOUS PRACTICES IN FUJIAN: A CASE STUDY

Kenneth Dean

Introduction

Just over 100 years ago J.J.M. de Groot lived in Amoy and took exhaustive notes on the customs and religious practices of the region.[1] There exists a considerable amount of material on religious developments in Fujian before that time, as Michel Strickmann has pointed out.[2] Still, de Groot remains a convenient point of reference, although Kristofer Schipper has recently demonstrated the inadequacies of de Groot's understanding of the relationship between Taoists and mediums.[3] Taoism in Taiwan has been the subject of a great deal of study as well, and provides another obvious point of comparison with the situation in Fujian and northern Guangdong. During 1985/86 I had the good fortune to do research in the Minnan region of southeastern Fujian.[4] I traveled extensively through this region, visiting the ancestral shrines of cults that have spread across Southeast Asia: the *Mazu zumiao* 媽祖祖廟 on Meizhou Island 湄洲, the *Guangze zunwang Fengshan si* 廣澤尊王鳳山寺 in Nan'an 南安, the *Baosheng dadi Ciji gong* 保生大帝慈濟宮 in Longhai 龍海 Baijiao 白礁, the *Sanshan guowang zumiao* 三山國王祖廟 in Jiexi xian 揭西縣 in Guangdong 廣東, the *Qingshui zushi zumiao* 清水祖師祖廟 in Fenglai 蓬萊 Anxi 安溪, the *Sanping zushi gong* 三坪祖師宮 in Pinghe xian 平和縣. In many cases I visited these temples on the birthday of the local deity and so was able to observe the religious ceremonies held to honor the god. Since 1978 an enormous number of temples for hundreds of local gods have been built throughout Fujian. The government has been partial to the Buddhist order, restoring over 2,000 of the 3,500 Buddhist monasteries and temples that had existed in Fujian prior to the Cultural Revolution.[5] But very little remains in Fujian of the *Quanzhen* order of Taoism, which imitated the Buddhists in the establishment of monasteries and the ordination of monks. Taoism in Fujian is overwhelmingly of the *Zhengyi* order. These Taoists are professional ritual specialists who live at home and tend to the spiritual and ritual needs of their community when requested to do so. They do not live in temples, which in China are really more like community centers, but are called upon to perform offerings in the local temples on the birthdays of the deity or on other important occasions. They are principally concerned with funeral rites and with minor rites of healing and purificatory exorcism. In many parts of Fujian local Buddhist monks also perform many of these functions. There is a great variety in the traditions of Taoist ritual practice but the underlying ritual structure and often the liturgical manuals and scriptures are nearly identical. There is also great variation with regard to the extent and nature of assimilation of the Taoist tradition with local cults in different regions.[6] *Zhengyi* Taoists pass down their tradition within their families. A priest is expected to hand-copy his father's liturgical manuscripts and scriptures before being raised from the status of acolyte. Few are chosen and most remain acolytes for life; they usually can play most of the accompanying instruments as well. A Taoist service is similar to an operatic performance. The music is greatly influenced by regional opera and instrumental music.

Since temples are really local community centers and Taoists in Fujian have no organizational framework, the building of temples and the conducting of rituals is a local affair. The role of Overseas Chinese with roots in local communities in supporting the reconstruction of temples and the revival of religious practices has been most substantial. Given the enormous role of Taoism in popular culture it would be possible to classify the tens of thousands of local temples in Fujian as primarily Taoist, but this is more a concern of Westerners who do not appreciate polytheism and of Chinese government officials in the Religious Affairs Bureau. Often temples in different localities are connected by division of incense to establish a subsidiary temple, or by regional ties or reciprocity. Any religious activity that gets much beyond the village level alarms the local authorities and requires sophisticated handling.

During the last eight years in Fujian, there has been a remarkable effort on the part of people committed to the restoration of important Chinese traditions before they are lost forever. The massive destruction of temples, religious artifacts, liturgical texts and traditions, regional music, local opera, traditional customs, etc., during the Cultural Revolution left the Taoist tradition in Fujian in shambles. Taoists have gradually pieced together the manuscripts, altar-hangings, vestments, etc., that they need. This process is continuing. But the break in the tradition, the uncertainties of the future, and the laborious process of picking up the pieces all suggest that this last generation of Taoist priests may not have the time or the means to pass on their tradition in its entirety to a decreasing pool of acolytes. If so, this would be a great loss for Chinese culture.

This essay presents a case study of a Taoist *jiao* 醮 offering performed in Fujian in 1986. A description of the setting, the preparations, the Taoist rituals, and the community rituals will be followed by some ideas on the significance of these events to the villagers.

The Village

Outside Zhangzhou 漳州 in southeast Fujian rice paddies separate nearby villages hidden from view behind dense thickets of sugar cane. In one such village a four-day Taoist *jiao* offering was held in January, 1986 for the first time in 37 years. The village has a population of approximately 1200 people divided into 250 families. Almost all share a common surname. This lineage has divided into 7 branches, as detailed on a sheet of red paper pasted to the back wall of the ancestral hall. At the founding of the village in the early 19th century, village leaders took incense from the Guandi Temple in a nearby corner of Zhangzhou to animate their ancestral hall, and so consider it a sort of ancestral temple. The observance of ancestral rites in the village is not organized by rotation through the branches but by the entire clan all together. People explained that some branches of the lineage have moved away, leaving only four. The ancestral sacrifice is celebrated once a year, in winter. Unfortunately, all the ancestral tablets of the lineage as well as the family genealogy *jiapu* 家譜 were burned during the Cultural Revolution.

The village is ideally located a few kilometres outside Zhangzhou near a major highway. Rice and sugar cane are readily transported into the city. Opportunities for household industry, small business ventures and labor within the city abound. Most families earn over 400 *yuan* a month, with individual incomes of 150 per month not uncommon. There is a brick factory, a construction company, a shipping company, three mushroom growing firms, and three very small stores in the village. In the last few years, almost half of the families in the village have built new homes. The village is divided into 13 neighborhoods, the largest of which consists of 38 households. The names of all the heads of the households, arranged according to the 13 neighborhoods, together with the number of male descendants (*ding* 丁), were posted on the outside wall of the ancestral hall throughout the ritual.

The village has three temples and an ancestral hall. The temples are dedicated to *Jialan* 伽藍 (Buddhist guardian similar to *Tudi gong* 土 地 公), to *Xuantian Shangdi* 玄天上帝 , and to *Taizi ye* 太 子 爺 , two martial gods, and *Zhuye gong* 朱 爺 公 , a deified brave general. These are all single room structures. The ancestral hall is one of the largest buildings in the village, with a gate-building consisting of two side chambers, a courtyard and a spacious back hall built completely out of wood. The ancestral hall also serves as the headquarters for the martial arts association, which is well known in the surrounding area. Each household has a domestic altar with an offering table and pictures of various gods (usually *Guandi* and/or *Guanyin*).

The village last held a *jiao* in 1949. Up to that time, a *jiao* had been celebrated once every 12 years. As many as 72 neighboring villages used to take part in the festivities, especially those which had subsidiary branches of the martial arts association. Some time in 1984 it was decided to proceed with preparations for a *jiao*. Prior to this time only one other village in the immediate vicinity had held a *jiao,* in the Fall of '84. During the Cultural Revolution all such practices had been violently suppressed but by no means extinguished. Many religious practices were carried on in secret on a small scale. The revival of more or less open religious practices began in some areas as early as 1978, but wide-scale construction of local temples did not take place until 1981. Proceeding hand in hand with economic reforms, growing personal prosperity, and greater freedom of conduct, religious rituals were revived throughout Fujian. But conditions varied from one locality to the next, depending on the attitude of local authorities and the determination of the villagers. The decision to hold a five-day *jiao* was therefore both ambitious and audacious. Such a ceremony would surpass anything performed in the entire district in the last several decades. Most communities in the last two or three years had been content with a one-day or at most a three-day *jiao* . Moreover, a celebration of such size would attract the attention of local authorities determined to keep such practices to a minimum.

Preparations

The decision to proceed with the *jiao* led to the decision to rebuild the *Xuantian Shangdi* temple, destroyed by a flood in 1960. Reconstruction commenced in December of 1984 and the temple was completed in February 1985. On the birthday of the Emperor of Heaven, the 9th day of the 1st lunar month (February 28, 1985), the community leaders gathered in the new temple to select, by casting the divination blocks, the *zongli* or manager of the *jiao* . As fate would have it, the gods selected the wealthiest

54

man in the community; his is the only three-storey house in the village. However, since he was only 45, and had been a child of six at the time of the last *jiao*, he urgently sought the aid of village elders to reconstruct the details of the ritual. Six days later, on *Yuanxiao Jie* 元 宵 節 , the Lantern Festival (15th day of the 1st lunar month: March 4, '85), community representatives (*huishou* 會 首) were selected from every family, again by casting the divination blocks. They were ranked according to the total number of successful casts of the blocks in succession. Further divination determined the date of the ritual, from the 5th to the 10th of the 12th lunar month (January 16-21, 1986). The next step was to select one of the three Taoist groups that had reorganized in the surrounding region to consecrate the newly built temple, to invite the gods to attend the ritual, and to perform the *jiao*. A group from southern Anxi was chosen by divination over the nearer groups in Zhangzhou and Shima.

The *Xuantien Shangdi* temple also housed a statue of Li Nocha, or Taizi ye, the child warrior. The martial arts association originally had its headquarters in this temple, and the connection between Li Nocha and martial arts groups can be found throughout Southeast Asia. Contributions to the costs of reconstruction were written on large red sheets and pasted on the inner walls of the temple. The list was arranged according to the 6 production teams that make up the village. The total came to 15,017 *yuan*, and total expenditure to 15,014 *yuan*. However, I was told that a great deal of labor and building materials had been freely donated and that the true construction costs exceeded 20,000 *yuan*. The *Zhugong* temple, a small single-room shrine, had collected 2 *yuan* from each of the 1195 inhabitants of the village. The total of 2400 *yuan* was used to repaint the shrine and to buy new altar hangings and lanterns. Costs for repairs to the *Jialan* temple came to 3322 *yuan*, but 4414 *yuan* had been collected from the heads of the 13 neighborhoods. The posters in that temple stated that the remaining 1092 *yuan* would be used towards the upkeep of the temple. A stone tablet about three feet high was set near the entrance of all three temples. The inscription read: "If anyone dares to violate the grounds of this temple, such a man will become ill and such a woman will encounter disaster". The inscriptions were dated winter of the *jiazi* year, 1984 or early '85.

Every family in the community began making preparations of one kind or another for the *jiao*, at that time still a year away. *Huishou* families selected or purchased a pig which they would raise with great care and sacrifice during the *jiao*. Although the villagers almost never eat mutton, they organized a trip to the

55

mountains nearby in order to purchase 45 kids to be raised for the sacrifice. Domestic altars were expanded, with new god paintings hung up, new embroidered altar hangings purchased. Often new incense burners, lacquer sweet boxes, candelabra, plastic flowers, etc., were added: everything to make the altars ready for the gods who would be displayed there from the 9th month to the end of the year. Large paper lanterns were ordered, bearing the inscription, "Jade Emperor Five-Day Offering: Peace to the family."

The Taoists were invited to come right away to consecrate the *Xuantian Shangdi* temple. This involved a full day of rituals to purify the temple, exorcise evil spirits, inform the local tutelary deity, and animate the gods by infusing the statues with the *yang* energy of the sun and the *yang* energy of a rooster's comb. Two months later, on the 10th day of the 3rd month, a delegation was organized to visit the largest temple in the district, the *Baijiao Ciji gong*, Temple of Merciful Salvation, the ancestral temple of the *Baosheng Dadi* cult. They presented offerings and collected spiritually charged incense from the temple to lend power to the new *Xuantian Shangdi* temple. Contributions for this pilgrimage were gathered from the six production teams and totaled 1040 *yuan*, with the martial arts association throwing in 100 *yuan* as well. Expenditures came to 907 *yuan*, primarily to rent two large buses for the trip. 200 *yuan* went to ritual offerings, and 80 *yuan* was offered to the *Ciji* gong.

Preparations began in earnest as the first day of the 9th lunar month drew near. This day had been selected by the Taoists for the invitation of the gods from the nearby *Guandi* temple. 2890 *yuan* was spent on the lanterns, which were hung up at this point, and which featured a painting of the trio: *Fu-Lu-Shou* along with the aforementioned inscription. A paper figure of the Jade Emperor, dressed in silk and seated on a tiger skin flung over a throne, was prepared for some 200 *yuan*. Another 120 *yuan* was spent to hire musicians, and 24 *yuan* went for a horse. When everything was in order, the procession set out from the ancestral hall. Some sixteen sedan chairs were carried out, seating all the god statues from the village's three temples, plus a few others from nearby temples. They were going out to greet the higher gods of the *Guandi* temple who would come to the temple to stay for three months until the *jiao*. A medium led the way, moving with skips and hops while holding sticks of incense. The Taoists also took part in the procession, along with the *huishou* and the martial arts association and several *huagu* drum troupes in costume. They walked a roundabout route of some 2 or 3 kilometres, marching through outlying neighborhoods of Zhangzhou before circling back to the *Guandi* temple. This temple stands on the banks of a branch of the *Jiulong* 九龍 river, which flows alongside Zhangzhou. Judging from extant stone inscriptions it dates back to the Ming but was most likely restored after the Taipings leveled Zhangzhou in 1861. Once the procession arrived, the gods were invited from the altar into the sedan chairs and escorted home. The Taoists performed a *kaiguang* 開光 ritual to animate the paper figure of the Jade Emperor. The other gods were distributed to the domestic altars of the *huishou*. The village was now ready to celebrate the *jiao*.

Taoist Rituals

The Taoists invited from Anxi made an imposing sight. The head priest was 62, a tall, thin man with four strong, tall sons. Only the eldest son was studying to become a Taoist, but the others came along to play musical instruments, help out and protect their family treasures which adorned the Taoist altar. There were a total of 22 men from Anxi in the group: 12 Taoist priests, 6 musicians and 4 helpers. They filled the back of a large pick-up truck. The family had been Taoists for many generations (they claimed 24!).[7] He claimed to have climbed a ladder of swords at his initiation to prove his worth. He collected talismans at the base and at the top of the ladder to distribute to petitioners afterwards. His description coincides in many respects with an initiation ceremony I observed in Tainan, Taiwan in the fall of '84. Although definitely a *Zhengyi* Taoist, he had no connection with Longhu shan in Jiangxi near the northwest corner of Fujian, for centuries seat of the *Zhengyi* Heavenly Masters. His ritual practice resembled that of the Northern Taiwanese *"hungtou* 紅頭 (Red-head)" tradition. This could be seen in his use of a metal "dragon-horn" in place of a buffalo horn used in the Quanzhou and Zhangzhou regions and in Tainan. The free-floating *sona* melodies, the swaying movements of the priests, the use of call and response songs, the incorporation of the *Jiejie* 解結 ritual into the *jiao*— all suggest some linkage with Northern Taiwanese Taoist ritual traditions. Unfortunately, it is difficult to go much further in the comparison as the bulk of the Taoist's manuscripts and robes had been destroyed. The surviving paintings resemble those used in a Northern Taiwanese *fachang* 法場 exorcism.[8] The chief priest had no embroidered square "robe of descent" to wear during important rites. His robe was distinguished from the others' red robes bordered with blue by a square piece of embroidery (resembling Qing insignia of rank), and a long bead rosary.

The *tan* (sacred area) was set up in the ancestral hall in four levels extending to the ceiling. The group from Anxi slept behind the *tan* in quilts on straw. The *tan* was set up as indicated in the Diagram at p. 78, explained as follows:

1,2,3: The Three Pure Ones, highest emanations of the Tao: *Yuanshi Tianzun, Lingbao tianzun* and *Daode Tianzun.* 元始天尊，靈寶天尊，道德天尊.

4: The paper figure of the Jade Emperor, 3 or 4 feet high.

5,6: 18 inch paper figurines, *Biaoguan* 表 官 Officials of the Memorial; they transmit the Memorial when burnt with it after the Presentation of the Memorial rite (see below).

7,8: Containers, or *dou* 斗, of rice, seats for the gods of the Southern and Northern Dippers.

9: A painting of *Zhusheng niangniang* 註生娘娘 which the Taoists called Empress *Huang furen,* 皇婦人 .

10: A painting of *Yanluo Wang* 閻羅王 , King of the Nether World.

11,12: Paintings of the four departments, Heaven, Earth, the Seas, the Nether World. (Other groupings include Humankind as one of the four.)

13: 11 emplacements for the most important gods.

14: Small wooden figure of *Tian Gong* 天公 , the Jade Emperor, with attendants.

15: Wooden statue of *Xuantian Shangdi* 玄天上帝 .

16,17: 4 foot paper figurines with flags reading, "Holding a *jiao* brings harmony to our homes", and "We pray for an abundance of the five grains".

18: A *dou* container with a flag stating "treasures of the Tao, the Master, and the Scriptures".

19: 13 god emplacements in front of 12 Lamps of Destiny belonging to various *huishou doudeng* (斗燈).

20: 12 more Lamps of Destiny.

21,22: 21 more Lamps of Destiny.

23: The Taoists working the altar, with another *dou* containing a sword, a red cloth with the spirit soldiers of the 5 Camps of the 5 Directions, 4 dragon horns, a wooden fish, gongs and cymbals, scriptures and occasionally a liturgical manuscript.

24: Emplacements for 11 deities, mainly stellar.

25: Emplacements for 11 more gods, mostly powers of mountains and waters and the soil.

26: Emplacements for 5 local gods of the earth.

27: Emplacements for ancestors of the lineage.

28: Lion-head and slogans of the martial arts association.

29: List of yearly festivals and the people in charge of them.
30: List of the branches of the lineage.
31: Life size papier mâché figure of *Kang* and *Zhao Yuanshuai*, 康 趙 元 帥 , Marshals who guard the doors of the temple.
32: Main doors.
33: List of names of all males in the village (on outside wall).
34: Proclamation (*bang* 榜).
35: Paper statue of *Dashi* 大士 .

The Taoists were not really capable of carrying out a full five-day *jiao* so they dragged things out and turned one or two days of rituals into five. It would be difficult at this time for any of the Taoist groups I have met in Fujian to perform a full five-day, not to mention a seven-day, or 49-day *jiao*. In some areas, seven-day funerals are still performed, but the Taoists from Anxi insisted that they rarely do funerals (in this regard they resemble Northern Taiwanese Taoists). The sequence of rituals was as follows:

Day 1

2:00 AM:	Opening Drum roll. 起鼓
3:00-5:00:	*Fabiao* 發 表 , Announcement of the Memorial, *Canxiang* 參 香 ,Invitation to Partake of Incense, *Zhaojun* 召 軍 , Invocation of the Soldiers.
8:00-10:00:	*Guotan* 過 壇 , a visit by a Taoist assemblage to every home.
10:00-11:00:	*Qingshui* 清 水 , Invitation of the Spirits of Water.
2:00-3:30:	*Guotan.*
5:30-6:30:	*Guotan* to the *huishou* homes.
7:30-8:30:	*Qingchan* 請 懺 , Invitation of the Scripture of Repentance, *Chaotien baochan,* 朝 天 寶 懺 in ten *juan.*
9:00-1:00AM:	Recitation of the first 3 *juan* of the *Chaotian baochan.*

Day 2

8:00-10:00:	*Guotan* of every home.
10:00-11:00:	Recitation of *juan* 4 of the *Chaotian baochan.*
12:00-1:00:	Qixian 七 獻 , Sevenfold Presentation.
3:00-4:00:	Recitation of *Chaotian baochan, juan* 5.
4:00-5:30:	*Guotan* to *huishou* homes.
7:00-8:30:	*Guotan* to *huishou* homes.

| 9:00-11:30: | Recitation of *Chaotian baochan, juan 6, 7, 8.* |

Day 3

8:00-10:00:	*Guotan* of every home.
9:00-11:00:	Recitation of *Chaotian baochan, juan* 9 and 10.
12:00-1:20:	*Baibiao* 拜 表 ,Worship (full presentation) of the Memorial.
2:00-3:00:	*Jiejie* 解 結 ,Untying of the Knots.
2:00-3:30:	*Guotan* of *huishou* homes.
3:45-4:30:	*Guandeng* 關 燈 ,Communication of the Lamps.
6:30-8:00:	*Guotan* of *huishou* homes.

Day 4

8:00-10:00:	*Guotan* of every home.
11:30-12:30:	*Qixian* , Sevenfold Presentation.
2:00:	Dismantle *Sanjie* 三 界 altar.
2:30-4:00:	*Guotan* of *huishou* homes.
6:00-7:00:	*Pudu* 普 度 , Feast of Universal Deliverance for disinherited souls.
7:30-8:10:	*Da huolu* Striking 打 火 路 , the Road of Fire.
8:30-10:00:	*Guotan* of *huishou* homes.

Day 5

8:00-10:00:	*Guotan* of every home.
11:00-12:00:	*Qixian* , Sevenfold Offering, Ritual visits (see below).
2:00-3:30:	*Guotan* of *huishou* homes.
6:00-7:30:	*Guotan* of *huishou* homes.
7:30:	Dismantling of the Taoist Altar.
11:00-12:00:	Sending home the Lamps of Destiny.

In the following paragraphs I will give a brief description of the 14 rituals performed by the Taoists during the *jiao* . Unfortunately, I was unable to attend the first sequence of rituals on day 1, but arrived in time to observe the *Qingchan.*

1. *Qingchan* : The Invitation of the Scriptures of Repentance was performed by a single Taoist on the stage before the altar of the Three Realms, across the square from the ancestral hall. The *huishou* lined up before the stage. The priest chanted, bowed and turned smoothly from side to side while gesturing from his heart to the altar with his free right hand, ringing a three-pronged bell with his left hand. His songs and movements were accompanied by a drum and a *sona,* the latter playing a free melody consisting of

many embellishments around the basic tune sung by the Taoist. Periodically the Taoist would blow his "dragon horn", a thin, flat curved horn made of tin with seven ridges or bumps just before the end. By this means, he summoned spirits to the altar. The songs, summons and recitative were all performed from memory. An incense burner, two candles, a plate of candies and one of fruit, the Taoist's *dou*, and the *Scripture of Repentance*, the *Chaotian baochan*, were placed on the altar table. There were another incense burner, more candles and fruit on the raised level of the altar table. After some 30 minutes of recitative, the Taoist took the scriptures and was followed back into the ancestral hall by the *huishou*. He placed the scriptures on the altar and began the lengthy recitation, first waiting for the *huishou* to light new sticks of incense at the altar.

2. *Song Chaotian baochan* : The recitation was performed before the altar with the entire group of *huishou* first standing, then kneeling, and finally sitting in attendance. The same young Taoist who had invited the Scriptures to the altar recited them. He never took part in the more important rituals and was a new recruit to the Taoist assemblage, although his father had been a *fashi* 法 師, or Ritual Master, all his life. Distinctions were drawn by the Taoists and elders between a *Taoshi* 道 士, or renowned Taoist priest, who would only perform *jiao* rituals and healings, a *saigong* 師 公 (*shigong* in Mandarin), who would also lead a funeral procession to a grave (*kaitong minglu*, 開 通 冥 路 Opening up the Dark Path), and a *fashi*, who had a dubious command of *fu*, talismanic arts for various purposes. Apparently, Anxi Taoists do not conduct entire funeral services except under very unusual circumstances.

3. *Guotan* : The ritual visiting of all the domestic altars of the community, or those of the community representatives selected by the god, took up the bulk of the ritual activity of the five-day *jiao*. The Taoists split up into two groups of five priests, each accompanied by two musicians, a drummer and a *sona* player. Before them marched several of the youngsters of the community who were taking the place of their elderly *huishou* parents. They had a great deal of fun carrying the banners and gongs and setting off firecrackers after each visit. The Taoists would enter each house, bow to the altar, and then make room for the *huishou* in their procession who would enter and place their sticks of incense into the home incense burner. Then the Taoists would sing while sanctifying the altar by sprinkling purificatory water with a sprig

of a peach-tree and sprays of rice. Then they would bow to the gods' statues on the altar, perform a brief criss-cross dance, and move on to the next household. On their rounds of the village the Taoists set out 23 *lujie* 路 界, 3 foot paper figurines that mark the spiritual boundaries and crossroads of the community.

4. *Qixian* : The Sevenfold Presentation was performed at noon outside the ancestral hall. An altar table was set up facing the hall, equipped with an incense burner, offerings of fruit and several *chuanlian* 串 聯, consisting of a few candies on sticks inserted into a rack at the center of the altar, several bowls of food, and the Taoists' tools: the *dou*, the wooden fish and metal bowl, the court tablets, the seals, etc. First the Taoists led the *huishou* onto the stage and around the Altar of the Three Realms. Then they went into the ancestral hall and circled the Altar of the Three Pure Emanations of the Tao. Finally the chief priest stood behind the outdoor altar with four acolytes behind him, two to a side. The chief priest picked up a hand-held incense burner with his left hand, and repeatedly formed a mudra with his right hand. He made a fist, then extended his index finger and his little finger, and made a flinging motion with this hand into the burner in his left hand. He used this mudra to transfer energy gathered within his body through visualizations into the burner. He repeated this movement three times, then performed it again to the south and the north (ritualistically speaking, the direction of the Altar of the Three Realms is to the South and that of the Three Pure Ones to the North). Again to the east and west. At this point he instructed assistants to burn paper money and to pour offerings of wine into cups on the inner and outer altars. Then he performed three series of three bows each to the south and began the presentation of offerings. First he danced with the hand-held burner back and forth up to the altar with quick steps and a great flinging of sleeves, weaving and bowing before passing the burner to the *zongli*, manager. He then retrieved it and replaced it on the altar. This process was repeated with candles, bowls of rice, flowers, fruit, tea, noodles, beans, pickled vegetables, cauliflower, and beans, greens, mushrooms, spring rolls, and last but not least, bananas. When all the offerings had been presented in this fashion, the priest picked up a pair of hand cymbals and joined in the spirited singing the acolytes had been doing throughout his dance. They were beating the time with *paiban* 拍 板, clackers made of five wooden slats about seven inches long tied together at one end. Then all five Taoists circled the table, bowed south, circled back counterclockwise and bowed south again. Then began an astonishing criss-cross dance in which the four acolytes would rush in front of the altar in pairs and fall violently to their knees and lean their bodies backwards to the ground and rise up again all in one motion. This went on for quite some time while the Taoist's sons and assistants tried to keep the mats on the ground from sliding out of place. Finally, the chief priest stopped them and purified the altar by sprinkling water with his sprig of peach. They then all bowed out with another great flourishing of sleeves.

5. *Baibiao* : The Worshipful Presentation of the Memorial on the third day was the climax of the entire ritual. Offering tables were set out in the square by the villagers facing the stage. These will be described below. The two chief *huishou* carried the paper and the wooden figures of the Jade Emperor out to the stage and placed them on top of the altar of the Three Realms. The Taoists brought over a yellow, rectangular, boxlike paper envelope addressed to the Jade Emperor, inside which there was a yellow envelope listing the Taoist's title, and the *shuwen* 疏 文 , or Memorial, and three sticks of incense. All the *huishou* then gathered before the stage and kowtowed. The Taoists entered the stage from the northeast side and brushed their court tables with their sleeves, fixed their flamelike hatpins, bowed to left and right, and bowed to the altar. An acolyte on the left handed the chief priest three sticks of incense and he said a prayer over them before handing them on to an acolyte on the right who placed them into the burner on the altar. The chief priest then made three deep bows, leaning backwards on his knees. Next the Taoists circled the altar. When they returned before it the four acolytes blew dragon horns, then twirled the horns in one hand while ringing their bells in the other, while the chief priest summoned spirits to the altar. Then he took the yellow envelope and bowed to the altar. He turned to the *huishou* who bowed three times. He returned their bows, then knelt before the altar and muttered *mijue* 密 訣 , secret incantations, over the envelope. The priest kowtowed north and south three times, retrieved the envelope, passed it over the fumes of the incense burner, and set it back on the altar. Then he picked up a hand-held burner from the altar and had wine poured out as an offering into cups set on the altar. He put down the burner, picked up the envelope, removed the *shuwen* and read it aloud. Additional names were read by a young Taoist from a separate list. The Taoist dropped a pair of small divinatory blocks connected by a string several times before the altar. The favorable response he finally received was the sign for the villagers to take the great strings of yellow paper money adorning their offering

63

tables to be burnt in a great bonfire. Meanwhile the chief priest was repeating the Presentation ritual, only with more elaborate dance steps. He passed incense, rice, flowers, and fruit to the *zongli*, who had been called up to kneel again at the edge of the stage. The Taoist held items using the *santai* 三 台 hand position, employing the thumb and first two fingers. He worked his feet back and forth, gliding from the altar to the *zongli*. A series of the acrobatic back-bending kneeling dives described above were performed by the acolytes to conclude the ritual. Great strings of firecrackers were set off.

6. *Jiejie*: The Untying of Knots was performed by the same young Taoist who had recited the *Chaotian baochan*. A single young *huishou* represented the village. The ritual was conducted inside the ancestral hall before the Altar of the Three Pure Ones. To the left of the altar a table had been set with candles, fruit, the Taoist *dou*, the metal incense stand, the metal bowl and an oil lamp. On the right of the entrance way was a bucket of water with a board across it. A Taoist court tablet lay across the *dou* like a bridge. Upon it were five piles of two one-*mao* (毛) notes (2 *mao* = U.S.$.05). The Taoist bowed to the *Sanqing* and then to the table of local earth gods (Diagram 1, #21). Then he took up a dragon horn in one hand and a fan with another 2 *mao* tied to it. He sang a long call and response recitative with the drummer accompanied by the *sona* and punctuated by blasts on his dragon horn. Then he read a document with some coaching from an older Taoist. Next he handed the fan to the *huishou* and circled the table while burning a roll of paper money and ringing his bell. As he came round the table he picked up a pile of money. Then he swooped down and flung the burning paper money onto the board across the bucket. He pretended to do the same with the money but in fact replaced the *mao* on the tablet bridge. When he came to the last pile he untied the money from the fan and circled in reverse direction around the table to conclude the ritual.

7. *Guandeng*: The Communication of the Lamps was very similar to the *Jiejie* ritual just described, except that it was performed by the chief Taoist before all the *huishou*. The Taoist performed the ritual before the *Sanqing*, using the side table more as a prop. On this table was the *dou* with a tablet bridge, this time equipped with seven piles of 5 × 2-*mao* notes. After lighting incense and bowing and kneeling before the altar, he began to sing and chant and blow on his dragon horn. Then he began to hop on one foot in a circle, twirling his horn in one hand and ringing his bell in time

with each hop. On his second time around he hopped with a roll of burning paper money and deposited it under the altar with a swooping motion. Then he collected one stack of bills and hopped happily around again. He repeated this process several times. Again there was no manuscript, all was recited and performed by memory.

8. *Pudu* : In preparation for the Feast for the Universal Deliverance of the Disinherited Souls or Hungry Ghosts, the villagers had set up offering tables (see below) facing east to an outdoor altar table. The Taoists first called the *huishou* to the altar, then led them around the tables of offerings, then on to the stage where the Altar of the Three Realms had been dismantled and a display of rice sculptures (see below) had been set up. Next the procession went to sing and bow and offer incense to the paper figure of *Dashi* 大 士 on the side of the ancestral hall. Finally they made their way to the outdoor altar table. Next to it stood a table covered with buns marked with red dye for distribution to the hungry ghosts. In addition to the usual tools of trade that the Taoists had brought out to the altar table there were paper umbrellas stuck into the *dou*. As the Taoists began the ritual a paper horseman and horse were burnt in front of the table. The Taoists read from an old manuscript by candlelight. The chief priest formed several mudras, sang several songs, recited sections of the text, and stuck ten sticks of incense into a bun, indicating that the hungry ghosts had all arrived. A document was read by the senior acolyte officially inviting the ghosts to dine. The document was then burnt. The chief priest flung sticks of incense, chopsticks, tea leaves, and buns into the area before the table. As he formed more mudras, the other Taoists and then the surrounding *huishou* all started flinging first, a plate of coins, then buns, fruit and other edibles, into a crowd of kids who were soon joined by older women, then by everyone in the village who scrambled for the blessed items. This despite occasional remonstrances from the Taoists that these things were for the hungry ghosts. At the conclusion of the ritual great piles of paper money were burnt.

9. *Da huolu* : The Striking of the Fire Road was performed in the inner courtyard of the ancestral hall. An altar table was set up to the right and a fire demon fortress to the left. This latter consisted of a bamboo and paper fort surmounted by eight paper figures. Two black flags hung down the front, and two flags stood on top, one of them round with one side red and bearing the character 'sun', the other side black with the character for 'moon'. The sec-

ond standing flag was square and black with a white circle in the middle. A great white paper 'ingot' was leant against the fortress. In front of the altar table was a bucket of water, on top of which rested a winnowing tray with five green talismans lying on bamboo fans on it. Each of these was addressed to a different direction and stamped with the chief priest's personal seal. The priest first sprinkled the talismans with purificatory water with his peach spring. Then the Taoists picked up the fans and swept with them this way and that around the table and then around the fortress skipping close in and skipping back while marking their steps with rings of their bells. Next, the talismans were placed on braziers set out in the 5 directions and fanned into flames. The chief priest formed mudras and pronounced *zhou* spells in each direction. Then the fortress, the paper ingot, the fans, winnowing tray and braziers were all burnt in a central bonfire. The priest circled the fire, forming more mudras, chanting mantras and spitting purifying water on the flames. When the flames died down, an acolyte (the chief priest's eldest son) poured water from the bucket over the ashes. The chief priest pressed down the smoke with his personal seal while his chief acolyte performed the corresponding mudra behind the altar table. In this ritual, talismans, seals, printing with seals, mudras and incantations are interchangeable aspects of one ritual act.[9]

10. *Song Tiangong*: The sending off of the Jade Emperor was done with little ceremony. The envelopes marking the emplacements of the gods were collected and stored away. Then five rectangular boxlike paper envelopes of five different colors that seemed to be unmarked (though they should have contained documents addressed to various celestial bureaucratic offices) were brought out from under the altar to be burnt. Candles were handed to three *huishou* from the altar and taken out together with the paper figures of the Jade Emperor. The *lujie* were collected and added to the bonfire. The great papier mâché marshalls were burnt too. Much later that evening, the *huishou* came to collect their *dou*, Lamps of Destiny, placing them carefully into baskets and taking them back to keep alight on their altars till the next morning.

11. *General Observation*: The *jiao* rituals which the Anxi Taoists performed differed to some extent from those I have seen other Taoist troupes perform in Nan'an, Zhangzhou, Tong'an, Taibei, Tainan, and the Penghu Islands. Several elements of a complete *jiao* were lacking in this five-day ceremony. Nonethe-

less, they put on an adequate performance in the eyes of the villagers. Few people under 50 had any experiences from which to evaluate them, but the Zhangzhou Taoists who came to look were very contemptuous. Those over 50 who might have noticed the Taoists' shortcomings were more than happy with the overall restoration of traditions in which they had an important role to play.

Community Rituals

The community had its own share of ritual specialists, elders who recalled the details of every aspect of religious practice. They organized themselves for religious purposes according to the ancient ritual models, first in their selection of *zongli* and *huishou* and again in their organization of the procession to invite the gods to the village and their subsequent distribution of the gods to the *huishou* domestic altars. The *zongli* kept a notebook filled with detailed lists of the rankings of the *huishou*, the distribution of the gods, the proper positioning of offering tables, *pudu* rice sculptures, gods in the processions, etc. More remarkable was the precision with which the three different tables of offerings set up in the course of the ritual were reconstructed. The first of these was for the *Baibiao*. Tables were set up facing the stage where the Taoist held an audience with the Jade Emperor, chief of the popular pantheon. During this audience, the Taoist reads once again the Memorial listing the names of the *huishou*, his prayers on behalf of the community and the sequence of rituals that he has performed. Thus it is a display of his power to hold an audience and transmit a message, as well as a demonstration of the merit he has reaped for himself and the community in the eyes of the gods. To this end he invites a large number of gods to the ritual. The villagers' offerings are presented to these gods by the priests' sevenfold dancing presentation at the altar. Each offering table was flanked by a sacrificial goat and pig. These animals had all been slaughtered the day before, many by the side of the new *Xuantian Shangdi* temple, where pools of blood could be seen for the remainder of the ritual. Gold lockets were placed around the necks of the carcasses, as the blessing accrued by participation in the offering could be transferred to the subsequent owners, the newborn of the village. On the table offerings were arranged in several rows. First came four empty teacups and a decanter of wine. Row 2 consisted of four teacups filled with rice. Row 3 had four teacups filled with dried fruits. Row 4, four cups of tea leaves. Two bowls of rice covered with circles of red paper with a stick of candies inserted in each rice bowl were set on either side of these sixteen

67

cups. Behind them came a lacquer box for offering sweets. Then came the family incense burner, flanked by candles, then a plate with a pig's head and tail next to a plate with five different kinds of meat. Behind these came a final row made up of large round flat brown glutinous rice cake, and a plate of small glutinous rice cakes made with a mold in the shape of turtle and imprinted with the character for 'longevity'.

On either side of the table strings of paper money were tied to bamboo poles; one variety consisted of long strips of serrated yellow paper tied to a pole with red strings, the second was made up of red and gold printed paper money folded into 'ingots' and grouped in wheels all up and down a string.

The second offering table was prepared facing east towards the *Pudu* altar table. The first row of offerings were ten empty bowls or spoons and ten pairs of chopsticks for the hungry ghosts. Next came empty cups with a container of wine. Next came four candy sticks inserted in the rice bowls flanking a lacquer box of sweets. Behind this were two candles and several bowls of dried foods, fruit and sweets. Then came a large container of rice, next to a plate of five different kinds of meat, next to a plate of pressed, molded rice-cakes. Behind these came the great rice cake and the pig's head and tail plate. Baskets of silvered sheets of paper money folded into ingots were kept in large supply in baskets next to the tables. The large bowl of rice on each table corresponded to the rice sculptures displayed on the stage.

The third offering table was set up on the afternoon of the fifth day at noon. The tables faced north, towards the ancestral hall. The priests emerged from the *tan*, walked around all the tables, and performed a final Sevenfold Presentation at an outdoor altar table. This offering seemed intended to honor the ancestors of the lineage, but there was some confusion evident in the arrangement of offerings. Some tables had ten bowls and chopsticks laid out, while others presented rows of wine cups, rice, candied fruits and tea.

Prior to the commencement of the ritual, community leaders throughout the area expressed their interest in participating by organizing temple delegations, carrying palanquins in the processions, and bringing martial arts troupes, drum and gong brigades, musicians, etc. However, the villagers feared attracting the disapproval of local authorities should the *jiao* turn into a regional event and they refused the offers. They did consent to ritual visits from five nearby villages. These visits took place on the afternoon of the fifth day. The offerings brought by the villages were displayed inside overturned tables on the stage. Then they were set in a U-shape around the outdoor altar where the Taoists had recently concluded their last Sevenfold Presentation. On the altar stood five antique wine decanters. The visitors lined up and knelt on the left, their hosts faced them kneeling on the right. Then each in turn went to face the altar and bow. They were handed incense with which they bowed several times before passing it back to the altar. Then they were handed one of the wine decanters. They knelt and poured a libation on the ground, passed back the decanter, and kowtowed before the altar. Then they returned to their place and made way for the next village headman.

No discussion of community rituals would be complete without mention of the ritual pieces acted out by the theatrical troupes before each afternoon and evening performance. These consisted of the sequence *Baxian zhuhe* 八 仙 祝 賀 , or "Eight Immortals' Longevity Blessing", *Tiaojia guan* 跳 加 官 , "Congratulations on Promotion", and *Tianfei songzi* 天 妃 送 子 ,"the Heavenly Maiden grants sons". Space does not permit a complete treatment here. Suffice it to say that numerous structural correspondences exist between the theatrical performance, community organization and the Taoist ritual. These are underlined when an actor dressed as a *zhuangyuan* 狀 元 and his his wife and maidservant carry a doll to the altar as an offering.[10] Performances of *Gezaixi* 歌 仔 戲 , regional opera, were held for twelve afternoons and evenings during and after the festival. Troupes included local and regionally known amateur troupes, as well as one of the best professional troupes available (thanks to good connections).

Yet another aspect of community ritual involved the visits by the Taoists to each household, as well as the ritualistic visits by the costumed drum troupes, *huagu* , to each household to conclude the five-day ritual. These troupes featured men in women's clothes and vice versa. The *huagu* troupes were part of the larger religious procession which involved a great deal of ritual organization on the part of the village.

Sending off the Gods

On the 13th day of the 12th lunar month, the gods of the village were collected from the *huishou* altars and placed on sedan chairs in the square before the ancestral hall. The highest gods, the *Tiangong* statue and statues of the City God and his wife, were escorted out of their homes with great fanfare by several *huishou* and musicians. Eleven *yaogu* 腰 鼓 , costumed drum and dance troupes, arrived from a village related by intermarriage. The martial arts troupe brought out the lion dance outfit and the temple

69

arms, which they had only recently recovered from the local authorities by claiming to be a physical fitness club, which of course they also are. The *yaogu* troupes performed in the square while the gods' sedan chairs were arranged in hierarchical order. The highest gods first stopped before the *Tiangong* temple where offerings had been displayed to greet them. Then the procession began moving in the following order: 1. The guardian of the ancestral hall, carrying the incense burner in a basket, flanked by youths pounding large gongs. Just behind them came four youths with red banners bearing the name of *Jialan wang, Jialan furen* 伽 儡 王 , 伽 儡 夫 人 , the Buddhist Guardian of the village and his wife. They were followed by two youths carrying lanterns on sticks inscribed with *Yuhuang Dadi, hejing pingan* 玉 皇 大 帝 , 合 境 平 安 , "The Jade Emperor Protects the Boundaries and Brings Peace." Two youths came behind with large embroidered umbrellas inscribed with the name of *Xuantian Shangdi*. 2. *Tudi gong*. 3. *Guandi*. 4. The 1st *yaogu* troupe. 5. *Li Nocha (Taizi ye)*. 6. 2nd *yaogu* troupe. 7. 3rd *yaogu* troupe. 8. *Tudi gong*. 9. *Xuantian Shangdi*. 10. Unidentified god. 11. *Zhuye gong*. 12. 4th *yaogu* troupe. 13. *Jialan gong*. 14. 5th *Yaogu* troupe. 15. *Yiyong jiangjun* 義 勇 將 軍 *(Chuye gong)*. 16. *Shuixian zunwang* 水 仙 尊 王 . 17. Unidentified god. 18. 6th *yaogu* troupe. 19. 7th *yaogu* troupe. 20. *Guandi*. 21. *Shangdi ye* 上 帝 爺 . 22. *Xuantian Shangdi*. 23. 8th *yaogu* troupe. 24. *Xuantian Shangdi*. 25. 9th *yaogu* troupe. 26. The Three Officials *(Sanguan Dadi)* 三 官 大 帝 . 27. *Bayin* 八 音 musicians. 28. *Tiangong*. 29. *Chenghuang ye, Chenghuang furen* 城 隍 爺 , 城 隍 夫 人 . 30. The *huishou* (會 首).

There were thus sixteen sedan chairs carrying deities and nine *yaogu* troupes in the procession, as well as two groups of *bayin* musicians who marched together and split up when the procession reached a new village. First the procession marched through the home village, then they made their way past rice paddies and sugarcane fields to five nearby villages, returning the ritual visit described above. Each village was circled by the procession while the last three sedan chairs lined up before the offerings set out in front of the village temple. Bows were made and incense was placed in the sedan chairs. Tremendous strings of firecrackers were exploded. Along the way at least every other home was burning paper money or incense or setting off firecrackers. Some had set up offering tables. When the procession entered the outskirts of Zhangzhou, the streets narrowed and the firecracker explosions grew even more intense. Finally the procession arrived before the *Guandi* temple. The martial arts group raced about and then lined the way into the temple. Each sedan chair rushed in, avoiding the lunging of the dancing lion. While the *yaogu* troupes were all dancing and drumming outside, the chairs were lined up inside and the borrowed gods returned to the altar. Incense was burnt and the *huishou* bowed before collecting their own gods and chairs and heading home. The gods were set on the stage for a while and then the *Tudigong* and *Taizi ye* were carried racing about the village by young lads. Finally all the gods were set before the stage to watch the evening performance before being escorted back to their respective temples. The *yaogu* troupes went from door to door before boarding buses at dusk to return to their village some 15 km. away.

The Revival of Yearly Festivals

The performance of this *jiao* underlines the renewal of obser-vances of annual festivals. A list of these was posted on the wall of the ancestral temple. The poster was entitled "List of the Head-men in Charge by Rotation of the Customary Festivals of This Village (*she* 社)". The festivals listed were as follows: 1. *Yuhuang Shangdi* (Jade Emperor's Birthday) 1st month 13th day. 2. *Zhongtan yuanshuai* 中 壇 元 帥 (*Li Nocha's* Birthday) 3rd month 10th day, Presentation of incense (to *Baijiao Ciji gong*). 3. *Jialan furen* (Birthday), 4th month 4th day. 4. *Shuixian zunwang* (Water Immortal's Festival) 5th month 5th day, Dragon Boat competition. 5. *Yulanshenghui* 盂 蘭 聖 會 (the Feast for the Hungry Ghosts), 7th month, 27th day. 6. *Fude zhengshen* 福 德 , 正 神 (the Earth spirit or *Tudi gong*), 8th month, 15th day. 7. *Xuantian Shangdi* , 9th month, 9th day. 8. *Jialan dawang* (Bud-dhist Guardian spirit) 10th month, 19th day. 9. *Yiyong jiangjun* (the Martial Spirit), 10th month, 19th day. 10. *Jialan dawang, furen* , 11th month, 10th day. One should add to this list the *Yuanxiao* festival on the 15th day of the 1st month which likewise involves parading the gods through the village.

An example of the cost of such festivals is the list of expenses for the *Duanwu* 端午 Dragon Boat festival (#4 above) posted on the walls of a temple in a small neighboring village visited during the procession:

1200 *yuan* for the boat, 400 for painting, 64 for decorations, 90 for a puppet show, 150 for a movie, 30 for the altars, 20 for food, additional costs, 693. Total: 2,597 *yuan* .

All the villages in the area compete in the dragon boat races in the canal before the Guandi temple. The preliminaries take two weeks from lunar calendar 5/5 to 5/19. The final race is given

71

added religious significance this year by inviting the image of the *Shuixian*, the Water Immortal, into a boat before the temple. The dragon boats race past and back to the god.

As for the costs of the entire revival of religious practices up to and including the five-day *jiao*, estimates hovered above 80,000 *yuan*. This breaks down as follows: rebuilding the *Xuantian Shangdi* temple: 20,000; repairing the other temples: 10,000; theater (amateur) for 11 days: 5,500; professional theater troupe, 34 days: 2,400; Taoists' fees (10 days, plus transportation, for 20 people): 4,400. This come to roughly 45,000 *yuan*, leaving 35,000 for goats, pigs, offerings, embroideries, sedan chairs, god carvings, lanterns, musicians' fees, *yaogu* troupes' fees, banquets, cigarettes, firecrackers, etc.

Conclusions

At this point I would like to explain my own involvement in the *jiao*. I had heard about the upcoming ritual from some Taoists in the area and had approached the village headmen about observing the ritual. They readily agreed, but asked that I first obtain permission from district and local authorities. Many felt that my presence would forestall any move on the part of the authorities to forbid the proceedings. Others thought that, with me there, they could invite all the villages that wanted to come and do it as in the old days. But voices of caution prevailed. I obtained permission through repeated visits to the authorities. During this process I came to realize that there was an unfortunate distance between the authorities and villagers restoring their traditions. Only in the past few years have economic reforms instituted from above caused the authorities at local levels to allow greater freedom of action to villagers. However, rather than participating at least to the extent of assisting with transportation problems, parking, etc., the authorities either turn a blind eye to the events or devise means to restrict or cancel them. Party members dare not involve themselves or show any support for religious activity, even though everyone else in the entire community is taking part in it. Party members fear that their own *belief* in Communism would be questioned if they showed any sympathy. This is one sense of the term "atheism", in Chinese, literally "no-spirit-theory" 無神論 . Official objections to religious activity usually take the following forms: 1.Local religious festivals lead to excessive economic wastage. 2. "Spiritual charlatans and witch-women" extort money from the people. 3. Religious events disturb public security. This latter complaint extends from problems with parking and parades to suspicions of political plotting.

I believe it is important to examine each of these objections in turn. In traditional Chinese society, religious festivals were occasions for great economic activity. Feasts were held in which the offerings were distributed amongst the villagers, down to the beggars. These large meals with their pork and other meat dishes were an important supplement to an otherwise subsistence diet of rice and a few vegetables. Meat sacrificed at the festival was distributed first to friends and relatives and then sold on the market. A tremendous amount of buying and selling would go on before and during these festivals. People would come from all over to do business. Thus these festivals brought about an increased regional economic flow and laid the groundwork for regional economic interconnections. Although standards of living have risen considerably in China, much of the above is still true. Not a single offering is wasted; feasting on such occasions still plays an important role in the nutritional life of the people. Moreover, a *jiao* is most often a celebration of thanksgiving after a bountiful harvest, and the spirit of community solidarity and economic progress are quite in accord with government exhortations to expand productivity. Economic ties continue to be established side by side with regional religious ties. In Fujian, the process spreads overseas, as many Overseas Chinese eagerly suport these events. It is not too far-fetched to suggest that if regional religious activity were allowed to develop, the network of economic exchanges would grow along with local revenues. However, China has just begun opening up its economic system and authorities experience considerable anxiety over the loss of direct power over economic affairs. Their inability to channel Overseas Chinese funds into state planned investments also frustrates them, but there can be little doubt as to the favorable impact on the local economy of such an influx of funds.

Occasionally, in the case of a renowned temple, which is under provincial government control while still being the locus of an active cult, local religious observances get caught in a tug of war about the temple revenues between local and provincial authorities. This can lead to sharp contradictions between the authorities and the community, in which Taoists and Buddhists end up the sacrificial victims. These confrontations usually do not last long, for the power of the people is formidable, and it would require a tremendous amount of force to completely suppress such religious activity. Such a show of force would only further alienate the people from the party just at a time when the party is hoping its guided economic reforms will restore the people's faith in the government. Should the government seek to prevent someone from performing a role in a ritual, there are always more than enough people to take his place. One can only hope that constructive ways can be found to resolve these problems.

Unfortunately, it is all too easy for the authorities to invoke the often repeated accusations against "spiritual charlatans" defrauding the people, when they accuse community leaders of forcing contributions from their fellow villagers. In all the nine *jiao* I have attended in the past few months, I never saw any evidence of forced participation or contributions. On the contrary, people felt it a great honor to participate in the ritual and an honor for the entire community to be holding one. The communities were entirely supportive and participation was virtually total. Old men and women were proud of their role in reconstructing the details of the ritual observances and very satisfied with the revival of their traditions. One young man in the village explained that he was certain participation in the ritual as a *huishou* would lead to prosperity even though it meant losing out on some short term deals. An old woman thanked me for attending the ritual, saying that since everyone had spent so much time and money raising goats and pigs, it would be terrible if the government were to call a mass meeting before the sacrifices were blessed. She felt this was unlikely to happen with me present. In fact, a meeting was called after the *jiao*, and local authorities demanded an end to "feudal superstitious practices", but said they would discount anything that had gone on up to the meeting. Similarly, the Taoists urged me to stay until the moment they had their altar, paintings and manuscripts packed and on the truck, for they felt insecure outside their home district and feared that the authorities might confiscate their family treasures. This fear may also account for the very small number of manuscripts they brought along, preferring to work from memory. In short, I have seen no evidence of fraud or extortion and instead discovered a great sense of common purpose and willingness to support the revival of religious traditions. No doubt charlatans do exist and prey upon the gullible, but probably no less so in the religious sphere than in business or even in political circles. Moreover, the experiences of the Cultural Revolution have given people plenty of perspective on religion. There is no longer an unbroken tradition of blind faith in anything.

This brings us to a consideration of the accusation that religious events constitute a threat to public security. This is another area where the government helps create a situation which it then accuses the people of fomenting. Instead of cooperating with the community on common issues like transportation, parking, parades, etc., they insist that simply having that many people in one place will lead inevitably to accidents, looting and crime. In fact, quite the opposite is the case. Cooperation and community organisation reach new levels through the planning and conducting of religious events. In one *jiao* I attended, a young tough had been selected through divination to be one of the *huishou*. When I asked him how he and his buddies felt about the ceremony, his answer indicated that his participation gave him a great deal of face, a new sense of belonging to the community, and hope of greater prosperity through social standing. His buddies also deeply respected the proceedings and instead of creating disorder would do their best to ensure that nothing went wrong.

As to the underlying uncertainty of secret societies forming out of Taoist rituals, this too arises from a misapprehension of the nature and function of Taoist ritual. Through his training and ordination, the registers of spirits at his command and his access to hand-copied manuscripts, the Taoist is able through his ritual to communicate with the Heavens on behalf of the community. During his ritual he embarks on an arduous visualized journey in which he first summons forth the gods of his body and mingles them with their counterparts in the Heavens. He presents a memorial on behalf of the community before the highest emanations of the Tao in the Nine Palaces of his cranium, now merged with the Palaces above the Dipper stars. Finally he merges with the ever and all transforming Tao in order to recreate the universe, speeding up cosmic cycles to generate cosmic merit and blessings for the community. The Taoist who performs rituals in the temple located at the center of the town, is at the hub of a wheel made up of the god's procession marking the spiritual boundaries of the community. That procession consists of practically the entire local popular culture; every form of the popular performing arts is represented. In these ways, Taoism at the local level in traditional society tended to legitimize the local community *vis-à-vis* the State. This is very different from attempting to overthrow the State. While Taoist ideology played an important role in peasant rebellions throughout Chinese history, in fact, Taoist social ideals are remarkably similar to those of the Chinese Communist party.

Rather than speculating on the meaning of these events to the authorities, let us turn to the question of the meaning of the rituals to the villagers. They do not understand exactly what the Taoist priest is doing in his rituals. They may not understand all the complex details of the community rituals they have reconstructed. A ritual act can have any number of meanings attached to it over time. Ritual may be primarily a *method* or *technique* of salva-

75

tion, but with many different levels of participation and understanding. Through these rituals the villagers re-enter ancient cycles. They reconnect with the ways of their ancestors, literally walking in their ancestors' footsteps in the prescribed movements of the ritual. For them, the ritual means first and foremost the restoration of local traditions. In the current political climate, it means a courageous statement of a degree of local autonomy. Village leaders assured me that the many shortcomings of the *jiao*, due to inexperience, would be corrected during the next *jiao* to be held 12 years later. That *jiao* would be even more magnificent.

In China, the importance of precedents is considerable. Although I have not traveled beyond Fujian, I understand that very few areas of China have restored religious traditions to the degree Fujian, Zhejiang, and Guangdong have. However, the economic reforms are under attack because of cases of corruption, which they are accused of having allowed or even fostered. The preferred weapon of attack on corruption within and beyond the party is "ideological reform". This can lead to cultural restrictions and these may be interpreted at the local levels as a call for reducing the revival of religious practices. Only time will tell.

NOTES

1 J.J. M. de Groot, *Les Fêtes annuelles célébrées à Emoui (Amoy), Etude concernant la Religion Populaire des Chinois*. Paris : Annales du Musée Guimet, 1886. Reprint: San Francisco, Chinese Materials Center, 1977. See also his *The Religious System of China, 6 volumes*. Leiden: E.J. Brill, 1911; Taipei Reprint: Southern Materials Center, 1982.

2 M. Strickmann, "History, Anthropology and Chinese Religion". *Harvard Journal of Asiatic Studies*, 40 (1980), 201-248.

3 K.M. Schipper, "Vernacular and Classical Ritual in Taoism". *Journal of Asian Studies, 45 (185), 21-57*.

4 K. Dean, "Taoism in Southern Fujian: Field Notes, Fall, 1985", to be published in a symposium volume by the Chinese University Press, Hong Kong. See also K. Dean, "Field Notes on two Taoist *Jiao* Observed in Zhangzhou in December 1985", *Cahiers d'Extrême-Asie*, No. 2, published by the Ecole Française d'Extrême-Orient, Kyoto, 1986.

5 *China Daily*, March, 1986.

6 R.A. Stein, "Religious Taoism and Popular Religion from the Second to Seventh Century", in H. Welch and A. Seidel, Editors. *Facets of Taoism*. (New Haven: Yale University Press, 1979), pp. 53-81.

7 On the grades and significance of Taoist ordination titles see M. Saso, "Red-head and Black-head: The Classification of the Taoists in Taiwan according to the Documents of the 61st Generation Heavenly Master", *Bulletin of the Institute of Ethnology* (Academia Sinica) no. 30 (1972), 69-82.

8 J. Keupers, "A Description of the *Fa-ch'ang* Ritual as practised by the *Lu-shan* Taoists of Northern Taiwan", in M. Saso and D. Chappell, Editors, *Buddhist and Taoist Studies, I*. (Hawaii: University of Hawaii Press, 1977), pp. 79-94. See also J. Lagerwey, "The Fa-ch'ang Ritual in Northern Taiwan", forthcoming, Hong Kong: Chinese University Press (see footnote 4 above).

9 M. Strickmann, "The Feel of the Law: Ritual Implement and Its Vicissitudes", in M. Strickmann, Editor, *Classical Asian Ritual and the Theory of Ritual*, forthcoming, Berlin: de Gruyter.

10 A. Kagan. "Eight Immortals' Longevity Blessings: Symbolic and Ritual Perspectives of the Music", forthcoming, Hong Kong: Chinese University Press (see footnote 4 above).

Diagram : Anxi Taoist *Jiao*

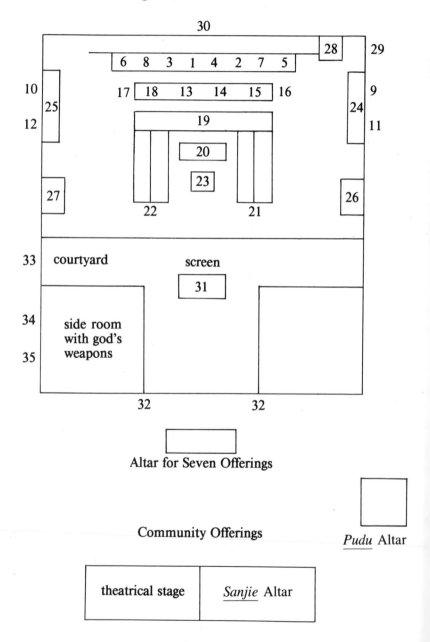

NEW DEVELOPMENTS CONCERNING BUDDHIST AND TAOIST MONASTERIES

Thomas H. Hahn

Prologue

This paper is mainly based on my fieldwork and recordings of excursions undertaken between September 1984 and September 1986. During these two years I visited around forty sites relevant to Taoist religious history (see the appendix for exact dates and places), climbed numerous mountains and talked to hundreds of different people engaged in or responsible for a social phenomenon that can be termed the rediscovering of a long abused religious consciousness. To thank all the people who helped me gather information I want to seize this opportunity and express my special gratitude to the following persons who may stand for the vast number of informants and helpers I have had to trouble with my many questions; foremost there are the monks of Qingchengshan 青 城 山 (Sichuan province) who were (and still are in fact) extremely helpful in making me comprehend modern Taoist history. There is furthermore Li Yangzheng 李 養 正 of the Baiyun Guan 白 雲 觀 (White Cloud Monastery) in Beijing who was helpful in clarifying a few problems concerning the new mode of education employed in modern religious institutions. The abbot Li Gaozhi 李 高 智 of the renowned Taiqing Gong 太 清 宮 in Shenyang 瀋 陽 (Liaoning province) must be thanked for his very cordial reception, which provided an unparalleled example of optimism and vigour drawn from a deep conviction in his faith. My feelings of gratitude also extend to the teachers and research workers at the Religious Study Department of the University of Sichuan who made valuable materials available to me. Finally, I have to thank Stephen Bokenkamp for reading through the draft, and preventing me from falling into the many pitfalls of English grammar and orthography.

Let me briefly describe the scope and limitations of this essay. I want to cover the time from 1980 until 1986. Many temples and monasteries have been restored on a nationwide level, many shrines have been erected at the "grassroots" level in the countryside. I do not attempt to give a socio-political explanation for these massive reconstruction projects. Rather, I will use a few examples as case studies to make a point about setting up organizations, religious "work units" etc. Three newspaper clippings from different Chinese publications used in the second section of this paper demonstrate the difficulty in dealing with numbers of monks and monasteries. The complexity of the problem of a future survival of religious life as keyed to the succession of old monks by young ones is the theme of the third section. Some afterthoughts represent my modest attempts to cope with the vast mass of data that have been piling up during these two years of extensive travelling. Logical deductions seem to be true and universal when they are verifiable by history. Yet history as it unfolded in China over the last one hundred years or so appears to be extremely hazardous. For this reason I refrain from giving definite statements about the meaning of religious activities in present day China, but rather advance some reflections about the religious potentiality that can be perceived by studying its outward manifestations.

Introduction

The religious policy of the People's Republic of China has been viewed with the utmost suspicion by foreign observers and scholars. Hasty as scholarly business tends to be nowadays, even the visiting experts are only able to catch a glimpse of a religious potential that sprouts forth throughout a China that is trying to redefine its traditional values. The feeling of suspicion is nourished by an authoritarian decision-making elite that tries hard to control the old creeds and customs as well as the new ideology and belief structure of the future. How far-reaching the radical narrowing down of socialist-realist "necessities" can be, was visible during the "Cultural Revolution" (1966-1976). China's religious heritage and practices were under accusation, and, having no spokesman of importance by its side, no belief system had the strength to withstand the onslaught of the Red Guards, who had elected a different god to be their master of destiny than, say, the Jade Emperor. Some of what was left in ruins for a decade or longer is now being rebuilt, and the generation of iconoclasts is left alone in pondering the damage done and the motivation that stood behind such a destructive annihilation of cultural diversity. Yet whatever the charges held against the spirit of the sixties and early seventies in China, mother nature is quickly covering over the scars of devastation with the lethargizing weeds of forgetfulness. The picture of a bright future has been distributed to everyone, and everyone is hurrying to get there. The ones who seldom forget are the victims of former days, in our case the monks, professional clergymen, priests and sorcerers, not to men-

80

tion the crowds of believers of every faith. Thus in the early years of this decade, voices were raised that propagated the reconstruction of destroyed monasteries and temples. We will see below that in some instances the recovery of religious activities is extending to "excessive" levels that have not been reached since late Qing times. I suggest that the surge of religious revival and reconstruction is keyed to economic growth and expansion, as well as to a degree of individual freedom never experienced before in China.

With the reconciliation efforts aimed at religious groups, a modest learning process by the authorities is becoming evident, with the effect that research centres for religious studies are being established.[1] At the same time monks and nuns are allowed to claim back their original habitats, the temples and monasteries. This paper deals exclusively with this phase of a religious recovery that belies the drastic and definite statements made only three or four years ago about the death of religion in China by Chinese and Western scholars alike (though with a completely different tone and arguing from opposing points of view)[2].

A New Start: The Revival of Guangdong Taoism as a Case Study

Based on information from the Hong Kong Taoist Association[3] and the Taoist monks at Luofu Shan[4] 羅浮山 , a list can be drawn up of the still existing (abbreviated 'E') and the still active (abbreviated 'A') Taoist temples in Guangdong province:

1. Sanyuan Gong 三元宮 (A), Guangzhou City 廣州
2. Yuanmiao Guan 元妙觀 (A), Huizhou City 惠州
3. Chongxu Guan 沖虛觀 (A), Luofu Shan 羅浮山 , Boluo Xian 博羅縣
4. Suliao Guan 酥醪觀 (A), Luofu Shan, 羅浮山 Boluo Xian 博羅縣
5. Jiutian Guan 九天觀 (A), Luofu Shan, 羅浮山 Boluo Xian 博羅縣
6. Dongzhen Guan 洞眞觀 (A), Zhonggu Yan 鐘鼓岩 , Nanxiong Xian 南雄縣
7. Yunquan Xian Guan 雲泉仙館 (E) (A?), Xijiao Shan 西樵山 , Nanhai Xian 南海縣
8. Feixia Dong 飛霞洞 (E) (A?), Qingyuan Xian 清遠縣
9. A temple or temples at Dinghu Shan 鼎湖山 (A?), Zhaoqing Xian 肇慶縣

There may be many more Taoist temples and shrines now under restoration in the province. The above nine names signify the main projects under restoration or completed already. The second on the list serves as the seat of the Guangdong Taoist Association[5] 廣 東 道 教 協 會 , and the first of the Guangzhou Taoist Association 廣 州 道 教 協 會

The number of resident monastic priests and novices is believed to be around fifty.[6] The process of reactivating any of the above monasteries may be divided into three stages:

1. The old monks are called back from the fields, the retired workers' hospitals or any kind of work unit to which they might have been assigned (at some places they had never left; see the examples of Hua Shan and Weiyu Shan given below). The religious affairs office 宗 教 事 務 處 at the district or the provincial level has lists of monks who were forced out during the Cultural Revolution.[7] Funds are now raised for acquiring building materials and the services of skilled workers who are able to execute the woodcarvings and stonemasonry on a more artistic level than the sloppy workmanship that can be noticed all over the rest of secular China. Officials sometimes take special pride in employing the best carpenters of the country, using the best materials[8] that can be found and attracting the largest crowds of tourists. The raising of funds, however, merits our closest attention, since it is closely linked to the initiative of the local people who at times are involved in the restoration process and at other times (and places) are not. Guangdong province seems to be a special case because it is the only province in China where money is being invested by foreign sources into Taoist temples. All over China the influence of Japanese money can be felt, but it usually only flows into Buddhist reconstruction projects. In Guangdong it is particularly Hong Kong money that is responsible for the fact that at least six of the above mentioned nine places will be opened to the public in the near future, or are open already. The reason for this investment is clear: the Hong Kong based Taoist institutions formerly came from such locations as Xijiao Shan (seventh on the list), but moved to the New Territories before 1949 (like the Qingsong Guan 青 松 觀 , which now runs a home for the aged and elderly, a junior high school and a hospital in downtown Kowloon). The old and well-established connections with authorities from the other side of the border pay off by allowing the Hong Kong institutions to be in full control of the restoration process and the assignments of monks. The stone tablets at the back of the Sanyuan Daidian 三 元 大 殿 of the Sanyuan Gong list the donors, all of them from Hong Kong.

2. While the temples are restored and the monks are called back, a Taoist association (on whatever level: local, district, or

provincial, depending on the importance of the monastery and its uniqueness in a given area)[9] is formed. This association consists of old and sometimes young monks, scholarly secretaries (jushi 居 士) and advisers assigned to the board by the government. The (fulltime) task of these advisers is to supervise the activities of the Taoists, implementing the religious policy of the hour (or the place).[10] They also supervise the contacts with other organisations and foreign visitors, acting as guardians and custodians of official opinion. The first Taoist association to be formed was on a nation-wide level, to be sure, but very soon provincial authorities were encouraging religious groups to enter into some form of organisational structure (Guangdong: 1985; Sichuan: 1983; Hunan: Feb. 1985; Hubei: May 1985; etc.) After a certain initial period, the Taoist organisations are allowed to manage not only the upkeep of the halls and shrines but also the financial side of their religious establishment.[11] Being in control of economic resources means being in the position to decide whether new monks (novices) should be taken into the monastery or not. This decision can only be made, however, in agreement with the local authorities and on the grounds that the faith in which the old monks believe, is indeed a living one.[12]

3. The next stage in our short survey is the founding of training centers for the young men and women who have decided to enter the monastery. Looking at the Sanyuan Gong we may draw up the following timetable:

1981 Funds are allocated for reconstruction
1982 Construction of the new halls starts
1983 The bell is cast; novices are admitted
1984 The drum is put into place, the *Kai guang* 開 光 ceremony is prepared
1985 Construction of the main halls is finished; a guesthouse is built and the Taoist association is establishing itself
1986 Printing of scriptures has started; pilgrims are able to stay in the guesthouse; a small-scale training school has been inaugurated.

Some of the more promising students are sent off to Beijing, where they participate in the lectures and seminars at the Baiyun Monastery. It is China's only "higher educational institution for Taoists" 道 教 知 識 專 修 班 , and enrols around forty students from all over the country each year. Lectures are given on Taoist history, *neidan* 內 丹 , politics and other topics. Thus the social, or at least the physical, mobility of the new generation of Taoists is very high. The training centres on a more regional level usually provide only one year's training, which, in comparison with the Buddhist high level research institutes, is very little time to digest 2000 years of history. Therefore, although the cultural level of the novices is higher than before 1949, the education still lacks a sense of identity, a substantial ingredient for a religion that is split into so many sects and subsects as Taoism. The same has to be said with some reservations about Buddhism. Included in this last stage of spiritual and administrative consolidation are the ordination ceremonies that should take place when the formal and required training of the novices is over. Ordination will be discussed in greater detail in part three of the study. Before we approach this specialised topic, I would like to draw attention to a problem of a more general nature.

The Difficulty of Counting Monks and Monasteries

The problem of how to estimate the number of monks in China nowadays is very complex. Obviously, the number of monasteries affluent enough to guarantee an economic standard of some significance is not high. In asking about the exact figures of religious professionals, we have to take into account the kind of priests who never leave home, who can marry and mingle with society.[13] We will see very soon that even government sources have great difficulties in sorting out who is and who is not a monk, a priest or a sorcerer. On a more superficial level we are confronted with monastic institutions of Buddhism and Taoism, and it appears that at least the active temples should be discernible and countable. Yet we are not only dealing with quantitative entities, but also with qualitative ones, as we will see below. Let me first point out that various sets of figures on monks and monasteries exist, and the qualitative distinction involved here does not concern the state of repair of certain temples, but rather the "state of mind" with which those statistics were put together. It is clear that at least three main sources of information must exist:

1. Government sources, based on registration tables etc.
2. Information gathered by the various research institutions that are either attached to the cultural affairs office of any level, or are affiliated with the Academy of Social Sciences.[14]
3. Taoist and Buddhist sources. The associations themselves undertake surveys involving religious activities in a given area.

For each of the above listed agencies which process information and data, I will give an example of printed evidence, some shorter, others more exhaustive. Before we enter into this maze of numbers, let me first give one definite starting point. The number of monasteries that attracted the attention of the Religious Affairs Bureau of the State Council in Beijing[15] is around 163. Lamaist temples are not included. Of these 163 religious institutions, 21 belong to the Taoist faith.[16] Thus Buddhism has a strong position throughout the country with 142 monasteries and shrines. That this figure can lead to absurd conclusions will be quickly revealed by the first newspaper clipping. Looking at the Taoist places only, it is evident that most of the temples are located in major cities or in areas where one would expect a Taoist shrine anyway.[17]

We may now proceed to our first example of "official" numbers and statistics.

Example 1

Under the heading "Fujian Sites restored to Buddhists" it is reported that "more than 2,400 Buddhist monasteries in Fujian Province have been returned to their original owners — monks and nuns — following repairs since China began normalizing its religious policy in 1979. Fujian in South-East China is one of [the]country's major Buddhist centres. It had 3,500 monasteries in the 1950's." Later in the same article we are informed that "there are 6,600 Buddhist monks and nuns in the 2,400 monasteries that have received repairs in recent years".[18]

Astounding as these figures seem, we will only know how to evaluate them by putting them into a contextual reference with the figures that are given in examples 2 and 3. Still, one is inclined to wonder whether the temples of the Goddess Tian Hou 天 后 (better known as Ma Zu 媽 祖) are not included here. Ma Zu shrines can be found in great numbers all along the southern coast of China.

Example 2

Under the title: "Visiting Taoist Temples in the Wenzhou and Taizhou Areas" we read:
> In September 1983 the [Taoist] association sent persons out to Zhejiang province, namely to inspect the state of affairs [of Taoist activities] in the city of Wenzhou 溫 州 , the counties of Pingyang 平 陽 , Cangnan 蒼 南 , Leqing 樂 清 , Ouhai 甌 海 , as well as to the counties of Wenling and others that belong to the prefecture of Taizhou 台 州 .[19] According to a preliminary statistic we have [counted] in the

above mentioned five districts and one city a total of 314 Quanzhen 全眞 Taoists (219 male, 95 female), and in respect to larger and smaller monasteries there are 104. In the Cangnan area alone there are 198 *daoshi* and 45 temples. Besides that, there are still *huoju daoshi* 火居道士 [20] who live among the common people, but because at the time the number was hard to determine we do not have a statistic.[21]

This internal report of the Taoist Association of China takes us very close to the stage that I have called the "grassroots" level. Looking at the two examples given so far, we are tempted to use example 2 to verify the figures given in example 1. Consequently one is led to the assumption that the number of nuns and monks in present day China must actually be much higher than the usually accepted 30,000 for Buddhism and 3,000 for Taoism. The problem enters its second qualitative stage when we have to ask the question: "Who, after all, can be considered a monk today?" Answers to this problem range from "A monk is someone who lives, works and dies in a monastery" to " A monk is someone who performs religious duties in society, i.e. a priest", to the last and probably most general statement: "A monk (or a priest) is someone who was ordained at some point in his religious upbringing and who is *still in the possession of his ordination documents* ". This last point has to be underlined because neither ordination registers (登眞籙) nor personal documents appear to have survived the turmoil of the Cultural Revolution.[22] Having no objective means of determining who is or who is not a monk or a priest or a magician (the latter seem to avoid the Chinese masses anyway and rather choose to perform among the minority groups to which they belong) is a further obstruction to our intention to evaluate our sources of information. Just to say that a monk should be a recognizable member of a given social group with religious inclinations and qualities like a monastic congregation is still too vague, because they may well live at home unrecognized and unwilling to show their spiritual (or physical) abilities in public. The problem is so complex, indeed, that I can only propose to cover the monastic side of religious life. One should be aware of the fact, however, that there is a certain number (be it high or low)[23] of religious professionals still performing (or again performing) services and ceremonies outside any monastery, an activity which is forbidden by law.

Again, one definite way of verifying a statistical figure is by comparing its peculiarity to any historical account that might contain the same irregularity. In our case this is the exceptionally high number of monks in the county of Cangnan (example 2 — I see no possibility to verify objectively the figures given in example 1). Cangnan is a new county created during the land reform instigated by communal cadres around 1951. It used to belong to the county of Pingyang. In 1923–24 the Zhejiang provincial government undertook a statistical survey of all the monks and nuns (Buddhist and Taoist) who were registered or at least recognizable (supposing that at that time the great majority did not have a reason to hide their true identity). Checking the figure of Taoist monks of Pingyang county we see that it is the fourth highest number in all Zhejiang with 702 monks and nuns. Only Shaoxing county (851), Hangzhou city (879) and Huangyan county (1893) range higher.[24] It is stated that most of the Zhejiang monks or nuns were of the Quanzhen order, subsect Longmen 龍 門 .[25] The total number was given as 15,530, a substantial number. Yet compared with the number of Buddhist monks and nuns (476,373!) the relationship stands at one Taoist for thirty Buddhists.

Of whatever nature our doubts with respect to the figures given in the two examples above, there is still a third newspaper clipping that will broaden our view and deepen our understanding of what actually is the "grassroots level" development I was hinting at above.

Example 3

Under the title of "The Practice of Excessive Restorations of Shrines and Temples in Shehong county" the newspaper informs the Chinese public about a certain religious expansion that is taking place at the same time as the country prepares to sell its own space shuttle services (the Einstein syndrome again). Since this third article is of considerable value, I translate it in full:

This year [1986] circumstances concerning the excessive renovation of temples and shrines in the villages of Shehong county are rather serious. According to the estimates of the authorities in charge, the number of *newly* erected temples and shrines in the 75 townships of Shehong county is not below one hundred. Within the first three months of the year altogether fourteen petitions for the erection of temples and shrines have been sent to the responsible government bureaus. Moreover, there is still a fair number of cases in which no application was forwarded but construction has been undertaken anyway. These temple restoration practices can be divided into two kinds: one is to build first and to report later. This year it was to a great extent the case that work had already begun or was near completion, and only then were the government offices in charge informed. At that time the "wood had already become a ship" (i.e. the temple was a reality). The other way is to build and not to report at all. A great number [of restoration or construction projects] are not reported to the authorities, and especially the township governments and the village party committees deal with the matter on the basis of "if we hear about it, we deal with it, otherwise we don't have anything to do with it". Therefore the real situation of the county cannot be grasped accurately. The process of building a temple is used by everyone as a pretext to give vent to one's pent-up feelings, and wood and materials are used [in great quantities] to the effect that the financial resources of the people are drawn together [for a disagreeable purpose]. Some people who have the intention of building a temple or a shrine collect money by offering to sell a bowl of auspicious "donation-rice" (消 災 飯) or something of this sort, and within a few short months they have collected some 80,000 *yuan* from everybody.[26]

Obviously these findings motivated the staff of the newspaper to write a criticism and discuss the problem on a more official level. That Shehong county 射 洪 縣 is not the only place where such phenomena abound begins to dawn on one as one reads through the review. Shehong county, it must be stated, serves as a warning example to other localities where cadres and officials seemingly have the same difficulties in obtaining information on what actually is going on in their region.[27] A reviewer analyses the situation in the following manner:

The excessive renovation and construction efforts of people building the temples and shrines in the villages of Shehong county deserve our closest attention. According to my understanding there are not a few places in the province [of Sichuan] where the situation is similar.

The circumstances of excessive repairs and construction of temples and shrines are quite complicated. Basically, there are three kinds of conditions. First, the Taoist or Buddhist monks or believers of either creed who live in the community or village[28] collect money out of their own initiative in order to build [or rebuild, the text is not clear here] a Taoist or Buddhist monastery or temple. In regard to these kinds of conditions, the responsible authorities of the local Bureau for Religious Affairs should negotiate with the followers of the faith and, on a reasonable scale, organize a religious activity centre, if in the past there has not been such a place. Against the disproportionate construction of temples and shrines, measures should be taken in the form of ideological work, and the founder of the temple should use the place for other purposes or tear it down by himself again.

Secondly, there are shaman-women and sorcerers who cheat the masses and extort money from the people by stimulating them to build Earth God Temples, Guandi Shrines, River God Temples, and so forth. These so-called "temples" can be found in abundance,[29] and according to political instructions should be forbidden.[30] Thirdly, some places are establishing a reputation as tourist spots, and in order to lure excursionists, new temples are built and statues of deities are put where there has never been a religious edifice before. The local authorities should take appropriate steps to counter these activities of erecting temples for the sake of financial profit. It is not allowed to build [new] temples and shrines.[31]

Summing up our last text, we have to come to the following conclusions:

1. There is, for whatever reason, at least a county-wide (if not province-wide or nation-wide) manifestation of old beliefs and practices.

2. Large parts of this manifestation are not in accordance with the laws. In fact, jurisdiction cannot keep up with the pace of events.

3. Authorities are hard pressed when dealing with excesses of any sort, but seem to turn their heads the other way when they hear of a case of "first build, then report". In at least one place, the author has stumbled across a case of embezzlement of funds allocated for a local Buddhist temple.

4. We have no reason to distrust the precise figures given. Adding all temples, shrines, and monasteries of today, we still do not reach the number of religious sites that Grootaers (together with Li Shiyu 李 世 瑜 and Zhang Jiwen 張 冀 文) counted in Wanquan in the 1940's.[32]

5. The revival of old traditions is tied to economic issues. The country is prospering, and so are its gods.

6. It still appears to be true that "modern cities as centres of secularization have only a limited effect on the common people's attitude towards religion."[33] With all the ideological training with which a citizen is confronted, it is clear that another set of values is implanted in the minds of the city folk.

Resistance is much stronger in such closely knit social units as single lineage villages. Thirty-seven years of communist rule have not quite quenched the common people's thirst for security, peace, wealth and health. If those basic quests remain unfulfilled, the turn to a supernatural authority is only natural and stringent. How "empty" these old beliefs are today is yet another complex notion that lies beyond the scope of this article.

Ordination: Fulfillment and Future Task

We enter now the third and last level of our survey. First I discussed the restoration process in detail, next I provided some information about the various levels of religious activity, mainly referring to developments taking place in the countryside. The maze of numbers that we had the impudence to enter lies behind us; we are moving on firmer ground again. Yet unfortunately, we are not quite through. Some rough figuring about the ordained monks in China today is still ahead of us. Turning to the official opening ceremony of one monastery,[34] we notice that the performing priests are in their sixties. This means that they themselves have been ordained before liberation or shortly after, in the early fifties. The generation gap that threatens Buddhist as well as Taoist survival results from those thirty years of uncertain religious policies of a Chinese communist party that was not necessarily united on the main issues of socialist progress. In order to analyse the succession crisis, let us first determine the respective age groups of monks; first Taoist, then Buddhist.

Starting from China's republican age (1912-49), we are hard pressed to resolve the question of the last ordination of Taoist monks. There used to be various places for ordination into monkhood, like the Baiyun Guan 白 雲 觀 in Beijing or the Erxian An 二 仙 庵 in Chengdu. Yoshioka Yoshitoyo gives the year of the last ordination in Beijing as 1927,[35] and in *Zongjiaoxue yanjiu* vol. 5 p. 35, we learn about an ordination that must have been conferred in Chengdu sometime between 1935 and 1937. The latest date for any large number of Taoist monks to be ordained is difficult to establish, yet there were numerous occasions when a smaller monastery would confer ordination titles after an informal (and unofficial) examination. These small-scale ceremonies were called *kou chuan* 口 傳, in contrast to the documented, full-fledged (and very costly) ceremonies held in the public monasteries. The Taoist associations of the time did not approve of those

90

practices, just as the Buddhist association nowadays does not approve of the unregistered, non-public ordinations that were going on in the early eighties on Wutaishan 五台山 in northern Shanxi 山西. With the takeover of the communist government, the Taoists were prohibited from performing any further ordinations. For many years a status quo was maintained, until the Cultural Revolution started. This ended officially in 1979, and we now see some younger monks with the traditional topknot in Taoist monasteries, but to the best of my knowledge no Taoist monk has been properly ordained to this day.[36] The reason for this fact is that there has not been a basis for proper training of novices until two years ago, when a class for advanced students was arranged at the Baiyun Guan. The classroom ironically lies directly opposite the ordination platform;[37] but to cross the court would also mean to bridge a sixty year timespan, which even in respect to the dimensions of Chinese history is a long time.

To state the situation clearly: neither from hearsay nor from written sources (I have access to the complete and newest *Daoxie huikan* materials) have I received notice of any ordination being performed by Taoists in the last few years. As an old and agile monk of the Tianshi Dong 天師洞 at the Qingcheng Shan put it, "We Taoists have to gain the rights for ordination through long self-examination and study (考戒), while the Buddhists receive their documents just like that." Also during talks held with two provincial-level Religious Affairs Bureau officials, I gathered that no Taoist association of whatever level has yet applied for the right to transmit the vows.[38] Still, I was told confidentially by monks on Qingcheng Shan that possibly next year (1987) there would be an ordination ceremony, but no official acknowledgement has been made of this rumour so far. To sum up the situation as we now can observe it, we have to say that the first novices were assigned to Taoist monasteries in 1981, and that at some places they even outnumber the old monks; but although they may have achieved a major rank in the administrative hierarchy of the monastery,[39] they have not been ordained. It follows that because in Taoist monastic institutions we are dealing with relatively older monks, the problem of succession is more serious than in Buddhist institutions, where ordinations were still going on some thirty years ago. An approximate statistic of the relation between young and old monks can be deduced from the table of places which I visited (see appendix).

Buddhism is a very different story. Not hampered by ignorant neglect nor by the accusation of being 80% superstitious, the Buddhist Association of China is following a straight course of expansion.[40] The toll of thirty years of deprivation and persecution was not as great as that of the numerically weaker Taoist community. What is of the greatest weight nowadays is the fact that Buddhism is an international faith, with Japan being the closest, and certainly the most generous, neighbour the Chinese could wish to have. Many temples would find it hard to support themselves without Japanese donations to cover the basic (and sometimes also some quite extravagant) needs and costs. The Xiangji Monastery 香 積 寺 near Xian is an example of such a "mummified" temple, that is artificially kept alive because the Japanese Jodo 淨 土 sect claims its origin from the founder of the temple; namely, Shandao Fashi 善 道 法 師 of Tang times.[41] Numerous as monks and monasteries are, I do not share the pessimism expressed in Jan Yün-hua's article about the future of Buddhism in China.[42]

Here is a short list of ordination ceremonies that were held since 1981, including and starting out from dates and figures given in Jan's article, which I take to be reliable:
1. 1981, New Year's Day, at the Guangji Si 廣 濟 寺 , Beijing: 47 monks.
2. 1982, January, at the Tiexiang Si 鐵 象 寺 Chengdu: 21 nuns.
3. 1983, October, at the Nanhua Si 南 華 寺 , Shaoguan: 250 monks and nuns.
4. 1983, October, at Qixia Si 棲 霞 寺 , Nanjing: 162 monks and nuns.
5. 1985, October, at the Baoguang Si 寶 光 寺 , Xindu: 119 monks and nuns.
6. 1986, January, at the Guiyuan Si 歸 元 寺 , Wuhan: more than 300 monks and nuns.
7. 1986, May, again at the Qixia Si 棲 霞 寺 , Nanjing: no numbers available.
8. 1986, September, at Jiuhua Shan 九 華 山 , Anhui: around 500 monks and nuns.

This list includes around 1,400 newly ordained monks, but it has two serious defects. First, it does not give a differentiated figure for the number of monks who were reordained (補 戒). Only for the Baoguang Si can the definite number of reordained, thus much older, monks be given, namely 11.[43] That is roughly one tenth of the whole. Presuming that at other places the average percentage of reordained monks ranges around the same 5 to 10%, we have to subtract around 140 monks from our list, leaving us with around 1,260 young ones between the age of 20 and 30 years. The second and more serious flaw of our little list is that it cannot be complete. Trivial as this may seem, we do have to take into account those very active Buddhist centres on the southern China coast like the Guoqing Si 國 淸 寺 (Tiantai county, Zhejiang), the Tiantong Si 天 童 寺 (near Ningbo, Zhejiang), Putuo Shan 普 陀 山 (Zhejiang), the Lingyin Si 靈 隱 寺 (Hangzhou). Going from Zhejiang further south, we are tempted to ask what is happening in Fujian province, which, according to *China Daily,* boasts 1,400 temples with 6,600 monks, supposedly some young ones among them. Important centres like the Nan Putuo Si 南 普 陀 寺 (Xiamen) or the Kaiyuan Si 開 元 寺 (Quanzhou) are clearly training grounds for young Buddhist élite students. Guangdong province also has a large Buddhist following (compare No. 3 on our list), while the Donglin Si 東 林 寺 of Lushan 廬 山 (Jiangxi) is responsible for the reprinting of a great number of popular Buddhist scriptures distributed throughout the country. Our list also does not include the scores of monks who were ordained on Wutaishan, where one would expect a great number of monks to be ordained because of its location. (It serves the whole northeast of China, and the stream of pilgrims and tourists never ceases.) Last but not least, the more than 500 lama-series in Western Sichuan and Qinghai, not to mention the ones in Tibet, opened recently (another *China Daily* figure, I am afraid, of March 1986) after years of repairs, have also to be staffed; otherwise they would not have reopened at all. Considering the numerical potential of the places just enlisted, we have to come to the conclusion that the number of 1,260 young monks could easily be tripled to maybe 4,000, and that this number represents only the development of the last three or four years. Taking into account the high number of students presently enrolled in the over 20 Buddhist colleges (佛 學 院) throughout the country, we can easily imagine that the number of ordained monks will reach around 10,000 by the end of this decade. How qualified these newly trained monks will be is yet another question.

I am inclined to accord (on a very general basis) the young monks of today a comparatively high cultural level (compared to republican times, when many monasteries were declared orphan-ages), because they come to the monastery of their own free will. In many instances they follow a family tradition, representing for example the fourth generation to shave the head (even as a single child). Around 65% are high school graduates, and most of the higher institutions of Buddhist learning take high school graduates only. The rest are of middle school level, but they all have to be 18 years of age to enter the monkhood. This creates a problem in itself, because the principal moulding and character shaping has been done by government-run socialization agencies like public schools (with their weekly study sessions on socialist theory). Another problem connected with this relatively advanced age of entry into a spiritual training school, is that a certain number of meditations cannot be performed or learned any more. They have to be trained and "physiologically locked" in the body and the mind of the apprentice at the age of 10, 11, or 12 at the latest, otherwise they are insufficiently and incorrectly performed. It appears to me that most of these late childhood meditations are already irretrievably lost.[44]

Government regulations set the eligible age for ordination at 20 years. Monks or nuns must have been in the temple for at least one year, and the monks who act as priests in the ceremonies must have made their vows at least ten years previously (the easiest requirement that I could think of). Monks who are ordained now can start teaching novices in five years. The ordination ceremony is organised by the Buddhist association on a very independent basis. Funds and fees have to be raised by the monks themselves, who have the duty to inform a government office of their intention. There is no limit to the number of monks that a monastery can house, as long as they can be fed appropriately. This means that the amount of grain required has to be taken into account as well as the general living conditions. The situation only becomes complicated when a monk wants to transfer to a temple in a large city like Shanghai, Beijing etc., but has lived so far in the countryside. Since the cities have their restrictions on new citizens, it may happen that the change of status poses a real obstacle to the monk's mobility. Otherwise, if the parents, and consequently government officials of the responsible departments, give their consent, there is no obstruction to the applicant becoming a religious professional (whether it is because a number of martial arts films made at the Shaolin Si have been shown on TV during the last couple of months or because the family really desires to uphold a long tradition). These requirements apply to Buddhists and Taoists alike.

General Remarks and Conclusions

The modest structural strengthening of China's religions has resulted in a consolidation of religious practices. Assuming that Chinese religious policy will not change over the next fifteen to twenty years, I foresee that we will not only have a number of very well written studies about a given religious topic, but also that the person who takes up the faith today will be perfectly able to embody the teachings and maybe even develop them in relation to the needs of the time.

A monastery may be a training place for the future generations of professional priests, it may become a museum or a pilgrim's target, or those qualities may be united in one larger temple. Very seldom have I had the luck to stay in a quiet and contemplative temple such as used to exist on Huashan and on many other mountains. Most of the monks in China today are engaged in some kind of production process like farming, tea planting, operating factories or other minor handicraft assignments. However positive such a development may be deemed, the problem of meaning remains. The traditional values of the Chinese people have been subjected to drastic changes, and the explosion of the birthrate starting in the 1950's (around 50 per cent of the population are 28 or younger) enhances the dramatic drain of historical consciousness. The services which a monastery, then, is able to render to the community outside, and to the community of monks inside its precincts, are (and will, for a long time to come, be), severely limited in scope and quality. It is exactly at this point that the young generation of monks has set to work, that is, to overcome the limitations, and to establish a meaningful social context for the monastery within its area of influence and within the Chinese state of tomorrow. How well this enormous task can be fulfilled, depends on many different factors; but certainly one of them, if not the most important, is the determination to objectively re-evaluate the balance of theist and non-theist spirituality. That this may sound a pathetic conclusion, is due to a few historical truths which the Chinese, as well as we, have to accept and to live with.

APPENDIX

1. List of Taoist Places Visited
by the author between September 1984 and September 1986

Some of the places have become Buddhist. In these cases I have added a B in parentheses behind the place name, and, where possible, give a date when the "conversion" took place. The letters DDT mean *Da Dongtian* 大 洞 天 ; XDT means *Xiao Dongtian* 小 洞 天 . I refer the reader to the list of Du Guangting in DZ 599/331. For a more detailed study see E. Chavannes, *Le Jet des Dragons. Memoires concernant l'Asie Orientale* III, 1919.

PLACE	TIME	NUMBER OF MONKS[45]
Hua Shan 華 山 (Shaanxi)	May 1985	around 35 + 20
Song Shan 嵩山 (Henan)	Sept. 1985	around 20 + 15
Heng Shan 衡 山 (Hunan)	Dec. 1985	5 + 4
Weiyu Shan 委 羽 山 , DDT 2 (Zhejiang)	Nov. 1984	2
Qingcheng Shan 青城山 , (Sichuan), DDT 5	five times so far, 1st visit in Jan. 1985	18 +23
Luofu Shan 羅 浮 山 , DDT 7 (Guangdong)	April 1985	2 + 4
Chicheng Shan 赤 城 山 , (Zhejiang) DDT 6 (B era of Qianlong 1736-1795)	Nov. 1984	none
Gouqu Shan 勾 曲 山 , DDT 8 (Jiangsu)	May 1985 and Sept. 1985	25 + 20
Linwu Shan 林 屋 山 , DDT 9 (Jiangsu)	Dec. 1984	none
Kuocang Shan 括 簪 山 , DDT 10 (Zhejiang)	Nov. 1984	none
Douquan Shan 竇 圈 山 (Sichuan) (B, Northern Song)	Sept. 1986	none
Yulei Shan 玉 壘 山 (Sichuan)	three times, first visit in Jan. 1985	4 + 3
Ge Ling 葛 嶺 (Hangzhou, Zhejiang)	Oct. 1984	12 + 5
Wu Shan 吳 山 (Hangzhou, Zhejiang)	Oct. 1984	30 (?)
Gaizhu Shan 蓋 竹 山, DDT 19 (Zhejiang)	Nov. 1984	1 + 1
Tongbo Shan 桐 柏 山 (Zhejiang)	Nov. 1984	1 + 1
Longhu Shan 龍 虎 山 (Jiangxi);	May 1985	3 +[46]
Wudang Shan 武 當 山 (Hubei)	Sept. 1985	55 + 40
Yaowang Shan 藥 王 山 (Shaanxi)	May 1986	none
Qian Shan 千 山 (Liaoning)	May 1986	18 + 14
Zigai Shan 紫 蓋 山 , XDT 33 (Hubei)	Feb. 1986	none
Emei Shan 峨 帽 山 , XDT 7 (Sichuan)	Sept. 1985	none[47]
Lao Shan 嶗 山 (Shandong)	May 1986	around 12 (?)

Zhonggu Yan (Guangdong)	鐘鼓岩	Apr. 1985	1
Heming Shan (Sichuan)	鶴鳴山	Sept. 1986	2
Pingdu Shan (Fengdu, Sichuan)	平都山	Dec. 1985	none
Zhongnan Shan (Louguan, Shaanxi)	終南山	Sept. 1985	20 + 15
Qingyuan Shan (Fujian)	清源山	March 1986	none

2. Monasteries in or near Cities

Xuanmiao Guan (Suzhou, Jiangsu) 玄妙觀	visited three times between Sept. 1984 and Sept. 1985	around 10 + 5
Changchun Guan (Wuhan, Hubei) 長春觀	Sept. 1985	over 50[48]
Taiqing Gong (Shenyang, Liaoning) 太清宮	May 1986	around 70
Qingyang Gong (Chengdu, Sichuan) 青羊宮	visited many times; first visit in Jan. 1985	15 + 12
Baiyun Guan (Beijing) 白雲觀	May and June 1986	around 70 (?)
Baiyun Guan (Shanghai) 白雲觀	visited many times between Sept. 1984 and Sept. 1985	11 + 6
Chunyang Guan (Xinjin 新津 , Sichuan) 純陽觀	Jan. 1986	none
Sanyuan Gong (Guangzhou) 三元宮	visited several times; first in April 1985	8 + 7
Chongyang Guan (Hu Xian 戶縣 , Shanxi) 重陽觀	May 1986	3
Baixian An (Xian, Shenxi) 八仙庵	visited several times	around 20 + 15

NOTES

1 For a description of the Religious Research Institute of Sichuan University, see my article "Fieldwork in Daoist Studies in the People's Republic", in *Cahiers d' Extrême-Asie* 2, Kyoto 1986.
2 Especially, K. M. Schipper's remarks in his *Le Corps Taoiste* (Paris, 1982), p. 32, have to be reconsidered in the light of very recent devel-

97

opments.

3 Visited in July 1985. The talk was with Leung Sing-chung 梁省松 , secretary general of the Hong Kong Taoist Association.

4 There were two at the time. Both were sent from the San Yuan Gong to act as overseers over repairs that had to be undertaken. One was a former teacher at Zhongshan University in Guangzhou, the other one held a *pinqingshu* 聘請書 from Qigong Yanjiu Shi 氣功研 究室 of the Zhongshan Yixueyuan 中山醫學院 (Zhongshan Medical College), also at Guangzhou. Both are in their early seventies and belonged to the Quanzhen/Longmen sect.

5 Established in the autumn of 1985.

6 Yu *daoshi* 余道士 of the Chongxu Monastery made this estimate in May 1985.

7 These lists are unfortunately not available in the Xinhua Bookstores, yet an old Taoist in Hangzhou told me about them and also stated that the ones that were considered first to return to their former state of monkhood were those belonging to the Quanzhen order, being the least harmful sect when compared to the costly ritual practices of the Zhengyi priests who do stress other forms of interhuman relationships through their rituals and funeral rites than the Marxist-inspired notion of equality of material resources permits.

8 People in charge of works at Longhu Shan 龍虎山 proudly state that the tiles used for the roofs of the Tianshi Fu Halls each cost one *yuan* (RMB, equal to one German Mark at the time), and that the only place where they were still being made was Xiamen in Fujian province.

9 For example, the Taoist Association of Hunan is located in the Xindu Guan on the Southern Hengshan 衡山 , simply because there is only one more Taoist temple in Hunan and a very small one at that (the Hetu Guan 河圖觀 in Changsha Xian 長沙縣 , not Changsha City!).

10 Some places tend to be more relaxed about their attitude towards "superstitious" practices like drawing lots and physiognomy.

11 Until only recently, the income from the entrance tickets of Qingcheng Shan went straight into the pockets of the local government.

12 A negative example is Xiangji Si 香積寺 near Xian. The local Jingtu 淨土 Sect is nearly extinct.

13 The appropriate term is *huo zai shehui shang* 活在社會上 . Some priests have gone "undercover" for thirty years or more, and find it hard (if they should desire to do so at all) to transmit their knowledge.

14 The Academy of Social Science of Shanghai issues two *neibu* 內部 papers, one called "Newsletter of Religious Studies" (宗教研 究通訊), the other is entitled "Investigation of Religious Problems" (宗教問題探索). The former is edited together with the Religious Study Society of Shanghai. Both papers deal with what I want to call "grassroots phenomena". This expression is meant to point to certain social occurrences of religious quality that are extremely limited and tied to a special place. For more information on these kinds of phenomena the reader is referred to the newspaper

clipping no. 3.

15 I held talks with officials of the Guowu Yuan 國 務 院 in September 1985 and in May 1986. One of their main problems seems to be why Einstein believed in God.

16 The list of 21 has been expanded to sixty. See my article "Fieldwork ...", cited in footnote 1, there especially footnote 8. So far I have been to around forty mountains and monasteries; see Appendix 1 for times, places and numbers of monks attending.

17 On the Five Sacred Mountains (五 嶽), for instance, or in the vicinity of a cave that counts as a "Grotto-Heaven" (洞 天). Taken altogether, we reach the figure of 51 places (five mountains and 46 grottoes), *all of them in the countryside.*

18 *China Daily,* March 25, 1986

19 I am preparing a major study on this prefecture during the Tang period.

20 I.e. the Zhengyi 正 乙 priests.

21 See DXHK No. 13, p. 4. On the same page the number of Taoists in Wuhan is listed as 162, 18 of them of Zhengyi *lineage.* 正 乙 道 士

22 The bonfire of sacred scriptures and documents lit in the court of the Xuanmiao Guan 玄 妙 觀 in Suzhou is said to have lasted for three days and nights.

23 It varies from area to area, but generally speaking the coastal provinces of Guangdong and especially Fujian have a high percentage of liturgical experts. As Kenneth Dean once pointed out to me, he had come across a community in the Jinjiang 晉 江 area where almost every person of a single lineage group could perform a complicated Taoist ritual, thus acting as a priest in charge of the ceremony (*fashi* 法 師). See Ken Dean's article in this volume about further research on religious practices in Fujian province.

24 I visited Huangyan county in November 1984 to examine the topography of the second big "Grotto-Heaven". I was received by two *daoshi* who were living in the Weiyu Shan Guan 委 羽 山 觀 . This inconspicuous little place has attracted the highest number of *daoshi* in all of Zhejiang, for reasons which the author has set out to explore in a forthcoming historical study.

25 Compare Zhejiang Tongzhi 浙 江 通 志 1926, Hangzhou, ch. 96. The relatively high number of Longmen monks is already emphasized by Yoshitoyo Yoshioka in his "Taoist Monastic Life" in Holmes Welch and Anna Seidel, *Facets of Taoism* (New Haven and London: Yale Univ. Press, 1979); see especially p. 233, bottom paragraph.

26 On the social priority levels "undermined" by religious ideas see Yang, *op. cit.,* p. 16

27 In some instances officials are supporting the construction of a temple for reasons of their own (one of them is listed under the third point of the translated newspaper review on pp.61-62 above).

28 A village with a very large (homogeneous) population of believers is called 居集村 .

29 On a recent field trip to Douquan Shan 竇 圀 山 in Jiangyou county

(Sichuan) I counted 14 little temples and shrines on a two-hour walk. Out of the 14, six were *Tuti Miao* 土 地 廟 i.e. Earth God temples.

30 As I was told in the Sichuan and Hunan Religious Affairs offices, the laws governing a number of items (pilgrim movements etc.) are still in the making.

31 Article and review from the *Sichuan Ribao* 四 川 日 報 , April 27, 1986.

32 Grootaers' fascinating materials have not been fully exploited. For an area of about 80 townships he counted 565 temples, roughly five times more than our Shehong county figures. See *Monumenta Serica* , 13 (1948).

33 Yang, *op. cit.,* p. 350.

34 Repairs on the Taiqing Gong 太 清 宮 in Shenyang will altogether take 6½ years. In a case like this, the monastery is advertising its services in the local newspaper before the official *Kai guang* ceremony has taken place.

35 *op. cit.,* p. 236 (Taoist Monastic Life).

36 Compare Jan Yun-hua's statement about two ordained young monks of Huashan and four more on Qingcheng shan. At least the latter must be an error of some sort. I have been five times to the mountain, and I was frequently (and with a kind of sad undertone) reminded that they had had no ordinations in Sichuan since the late thirties. See Jan, "The Religious Situation . . . ". *Journal of Chinese Religions* , 12, (1985) pages 46 and 49 respectively.

37 Yoshioka's plan of the monastery, *op. cit.* p. 250. Compare the position of the "Room of the mountain where many monks gather" (No. 71) with the ordination platform (No. 64).

38 There is a slight possibility that at this year's meeting of the national Taoist Association (terminated on September 18) the problem of ordination was discussed, and maybe even some definite steps were taken in the direction of a small ceremony some time within the next two years.

39 For ranks and titles see Yang in ZJXYJ 5, 1985, pp. 38-39. According to enquiries they have not changed since the 1930's.

40 There was a curious report about establishing a "World Buddhist Dharma-King Hall" in Seoul this year in the RMRB/HWB (September 16, 1986). The "Hall" was not recognized by the Buddhist Association of China on the grounds that it resembles the Vatican in Rome, which remains a very touchy subject for the Roman Catholic followers in China.

41 About Shan-dao Fashi, see the article by Julian F. Pas, "Dimensions in the Life and Teachings of Shan-tao', in *Taoist and Buddhist Studies* , vol. 2, edited by David Chappell. (University of Hawaii Press, 1987). pp. 65-84.

42 *Op. cit.,* p. 44. Concerning the Taoist faith, however, I find myself in total agreement with Jan (*op. cit.* p. 51).

43 My more specialised article on the ordination ceremony of Baoguang Si 寶 光 寺 is in preparation.

44 This applies even more to Taoist meditation practices.
45 The first figure indicates the number of old monks or nuns, the latter gives the number of young ones. A number followed by a question mark means the total cannot precisely be broken up into old and young monks.
46 I am presently compiling a separate study on Longhu Shan in Republican and later times up to the present.
47 Especially interesting are the two small Emei mountains some distance to the north of the main peak. All the place names have Taoist connotations. Unfortunately, nothing is left of the original structures except the foundations.
48 Compare my own recordings with the figure given by the Taoist Association, see footnote 21.

SYMBOLIC AMULETS AND JEWELRY IN CHINESE POPULAR CULTURE

Alvin P. Cohen

Jewelry is often seen in museum exhibits and books concerning the minor arts of China. This jewelry was usually manufactured for and worn by members of the imperial household or the upper classes, whereas that used by ordinary people is seldom displayed or discussed. Although the latter is seldom seen in museum exhibitions, the craftsmanship is often of high quality and the designs are delightful to behold. The designs on jewelry made for and worn by ordinary people very frequently contain motifs that are closely related to the symbols commonly used in popular religion.

During the Cultural Revolution (1966-1976) large quantities of folk jewelry were collected by the Chinese authorities (or Red Guards) — they said the pieces were "donated by the people". These jewelry pieces were regarded as artistically insignificant since they were outside the range of fine jewelry, as defined by the products manufactured for use by the upper classes. Consequently the pieces were all scrambled together without any regard for provenance. Miscellaneous large bulk lots of this jewelry were sold to foreign buyers during the early 1970's. A short time later, the Chinese authorities terminated these bulk sales, possibly because they could get higher prices by selling only small lots or because there was some realization of the cultural value of the pieces as folk art.

This essay is a brief introduction to several types of folk jewelry drawn from a collection[1] purchased in the People's Republic of China from 1971 to 1976. Many design types are represented in the collection, some types by numerous examples while others have only a few examples, and a few pieces are unique.

As far as I know the pieces were all collected within the P.R.C., but otherwise their provenance is unknown. The reverse side of some pieces bears what appear to be craftsmen's hallmarks engraved on tiny metal strips soldered to the back. The hallmarks are comprised of two or three characters — apparently names — but there is no way to identify these provincial craftsmen or their locations.[2] I have discussed many of the jewelry pieces with people in the P.R.C., Taiwan, and the U.S. who are knowledgeable about Chinese folk customs. The consensus of opinion is that they were most likely manufactured during the nineteenth and the early twentieth centuries, i.e., the latter years of the Qing dynasty (1644-1911) and the early years of the Republic of China.

Although I have not yet subjected the jewelry pieces to chemical analysis, it appears that most of them are made from silver alloys. Some also seem to be made of various types of "white metal" — silver colored alloys of lead or tin with copper and zinc. A small minority of pieces are made of brass. Most of the pieces are made of very thin sheet metal (about 0.5 mm or less thick) and are either single-sided or hollow when double-sided. The latter are usually made of two halves soldered together. The designs are mostly produced by hammering or engraving, while some pieces appear to have been cast and polished, with some detail added by engraving.[3] Many pieces have additional decor made of flat sheet or wire soldered onto the main sections. Chains and jump-rings seem to be made from drawn silver wire.

The jewelry shows varying degrees of wear, from very badly worn to almost none at all. The silver pieces all have varying amounts of black silver-sulfide corrosion, especially in the indented areas. The protruding surfaces are generally brightly polished from exposure to wear. Many of the pieces were quite dirty when acquired and have therefore been cleaned in an ultrasonic cleaner with soap and water. Beyond this, I have not attempted to clean or polish the pieces.

The manufacture of the pieces reveals some very skillful craftsmanship. Many pieces show intricate engraving, embossing, and filigree work. Some even have moving parts — there are many examples of fish designs (not shown here) with twenty to thirty moving parts connected by pivot posts, rivets, and jump-rings. It is very clear that the more intricate of these pieces required many hours of patient skilled craftsmanship for their manufacture.

Because folk jewelry has barely been studied in its social or cultural contexts, there is no basis for discussing specific uses of the pieces beyond a few obvious cases. It is necessary to leave many important questions open: for example, by whom were they worn (men, women, adults, children, babies, boys, girls)? How do the designs or motifs vary with locality? When were they worn (daily, for special occasions, on festival or ritual days)? To what extent were they worn solely as decorative jewelry or as amulets to invoke good fortune or to ward off evil influences? Until studies of folk jewelry in its cultural contexts become available, the answers to these issues will have to remain speculative and tentative.

The only extensive study related to this type of jewelry is Cammann's analysis and catalog of a collection of Chinese belt toggles worn by men.[4] Some of the pendants discussed below might have been used as toggles (e.g., Figs. 12 to 16, 20, 24) since the toggles were also made of metal.[5] A large proportion of the symbolic design motifs found on the toggles are also used in the jewelry designs discussed here. These symbolic motifs were drawn from a common body of symbols that were, and frequently still are, widely used throughout Chinese culture.

The motifs within the designs are mostly well-known symbols associated with popular religion, folklore, and the culture of ordinary people. Many of these motifs have antecedents in literature and art as old as the Eastern Zhou dynasty (770-256 B.C.). Some of the pieces were clearly used as amulets or votive devices, while others could provide double service as both amulets and decorative jewelry. The symbols invoke desires for long life, male offspring, wealth, protection from harm, or general good fortune, and sometimes homage to various deities. Some of the amulets are containers for charms or what were regarded as magically efficacious substances. Most of these contained the remnants of hair or decomposed (organic?) substances when they were acquired. Some of the pieces bear inscriptions with auspicious slogans. The inscriptions are usually in the traditional full-form characters, however abbreviated characters are not at all uncommon, as can be seen in Figs. 2, 4, 5, and 20A. Although most of the inscriptions read from right to left, some read from left to right.

Because relevant scholarship and research materials for the study of Chinese folk jewelry are very sparse, my discussion of the pieces will largely be confined to descriptions of the symbolic content of the designs and motifs. Comments relating to the religious and cultural contexts of the pieces will be largely speculative and preliminary.

I will describe several examples from six categories of jewelry pieces. This small set should not be taken as fully representative of the entire range of categories. Space limitations and the availability of satisfactory photographs have determined the selection presented here. Suffice it to say that there are additional categories of designs, and nearly all are represented by numerous variations. For purposes of introduction and description, I have divided the pieces into categories more or less on the basis of function and design rather than on any clearly specified classification in accordance with well-defined criteria.

The six categories are: 1. pendants in the form of "baby locks", 2. pendants on the theme "the *qilin* brings sons ", 3. pendants in the form of charm cases, 4. pendants with other symbolic motifs, 5. attachments for clothing, 6. hairpins.

Pendants in the Form of Baby Locks

This is a type of amulet worn by babies, but whether by both males and females and in what age range is not clear.[6] The basic theme of the amulets is protection for the baby's life, while a subsidiary theme is material success in later years. The design of these pieces is based upon the old type of Chinese lock which is very similar to the baby lock in Fig. 1A. Most of the baby locks were pendants with a fixed bar for attaching a chain. Some of the baby locks strive for authenticity[7] to the extent that one end is actually detachable as shown in Fig. 1B. The sections of the baby lock are held together by the friction between the lower split bar and an internal receptacle plus the friction between the end piece and the body of the case.

The baby locks are often referred to as "hundred family locks" *baijia suo* 百 家 鎖 (or 肖), a phrase which is often inscribed on the locks (see Figs. 2 and 4). The most common explanation that I have heard is that the family of the baby collected small donations for the lock from all their neighbours, thus ensuring the auspicious influences of all the families ("the hundred families") in the community.[8] A family that could not afford a silver lock might even embroider a fascimile of a lock on the baby's bib or apron.[9]

Figs. 1A and **1B** (piece no. 272).[10] Baby lock, silver. Engraved on the obverse (Fig. 1A) is the inscription: *changming fugui* 長 命 富 貴 "Long life, wealth and honors". The reverse (Fig. 1B) has a floral design. This is a good example of the type with a detachable suspension bar in direct imitation of the old type of Chinese lock.

length 44 mm, height 23 mm, thickness 14 mm.

Fig. 2 (nos. 105 to 110). Baby locks, silver. These represent a variety of baby lock shapes and inscriptions. All are hollow. No. 105 is cast, Nos. 109 and 110 are hammered. Nos. 106, 107, and 108 are engraved. The suspension bars are not detachable. No. 105 is inscribed is seal-type characters (R-L): *yutang* 玉 堂 "[May] jade [fill your] hall". Nos. 106, 107, and 109 are inscribed (R-L): *baijia suo* 百 家 鎖 "Hundred families lock" (note the abbreviated form of *suo*). No. 108 is inscribed (R-L): *songzi* 送 子 "[May you] get a son". Could this be a lock worn by a prospective mother? No. 110 is crudely inscribed (R-L): *baisuo* 百 鎖 "Hundred [families] lock" (again the abbreviated *suo*).

length 25 mm to 37 mm.

Fig. 3 (no. 263). Baby lock, silver. This is hammered from thin sheet metal. Additional small decor pieces can be attached through the three holes in the bottom rim. The inscription reads (R-L): *changming fugui* 長命富貴 "Long life, wealth and honors". On bosses at the top right and left are the characters for *ri* 日 "sun" and *yue* 月 "moon", respectively, with what may be three light rays beneath each of them. The moon and sun symbolize the *yin* and *yang* respectively, the primal creative forces of the universe (see also Fig. 19). A bat is depicted at the bottom center. The word for bat is a two-syllable word, *bianfu* 蝙蝠 , the second syllable of which is homophonous with *fu* 福 "blessings".[11] The bat is therefore a very common motif in both jewelry and embroidery designs. At the top-center and bottom-left and right are plum flowers, *meihua* 梅花 , usually depicted with five petals to symbolize the five deities of good fortune. The plum tree is the very first to bloom in the spring.[12] Therefore it is symbolic of the survival of hardship (and the "death" of winter) and the renewal of life and youthful vigor. The plum flower is also a very common motif in jewelry design.

length 69 mm, height 24 mm.

Fig. 4 (no. 126A). Baby lock, silver. An unusual variation on the lock design. It is hollow, with an embossed design. The central rod is removable (for what purpose?). The thick suspension bar and central rod make this piece quite heavy. The inscription reads (T-B-R-L): *baijia baosuo* 百家保鎖 "Hundred families protecting lock" (again the abbreviated *suo*). In the center is the character *shou* 壽 "Longevity", which is undoubtedly the most common decorative character in all Chinese popular arts and crafts.[13] A plum flower and leaf separate each of the characters of the inscription. The reverse side has a floral design.

overall length 71 mm, width 40 mm, thickness 15 mm.

Fig. 5 (no. 262). Baby lock, silver. This piece has a hammered design on one side only. It is made of thin sheet metal and has a large flange that provides the illusion of thickness. The inscription reads (R-L): *changming baisui* 長命百歲 "Long life for a hundred years" (note the abbreviated form of *sui*). Plum flowers surround the inscription.

length 84 mm, height 65 mm, thickness (width of flange) 9 mm.

Fig. 6 (no. 50). Baby lock, silver. This large piece has a hammered design on one side only. It is made of thin sheet metal and has a large flange that provides the illusion of thickness. Five wire loops soldered to the bottom part of the flange (the center loop is broken off) provide attachments for chains or additional decor pieces. This piece represents a sub-type that has an anecdotal scene without an inscription.[14] This scene depicts Wu Song 武松 killing a tiger with his bare hands, which alludes to an episode in the fourteenth century novel *Shuihu zhuan*.[15] Apparently Wu Song's courage is being evoked as a wish for the wearer of the lock.

length 102 mm, width 81 mm, width of flange 8 mm.

Pendants on the Theme "the *qilin* brings sons"

This type of amulet is represented in numerous variations in the collection. The basic design is a male child riding on the back of a *qilin* 麒麟, a mythical animal with the body of a deer covered with scales, the hooves of a horse, the tail of an ox, and a single horn on its head. This beast, often termed "unicorn" in English, supposedly appears only once every five hundred years and heralds the appearance of a great sage. Therefore the amulet implies the wish for a son who will be a great paragon of wisdom and virtue.[16] Doré implies that this type of amulet is worn by women who wish to give birth to a son.[17]

These pieces are usually two-sided and show the front and back of the child and the *qilin* (rather than having identical sides). The child and *qilin* both face in the same direction, which may be either right or left. The child holds onto the *qilin* with one hand and holds an object in the other hand, such as a peach, a lotus flower, or a *ruyi* 如意 scepter. The peach is a very common symbol of longevity, and its wood will destroy evil spirits.[18] The lotus flower is a symbol of moral purity, and also, because of the many seeds in its pod, implies numerous offspring, especially sons. The lotus (*lian* 蓮) also symbolizes the wish to continue begetting sons one after another — through homophony with *lian* 連 "continuous".[19] The *ruyi* is a scepter-like baton which implies *wanshi ruyi* 萬事如意 "May all things be as you wish", and in this design symbolizes the wish for professional and social success and advancement.[20]

Fig. 7 (no. 206). "The *qilin* brings sons" pendant, silver. The piece is hammered from thin sheet metal and is hollow. The suspension rings are on the *qilin's* head and tail. Four holes in the plane beneath the *qilin's* legs permit the attachment of additional decor. The child, with his hair tied in tufts on the sides of his head, holds a peach in his right hand. The character *shou* 壽 "longevity" decorates the child's sleeves and trousers. The pockmarks on the *qilin* evidently represent the scales that are supposed to cover its body.

length 81 mm, height (T-B) 80 mm, maximum thickness (T-B) 14 mm.

Fig. 8 (no. 202). "The *qilin* brings sons" pendant, silver. The piece is also hammered from thin sheet metal and is hollow. Four holes in the *qilin's* hooves permit the attachment of additional decor. The child, wearing a hat, holds a *ruyi* scepter in his left hand, the head of the scepter lying on his shoulder. Behind him is a bat. His trousers are engraved with a floral design. Around the *qilin's* neck and flanks is a harness, from which a pendant hangs between its forelegs.

length 70 mm, height 67 mm, maximum thickness 13 mm.

Fig. 9 (nos. 297 and 298). "The *qilin* brings sons" pendants, silver. Both pieces are two-sided, hammered from thin sheet metal, and hollow. In both pieces the child's hair is tied in tufts at the side of the head. In No. 297 the child holds an object (perhaps a fly-whisk, a symbol of authority)[21] in its left hand. Plum flowers decorate his trousers. In No. 298 the child holds a lotus flower in his left hand and wears what appears to be an elaborate belt buckle.

length (L-R) 40 mm (no. 297), 39 mm (no. 298).

Fig. 10 (no. 173). "The *qilin* brings sons" pendant, silver. This piece is hammered from thin sheet metal, and is one-sided. The design is set into an openwork background and completely surrounded by a raised border. The child, with his hair in tufts at the temples, holds a peach in his left hand. His right hand appears to be holding a lotus flower on a stalk over his shoulder. A plum flower occupies the center of the *qilin's* tail.

length (L-R) 88 mm, height (T-B) 72 mm, thickness 5 mm.

Fig. 11 (no. 172). "The *qilin* brings sons" pendant, silver. This piece is hammered from thin sheet metal, and is one-sided. Four holes at the bottom permit the attachment of additional decor. The design is raised from a solid background without a surrounding border. The triangular motif at the bottom represents mountains, symbolic of cosmic order and permanence.[22] The three-lobed motifs at 1:00, 4:00, and 8:00 o'clock may be bats. The child, wearing an elaborate hat, holds the stalk of a lotus flower in his right hand, with the stalk resting on his shoulder. The *qilin's* scales are clearly delineated, and around its neck is a ring of bells.

length (L-R) 91 mm, height (T-B) 72 mm, thickness 5 mm.

Pendants in the Form of Charm Holders

This type of pendant is a holder or case for some sort of charm. The charm might be either an inscription on paper or some substance that was supposed to have the power to ward off evil influences and protect the wearer from disease and ill fortune. Many of these pieces in the collection contained hair or the residue of some (organic?) matter when acquired. Human hair was used as a charm to ward off evil spirits, as well as the representative of a victim in black magic.[23]

Figs. 12A and **B** (no. 258). Charm case, silver. This piece depicts a male child[24] wearing an elaborate crown decorated with a plum flower, and holding a *ruyi* scepter (or a lotus flower?) over his right shoulder. The body separates at the waist for the insertion of the charm (Fig. 12B). The case is held together by a chain attached to the bottom section, passing through the head where it is attached to a large jump-ring behind the hat. A small jewelry piece decorated with a plum flower is attached with a jump-ring through a hole at the bottom.

length (closed) 74 mm, maximum width 27 mm.

Figs. 13A and **B** (no. 055). Charm case, silver. This piece depicts a female in the typical costume of the Hakka women of southern Guangdong Province, which suggests the provenance of this charm case. The hat is decorated with rings to simulate the long fringes on the Hakka hat. In her ears are movable earrings, and around her neck is either a collar or a large necklace. The ends of the ribbons around her waist hang down in front of her skirt and over her right arm. The device over her left shoulder (see also 13B) is a basket that may have originally been suspended from her left hand. The axial chain holding the case together may be seen clearly in Fig. 13B. The attachment at the bottom of the amulet holder is decorated with a plum flower. The careful workmanship and attention to detail seen here is also found in many other jewelry pieces.

length (closed) 56 mm.

Figs. 14A and **B** (no. 068). Charm case, silver. This case is in the form of a cicada (cf. Fig. 15), an ancient symbol of immortality. Its reproductive habit that gives the impression of rebirth after burial suggests that the cicada symbolizes the continuation of the family line generation after generation.[25] This piece separates into three sections, with the cage-like body being the charm holder (see Fig.

14B). A ring at the bottom permits the attachment of additional decor. A peach is attached to the suspension chain.

length (closed) 49 mm, width 24 mm.

Fig. 15 (no. 077). Charm case, silver. This case depicts a cicada (cf. Figs. 14A and B) that separates into two sections. Both sections are decorated with plum flowers. The charm is inserted into the body. A ring at the bottom permits the attachment of additional decor. The small hook at the end of the chain appears to be an imitation of the *ruyi* scepter.[26]

length (closed) 77 mm.

Fig. 16 (no. 067). Charm case, silver. This case, in the form of a fish, separates into two sections. The fish (*yu* 魚) is a common symbol since it is homophonous with the word *yu* 餘 "surplus, more than sufficient". Thus the fish symbolizes *fugui youyu* 富貴有餘 "May there be more than enough wealth and honors" [than is merely sufficient].[27] A ring at the bottom permits the attachment of the small bell.

length (closed) 59 mm.

Pendants with Other Symbolic Motifs

This is a large miscellaneous class of jewelry pieces and encompasses designs that have overt symbolic implications as well those that seem to be just decorative jewelry. The range of designs includes human, animal, and vegetable motifs, as well as symbolic plaques with inscriptions.

Fig. 17 (no. 023). Human figure pendant, silver. This piece depicts the well-known wealth deity, Liu Hai 劉 海 . He typically appears as a male child swinging a string of coins around his head and standing with one foot on a three-legged toad.[28] This alludes to the phrase *Liu Hai xi chan* 劉海戲蟾 "Liu Hai plays with the [moon] toad", a configuration that is supposed to bring good fortune. The three-legged toad lives in the moon and is a symbol of wealth and longevity.[29] The long double-edged projection at 5:00 o'clock represents the two ends of Liu Hai's belt. This is a single-sided casting with additional engraving work.

length 69 mm, max. width 30 mm, max. thickness 5 mm.

Fig. 18 (no. 273). Plaque pendant, silver. This design is very similar to the baby locks, both in its overall configuration and the implication of its inscription, and should perhaps be classified as a sub-type of the lock. The inscription reads: *zhuangyuan lang* 狀 元 郞 "A boy who attains first place in the examinations". *Zhuangyuan* is the title that was conferred upon the person who attained first place in the final (i.e. palace) civil service examinations under the imperial system.[30] In the top-right and left corners are dragons, a very common symbol of male vigor and fertility as well as of the Son of Heaven.[31] Bats flank the inscription to the left and right. A flower (peach?) occupies the bottom-center. I cannot identify the two symbols at the bottom-left and right. This piece is hammered from thin sheet metal.

length (L-R) 96 mm, width (T-B) 68 mm.

Fig. 19 (no. 130A). Plaque pendant, silver. There are three motif sets on this piece. In the outer circle are the Eight Trigrams (*bagua* 八 卦), and in the adjacent circle are the characters for their names. The *bagua* are the basic building blocks for the hexagrams of the famous oracle text, the *Book of Changes (Yijing* 易 經), and therefore provide cosmic protection to the wearer.[32] The *yin-yang* 陰 陽 interlocked comma-like symbol occupies the center of the plaque, signifying the primal generative forces of the cosmos.[33] This pendant is hammered from thin sheet metal.

diameter 53 mm.

Figs. 20A and **B** (no. 378). Pendant, silver, This design depicts a peach. On one side (Fig. 20A) is the inscription (T-B-R-L): *fushou shuangquan* 福 壽 雙 全 "Blessings and longevity shall both be completely [fulfilled]" (note the abbreviated character for *shuang*). On the other side (Fig. 20B) are plum flowers. This piece is hollow and seems to be a casting.

height (T-B) 47 mm, thickness 11 mm.

Fig. 21 (no. 392). Pendant, silver. Two "Buddha hands" (*foshou* 佛 手) are depicted on this piece. The Buddha hand is an inedible sweet-smelling yellow citrus fruit with long projections that look like fingers — the entire fruit resembling a hand. Through phonetic similarity, the *foshou* symbolizes *fushou* 福 壽 "blessings and longevity".[34] This piece is hammered from thin sheet metal.

length (L-R) 66 mm, width (T-B) 47mm, thickness 5 mm.

Fig. 22 (no. 196). Pendant, silver. This design depicts a bat, which we have already encountered as a common symbol of good fortune. The piece is hammered from thin sheet metal.

length (L-R) 70 mm, width (T-B) 56 mm.

Fig. 23 (no. 145). Pendant, silver. The creature depicted on this piece is a lion (*shizi* 獅 子), an animal which is not native to China. Therefore Chinese lions are very stylized figures derived from descriptions. The embroidered ball between the lion's forepaws indicates that it is male (a female lion will have a lion cub between its forelegs). The lion symbolizes valor and energy, and is a common guardian figure that will scare away evil spirits. Therefore a pair of lions frequently flanks the entrances to temples and other buildings. This stylized beast is often called a "fu-dog" or "fo-dog" by antique dealers.[35] This piece is hammered from thin sheet metal. It has five rings at the bottom for additional decor. The two hooks on the top rings have floral designs.

length (L-R) 89 mm.

Fig. 24 (nos. 236 and 237). Pendants, silver. These pieces depict donkeys equipped with saddles and stirrups. The collection contains many donkey pendants, and all of them depict the donkey hanging upside-down with its four hooves together. Some, such as no. 237, have an additional ring at the bottom, but the orientation of the figure is always upside-down. I have not been able to find out why the donkey is depicted in this posture. Some informants suggested that it is an allusion to: *Zhang Guolao daoqi maolü* 張 果 老 倒 騎 毛 驢 "Zhang Guolao rides his donkey backwards." Zhang Guolao is one of the Eight Immortals, and is said to ride on his donkey facing backwards. When he finishes riding, he can fold up the donkey and put it into his pack.[36] But Zhang Guolao rides his donkey backwards, not upside-down. The donkey in China is such a humble animal that it is seldom mentioned. However, there are an enormous number of donkeys distributed all over northern and central China where they have been a major source of power for agriculture, mechanical devices, and drayage.[37] Although the upside-down donkey is a common pendant design in the collection, I cannot explain its symbolism. These pieces (nos. 236 and 237) are both hammered from thin sheet metal; both are two sided and hollow.

length (L-R) 42 mm (no. 236) and 36 mm (no. 237)

Attachments for Clothing

This is another large class of jewelry pieces with a seemingly endless number of varieties. These pieces do not have rings or hooks that would easily permit their use as pendants. They were apparently attached to clothing, hats, and belts with threads sewn through the openwork holes in the designs (e.g., Fig. 25) or through mounting holes that are not part of the designs (e.g., Fig. 26). These are all hammered from thin sheet metal. Designs cover the entire range of animal and vegetable symbols, plus inscriptions and human figures. Two sub-classes of this category are buttons (not included here) for clothing and very small attachments for boys' hats. The former have designs that include floral patterns as well as single auspicious characters. There is sufficient variety among the characters to suggest that an appropriate sequence of buttons on a garment could produce an auspicious slogan. The attachments for boys' hats are discussed and illustrated by Stevens.[38] Many of these are round or oblong plaques that bear single character or character-like inscriptions (see Fig. 30), or images of deities or lions, and combine with other decor on a hat to form an auspicious slogan and symbolic configuration that portends a successful future for a boy. Another variety of these hat attachments is the semi-three dimensional figure such as Fig. 29. This type is hammered from thin sheet metal, but the upper part of the body is three dimensional while the lower part is intended to attach flat onto a surface.[39] Mounting is accomplished by sewing through small holes in the lower back and front edges. These pieces depict various deities who are expected to protect the boy and guide him to success.

Fig. 25 (nos. 090 and 091). Clothing attachments, silver. No. 090 depicts four bats flying counterclockwise. It is more common to find a configuration of five bats representing the "five happinesses": longevity, wealth, health, love of virtue, and a natural death.[40] I cannot identify the 43-serration motif on the outer rim − if it is anything more than a decorative edge. I cannot identify the eight-sectioned motif encompassing the outer rim (perhaps flowers?).

diameter of both pieces 47 mm.

Fig. 26 (nos. 070 to 074). Clothing attachments (nos. 070, 072 to 074) and a pendant (No. 071), silver. These pieces all depict four-legged frogs/toads (cf. the three-legged toad in Fig. 17). Frogs and toads (which are not clearly distinguished) are common motifs in both jewelry and embroidery design, however, their symbolism is not clear. The frog/toad is sometimes listed as one of the Five Poisonous Creatures (*wudu* 五 毒): snake, scorpion, centipede, spider, toad/frog. A mixture of their poisons is used to counteract evil influences.[41] Perhaps the frog/toad's reputed long lifespan makes it a symbol of longevity.[42] Its prolific reproductive habits suggest that it also may be a fertility symbol.[43] Two holes next to

each leg permit the pieces to be sewn on. The pendant (no. 071) can be suspended from the two rings between its head and forelegs.
length (T-B) 27 to 37 mm.

Fig. 27 (no. 252). Clothing attachment, silver. This is a male lion with its left forepaw resting on an embroidered ball (see also Fig. 23). It is attached by sewing through the openwork around the edges.
length (L-R) 52 mm, width (T-B) 36 mm.

Fig. 28 (no. 228). Clothing attachment, silver. The design has a chrysanthemum flower with leaves and three buds. The word for chrysanthemum (*ju* 菊) is phonetically similar to *ju* 居 "to reside, remain" and to *jiu* 久 "long time, long duration", which makes it a symbol for longevity, ie., remaining for a long time. Since the chrysanthemum was used in a medicinal wine for warding off evil influences, it may also function as protective amulet.[44] This piece was possibly attached to clothing by sewing through the openwork of the design. On the other hand, its construction also suggests that it might originally have been the top of a hairpin (cf. Figs. 31 and 32).
length (L-R) 62 mm, width (T-B) 51 mm.

Fig. 29 (no. 047). Clothing attachment, silver. This semi-three-dimensional piece is quite likely an attachment for a boy's hat (as mentioned above). The upper part of the body is three dimensional while the lower part is spread out to rest upon the cloth. Two pairs of holes in the back edge plus one pair at the bottom of the front edge allow it to be sewn on.[45] The design is of a male figure wrapped in a cloak with his chest and abdomen exposed. He holds a lotus stalk (or a *ruyi* scepter?) in his left hand and a sword or scepter or fly-whisk (?) with a curved hook on the end (perhaps a *vajra*, the Buddhist thunderbolt scepter)[46] in his right hand. His left leg is bent at the knee and lies flat on the plane of the base, with his foot resting sideways under his abdomen. The right leg and foot face out, and his right hand rests upon his knee. Long ear lobes are formed from wire soldered to the head. He wears a long beard formed into three tufts. At first glance this may seem like a Buddhist figure (such as a *lohan*, an enlightened disciple of the Buddha)[47] or even one of the Eight Immortals, but the beard and the two horn-like projections on the head suggest otherwise. So far, I have not been able to identify the figure.
height of front 64 mm, height of back 33 mm, maximum width 61 mm.

Fig. 30 (no. 380). Clothing attachment, silver. The general construction of this high-domed thin metal plaque, attached by sewing through the four pairs of holes at 3:00, 6:00, 9:00, and 12:00 o'clock on the rim is the same as the many other clothing attachments in the collection. They are quite possibly attachments for boys' hats (as mentioned above). The designs on most of these pieces are comprised of a Chinese character surrounded by a decorative border. In contrast, the design on this piece (no. 380) is very unusual, since it is comprised of ancient Christian symbols. The configuration in the top half is made up of a Latin cross above the letters IHS. IHS denotes the first letters of the Greek name of Jesus, and has also been popularly interpreted as an abbreviation for either *Iesus Hominum Salvator* (by the Jesuit order) or for *In hoc signo* ("In this sign", i.e., the sign of the cross of Christ).[48] The star-like configuration at the bottom is a stylized abbreviation for the first two letters of the Greek word for Christ, ΧΡΙΣΤΟΣ, and is known as the *Chi-Rho* symbol from the names of these two letters. The two Greek letters are often superimposed upon each other to form the symbol 朱 , but the simplified form on this jewelry piece is also found. The *Chi-Rho* symbol was also used as an amulet to ward off sickness or danger from evil spirits.[49]

There are only two other clothing attachments in the collection that contain Christian symbols, but neither is included in the present description. One (no. 382) is very similar to no. 380, except for variations in the cross, IHS, and *Chi-Rho* signs. The other (no. 405) has only a Latin cross surrounded by six plum flowers. Since the construction of these jewelry pieces is the same as the common hat attachments, they were most likely made by or for Chinese Christians. This is an excellent example of an innovation within the framework of tradition. Christian boys could wear traditional hat decor with symbols appropriate to their religion — symbols which, at first glance, could appear to be abbreviated forms of *shou* 壽 "longevity".

height (T-B) 48 mm, width 42 mm, thickness: 11 mm.

Hairpins

Most hairpins have a decorative head piece plus one or two long prongs which are inserted into a chignon or other kind of coiffure. Some have the decorative section in the middle of a single bar that is used to support a braid or other tightly pulled hair style. This latter type is not included here. The designs usually have vegetable or animal motifs. Small chains and other decor pieces may sometimes be attached to a hairpin, and some have decorative pieces attached by long coil springs that allow the piece to vibrate as the head moves.

Fig. 31 (no. 315). Hairpin, silver. This piece is double-sided. The two sides are hammered from thin sheet metal and soldered together. The two-pronged pin is soldered to the back side. The design shows a bat hovering over a bowl in which are two plum flowers (B-L and B-R), a peach (C-R), a Buddha hand (C), and a pomegranate (C-L). The pomegranate (*shiliu* 石榴) has many "seeds" (*zi* 子) which is the same word for "sons" (*zi* 子). Therefore this fruit symbolizes the wish for many sons. The three kinds of fruit in this design — the peach, Buddha hand, and pomegranate — are the three fruits of good fortune and symbolize longevity, blessings. and progeny.[50]
length (of decor section) 67 mm; width 58 mm.

Fig. 32 (no. 372). Hairpin, silver. This piece is single-sided and is hammered from thin sheet metal. The two hairpin prongs have been removed. The motifs are almost identical with those of the hairpin in Fig. 31, although their configuration is different. The entire right side is occupied by a large bat — its body is at the lower right — with plum flowers decorating its wings. At 12:00 o'clock is a Buddha hand, at 10:00 is a peach, and at 8:00 is a pomegranate.
length 56 mm, width 45 mm, thickness 4 mm.

It is not reasonable to draw any conclusions from this brief preliminary description of the jewelry pieces. They are clearly a source for the study of symbolic motifs in folk art and religion, and of their relations to popular culture. Their fine craftsmanship should be an inspiration for the study of local craftsmen and their products.[51]

NOTES

1 The collection belongs to my wife, Dade Singapuri, to whom I am very grateful for helping me prepare this essay. My sincere thanks are due to Professor Ching-mao Cheng for helping me decipher the abstruse inscriptions on some of the pieces. I am especially grateful to Mr. Louis J. Musante of the University of Massachusetts Audio-Visual Services for teaching me how to photograph these small jewelry pieces whose surfaces are very uneven in texture and light reflectivity. Whatever good qualities there are in my photographs are the results of Mr. Musante's patient instruction and criticism. My thanks are also due to the University of Massachusetts for an Educational Needs Grant to print the photographs used here.

2 For an example of this type of hallmark see Keith G. Stevens, "Chinese Miniature Silver Deities," *Arts of Asia*, 11.1(Jan.-Feb.1981), 94 right-center. My thanks to Ms Carol Stepanchuk for bringing this article to my attention.

3 Schuyler Cammann, *Substance and Symbol in Chinese Toggles: Chinese Belt Toggles from the C.F. Bieber Collection* (Philadelphia: University of Pennsylvania Press, 1962), p. 84 and Toggles Nos. 36, 37, 38.

4 Cammann, *Substance*; see also S. Cammann, "Toggles and Toggle-Wearing," *Southwest Journal of Anthropology*, 16 (1960), 463-475.

5 Cammann, *Substance*, pp. 83-91.

6 For examples of baby locks, including some suspended from necklaces, see Tseng Yu-ho Ecke, *Chinese Folk Art II: In American Collections, from Early 15th Century to Early 20th Century* (Honolulu: University of Hawaii Press, 1977), Figs. 71 to 74. An unusual variation of the baby lock design is shown in Fig. 70. See also Cammann, *Substance*, p. 134; and Henry Doré, *Researches into Chinese Superstitions* (trans. M. Kennelly; first pub. Shanghai 1914-20; rpt. Taipei: Ch'engwen Publishing Company, 1966), I.13, Fig. 10.

7 For an example of an old style Chinese lock see Rudolf P. Hommel, *China at Work* (1937; rpt. Cambridge, MA: M.I.T. Press, 1969), Fig. 447; the same illustration is reprinted in Joseph Needham, *Science and Civilisation in China* (Cambridge: Cambridge University Press, 1965), vol. IV: 2, Fig. 490.

8 This custom seems quite similar to the collection of scraps of cloth and thread from the neighboring families to produce the "Hundred families dress" *baijia yi* 百 家 衣) and "Hundred families tassel" (*baijia xian* 百 家 線) described by Doré, *Researches*, I.21.

9 For example, see Ecke, *Chinese Folk Art II*, Fig. 14.

10 The piece number in the photograph denotes the piece number in the jewelry collection.

11 Wolfram Eberhard, *A Dictionary of Chinese Symbols* (London: Routledge & Kegan Paul, 1986, first published Cologne 1983, in German), pp. 32-33; C. A. S. Williams, *Outlines of Chinese Sybolism and Art Motives* (1931; third ed. 1941; rpt. of third ed. Rutland, VT: Charles E. Tuttle, 1974), pp. 34-35.

12 Eberhard, *Dictionary*, pp. 239-240; Williams, *Outlines*, pp. 330-332;

Doré, *Researches*, V.726.

13 For one hundred variations on the forms of the characters *shou* and *fu* see J. J. M. de Groot, *The Religious System of China* (1892-1910; rpt. Taipei: Literature House, 1964), vol. VI, Plate XIV, following p. 1032.

14 Anecdotal designs are also found on some toggles; Cammann, *Substance*, pp. 150-154.

15 *Shuihu zhuan* 水 滸 傳 attributed to Shi Nai'an 施 耐 菴 (14th c.) and Luo Guanzhong 羅 貫 中 (ca. 1330-1400) (Hong Kong: Guanzhi shuju 廣 智 書 局 , n.d.), chapter 23. For an analysis of this episode in the novel see John C. Y. Wang, *Chin Sheng-t'an* (New York: Twayne Pub., 1972), pp. 75-81.

16 On the *qilin* see Eberhard, *Dictionary*, pp. 168-170; Doré, *Researches*, V.672-676.

17 Doré, *Researches*, V.675.

18 Eberhard, *Dictionary*, pp. 227-229; Williams, *Outlines*, pp. 315-317; Doré, *Researches*, V.717-721.

19 Eberhard, *Dictionary*, pp. 168-170; Williams, *Outlines*, pp. 255-258; Doré, *Researches*, V.273.

20 Eberhard, *Dictionary*, pp. 258-259; Williams, *Outlines*, pp.238-239.

21 On the fly-whisk, *fo-ying* 拂 蠅 , see Eberhard, *Dictionary*, pp. 112-113; Williams, *Outlines*, pp. 193-194.

22 Eberhard, *Dictionary*, pp. 194-197.

23 For lore concerning hair see Jiang Shaoyuan江 紹 原. *Fa-xu-zhua : Guanyu tamen de mixin*髮 鬚 爪 ： 關 於 他 們 的 迷 信 928; rpt. Taipei: Orient Cultural Service, 1971), chapts. 3 and 4. Doré, *Researches*, I.12, notes that dog hair will ward off evil influences. Wolfram Eberhard, *The Local Cultures of South and East China* (1943; revised and translated Leiden: E.J. Brill, 1968), p. 153, notes the use of human hair as a love charm. Wolfram Eberhard, "Chinese Building Magic" (first published 1940) in *Studies in Chinese Folklore and Related Essays*, by W. Eberhard (Bloomington: Indiana University Research Center for the Language Sciences, 1970), p. 64, notes the use of human hair in black magic.

24 For toggles shaped like human figures see Cammann, *Substance*, p. 36.

25 Eberhard, *Dictionary,* p.66; Williams, *Outlines*, pp. 70-71; Cammann, *Substance*, pp. 131-132; W. Eberhard, "Chinese Toggles" (first published 1942), translated and revised in Eberhard, *Studies in Chinese Folklore*, p.214.

26 Cf. the illustration in Williams, *Outlines*, p.238.

27 Eberhard, *Dictionary,* pp. 106-107; Williams, *Outlines*, pp. 183-186; Doré, *Researches*, V. 716.

28 Eberhard, *Dictionary*, p. 166; Williams, *Outlines*, pp. 402-403; Cammann, *Substance*, pp. 130, 148, Toggle No. 225; Dore, *Researches*, IX.64-68; Eberhard, "Chinese Toggles", pp. 211, 212, 214. For a contemporary depiction from the P.R.C. of Liu Hai holding a string of coins over his head and standing on a three-legged toad, see the painting titled "New Kitchen" by Zhang Xinying in *Chinese Literature*

(Beijing), Spring 1986, following p.60.

29 Eberhard, *Dictionary* , pp.292-293; Eberhard, *Local Cultures* , p. 205; V.R. Burkhardt, *Chinese Creeds and Customs* (Hong Kong: South China Morning Post, 1953-60), I.121, 148-149, II.46-47.

30 Charles O. Hucker, *A Dictionary of Official Titles in Imperial China* (Stanford: Stanford University Press, 1985), no. 1515.

31 Eberhard, *Dictionary* , pp. 83-86.

32 Eberhard, *Dictionary* , pp. 298-299; Williams, *Outlines* , pp. 148-151. Designs similar to this pendant are illustrated in Williams, *Outlines* , p.150 and Doré, *Researches* I.18 (as a necklace) and vol. IX, Fig. 153, following p.64.

33 Eberhard, *Dictionary* , pp. 321-322,323.

34 Eberhard, *Dictionary* , pp. 104-105.

35 Eberhard, *Dictionary* , pp. 164-165; Williams, *Outlines* , pp. 253-255; Doré, *Researches* , V. 713; Cammann, *Substance* , pp. 128-129; Eberhard, "Chinese Toggles", p. 213.

36 Eberhard, *Dictionary* , pp. 82, 324; Williams, *Outlines* , pp. 127, 153; Doré, *Researches* , IX. 44-45.

37 For the distribution of donkeys in China see *Zhonghua minguo ditu ji* 中華民國地圖集 (Taipei: Guofang yanjiu yuan 國防研究院, 1959-62), vol. V, p. A16 (the data are probably from the 1930's).

38 Stevens, "Silver Deities", pp. 92-93.

39 These are well-illustrated in Stevens, "Silver Deities", pp. 94-95.

40 Eberhard, *Dictonary,* p. 32, Williams, *Outlines* , pp. 35 and 450; Doré, *Researches* , V. 714-715.

41 Eberhard, *Dictionary,* pp. 208-209; Williams, *Outlines,* p. 187-188; Eberhard, *Local Cultures* , pp. 202-204; Wolfram Eberhard, *Guilt and Sin in Traditional China* (Berkeley: University of California Press, 1967), p. 36. The members of the *wudu* set varies; see *Daikanwa jiten* 257.899.4. The *wudu* creatures are also found on embroidered Chinese charm pouches.

42 Eberhard, *Dictionary* , pp. 119-120; Williams, *Outlines,* p. 401-403.

43 Cammann, *Substance* , p. 130; Eberhard, "Chinese Toggles", p. 214.

44 Eberhard, *Dictionary,* pp. 63-65; Williams, *Outlines,* p. 12, 69-70; Cammann, *Substance* , p. 109.

45 For an example of this construction see the illustration in Stevens, "Silver Deities", p. 94 right-center.

46 For various scepter, sword, and *vajra* designs see Williams, *Outlines* , pp. 248-249.

47 Eberhard, *Dictionary,* pp. 171-172; Williams, *Outlines* , pp. 152-168.

48 *New Catholic Encyclopedia* (New York: McGraw-Hill, 1967), VII.77 and IX.1064; *Dictionnaire d'Archéologie Chrétienne et de Liturgie* , eds. Fernand Cabrol and Henri Leclerq (Paris: Letouzey, vol. 7, 1926), VII.2468. The identical Latin cross over an IHS configuration is found on the coat-of-arms of the Society of Jesus (*New Catholic Encyclopedia*) VI.1050, illus. f). I am grateful to Professor Julian F. Pas for calling my attention to the appropriate interpretations of the symbols on this piece and for guiding me to the relevant references.

49 *New Catholic Encyclopedia*, IX.1064 and VII.78, fig. 1; *Dictionnaire d'Archéologie*, VII.265-267.
50 Eberhard, *Dictionary*, pp. 240-241; Williams, *Outlines*, p. 332-333; Cammann, *Substance*, p. 113; Doré, *Researches*, V.722.
51 I plan to continue photographing the jewelry pieces and to eventually prepare a descriptive catalog of the collection. This should provide a workable corpus of data for future research on Chinese folk jewelry and symbolic motifs.

Fig. 1A

272

121

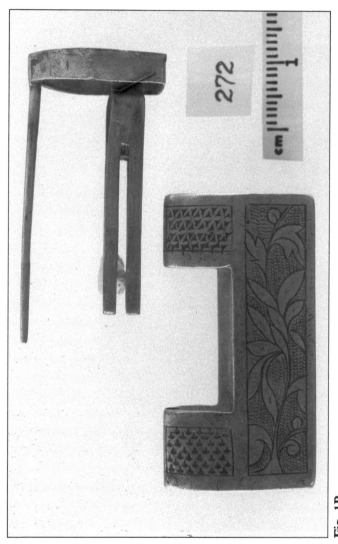

Fig. 1B

122

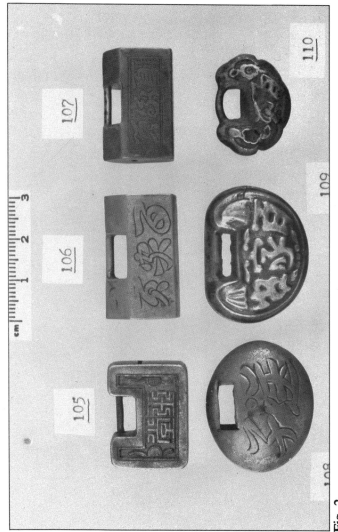

Fig. 2

123

Fig. 3

126A

Fig. 4

Fig. 5

Fig. 6

127

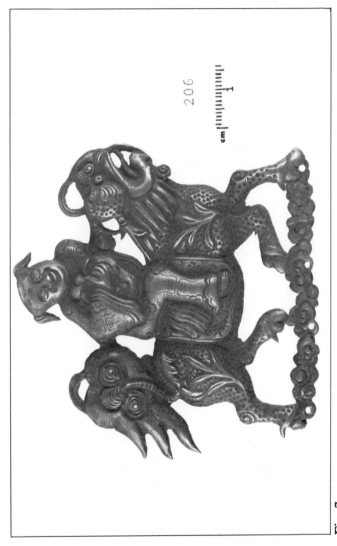

206

Fig. 7

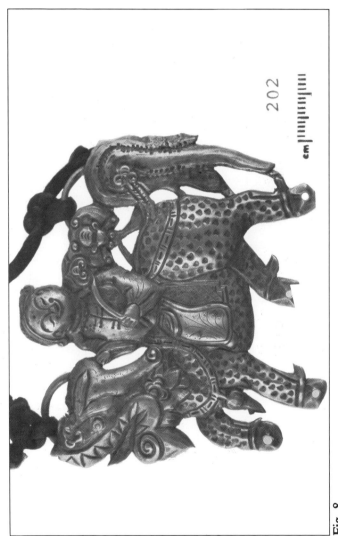

Fig. 8

Fig. 9

130

Fig. 10

131

Fig. 11

Fig. 12A

258

cm |||||||||||||||||||||||||||||||||||||||
1   2   3

Fig. 12B

Fig. 13A

Fig. 13B

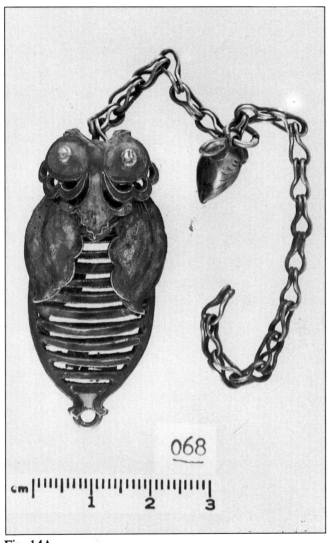

Fig. 14A

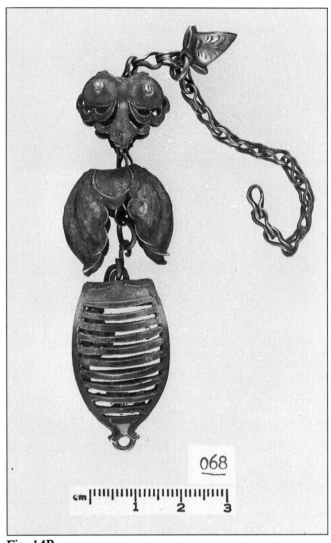

Fig. 14B

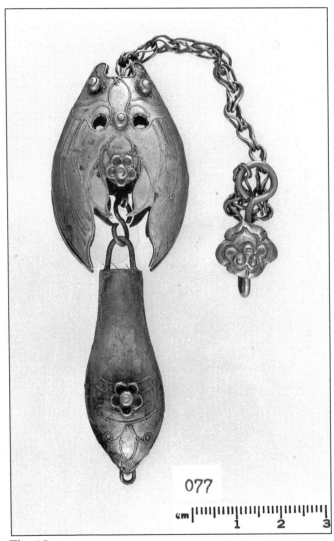

Fig. 15

139

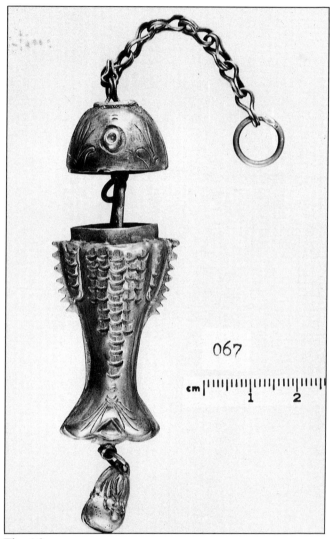

067

Fig. 16

Fig. 17

Fig. 18

130A

Fig. 19

143

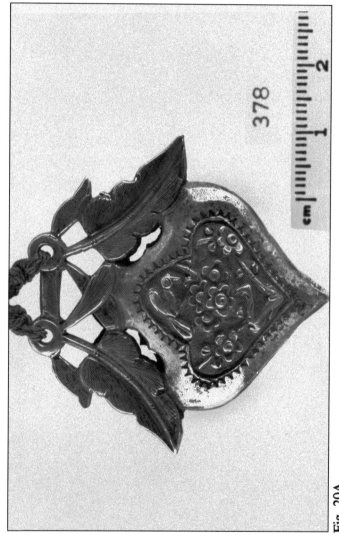

Fig. 20A

144

Fig. 20B

145

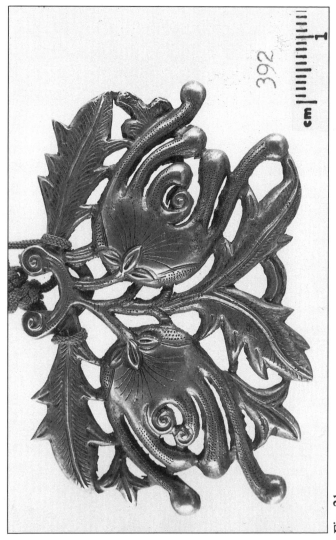

Fig. 21

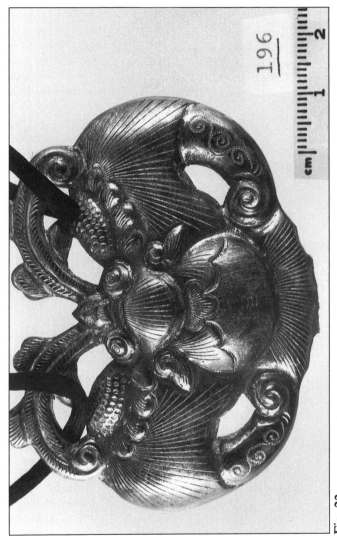

Fig. 22

147

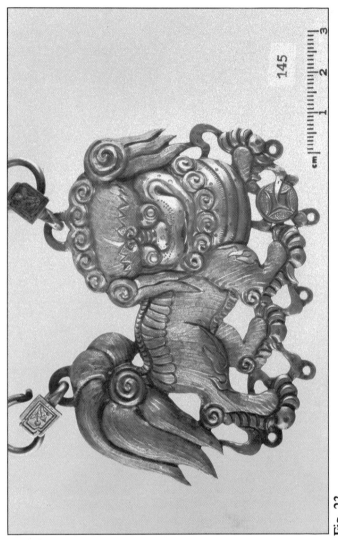

145

Fig. 23

Fig. 24

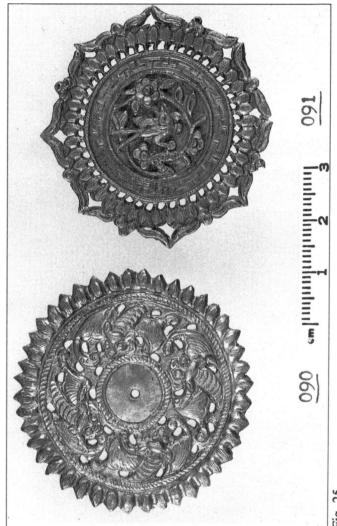

091   090

Fig. 25

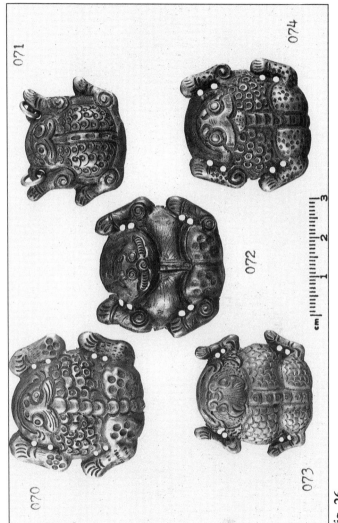

Fig. 26

151

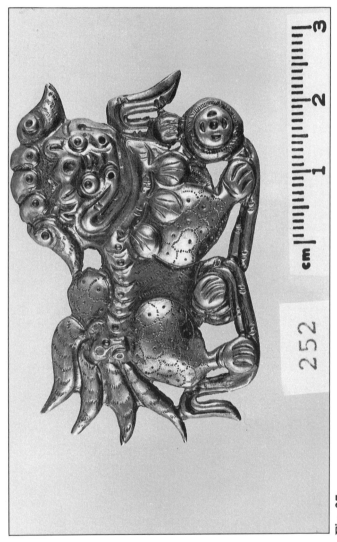

Fig. 27

152

Fig. 28

153

Fig. 29

154

Fig. 30

155

315

Fig. 31

Fig. 32

REVIVAL OF TEMPLE WORSHIP AND POPULAR
RELIGIOUS TRADITIONS

Julian F. Pas

In a great number of discussions of Chinese religion, articles, textbooks, monographs, the emphasis seems to be on historical developments, on the textual traditions, on the philosophical issues. Yet what I consider to be the most vital dimension of any religious tradition has often been missing: ritual and devotional expressions of faith, the mystical experience. There has been a change, however; more attention has been given in recent years to the many sides of the Chinese folk religion, to popular and ritual Taoism, to popular Chinese Buddhism. Although the mutual relationships of all these streams and tendencies is not very clear, yet it is a fact that popular expressions of faith start to demand a greater attention from scholarly observers and experts.

During my recent China trip (January – April 1985), I had the opportunity to attend a moderate number of rituals, mostly in temples within city boundaries, and moreover, was also struck by the gradual return of some folk religious customs in the countryside. Most visitors to China stay in the major cities and do not often have a chance to see what goes on in the rural areas. Yet, what is happening there among the peasant population is very fascinating.

In this essay, I shall discuss these two quite different aspects of the religious traditions in China; first, some of the Taoist and Buddhist rituals in the cities; secondly, the return of folk religious customs in rural China, such as the door spirits and New Year's scrolls, and protective talismans.

Taoist and Buddhist Rituals

In the Hong Kong publication *Chung-kuo Hsin-wen* 中國新聞 *(China News)* of December 8, 1984, it was mentioned that a special ceremony had taken place on December 7 in the Beijing Baiyun Temple 白雲觀 to celebrate the birth of the Ruler of the Water Realm (*Shui Guan* 水官), one of a trinity of Taoist deities: Heaven, Earth, Water. As a result of this announcement, I looked forward to seeing similar rituals being held on other traditional festival days in China. In fact, I witnessed ritual ceremonies at the Baiyun monastery on three different occasions: first on the

1st day of the new lunar year (or Spring Festival, February 20, 1985). This was nothing more than an ordinary morning service (*zaotan gongke* 早 壇 功 課) chanted every month on the 1st and 15th days. Seven Taoist priests in colourful embroidered robes recited the texts, using a 1983 reprinted edition of a prayer book titled *Taishang quanzhen zaotan gonke jing* 太 上 全 眞 早 壇 功 課 經 . There is a second volume for night recitation as well, *wantan gongke jing* 晚 壇 功 課 經 . The chanting is interesting, and (as I wrote in my notebook) "the singing sounds better than the rather poor performances of Taoist priests in Taiwan". This is perhaps due to the fact that the Baiyun guan Taoists are living a monastic life and take special care of their chanting. There were several lay-visitors, all standing in the recitation hall but none participating in the rituals, not even burning incense.

My second attendance at a Taoist ritual was on the 9th day of the first lunar month (Feb 28, 1985): the traditional birthday of the Jade Emperor. The set-up was quite different from New Year's Day. The ritual took place in the Jade Emperor's hall and extended to the open air platform connected with it. On the terrace floor a young monk had laid out rice grains to design a large Eight Diagrams picture. The ceremony started soon after 10:00 am: seven priests in embroidered robes performed the ritual of chanting the *Yuhuangjing* 玉 皇 經 . The main celebrant was a young monk, not more than 30, who performed the ritual prostrations and a sort of "Steps of Yu" on the outside terrace with grace and fluency. Part of the ritual consists of the celebrant's walking, even running, in large circles from inside to the platform outside, while brandishing his ritual sword, as if performing an exorcism. Musical accompaniment is done with cymbals and gongs and a drum by the assisting priests; there were no separate musicians as is the case in Taiwan's Zhengyi 正 乙 sect rituals. A large crowd of people were present, with an exceptional number of photographer-reporters. The ritual had been advertised in the newspapers, and visitors were most welcome.

The third ritual was held to commemorate the birthday of Qiu Changchun 邱 長 春 , one of the earliest patriarchs of the Quanzhen order, and the founder of a subsect called *Longhua pai* 龍 華 派 (March 10, 1985). It was a sunny day as well as a Sunday, which explains why a crowd of at least a thousand spectators was present, with again, a large number of photographers. Because of the crowd I could not get close to the hall of worship, which is situated deeper in the temple complex, specially dedicated to the founder Qiu. I presume the ritual was similar to the one on Febru-

159

ary 28, with the difference of the texts recited. The majority of the people did not seem to understand much of Taoism; few came forward, before or after the ritual, to prostrate and burn incense.

During my stay in Shanghai I had a chance to witness a Taoist ritual performed by the Zhengyi sect at the temple situated on the peninsula north of the city. The occasion was the birthday of the bodhisattva Guanyin (19th of the 2nd month or April 8), amazingly also worshipped in a Taoist temple: her name given by Taoism is Cihang daoren 慈航道人 "Taoist sage of the Boat of Compassion". The ritual was executed by five Taoist priests, all older men, especially the abbot who was in his late seventies. They were wearing beautifully brocaded garments, very similar to the Taoist robes I had seen in Taiwan. The ritual was a very stereotyped recitation of a sutra, with prostrations and bowings. A large group of musicians accompanied the chanting; these also were all older men. I was told that this ritual had not been planned but was performed because of urgent requests of the community. It was the first time such a ritual had taken place in many years.

The temple had suffered a lot during the Ten Years of Chaos and was only returned to Taoism in 1982. About 215 priests are connected with this temple, although the majority do not live here. Besides, since this is the *zhengyi* sect, the priests are married and live at home; they come to the temple to perform rituals or receive visitors. The central image of the Jade Emperor is brand new, and is not consecrated yet. The official consecration *kai guang* 開光 (lit.: opening of the eyes to light) will take place later in April.

To conclude this short report on Taoist rituals, I must add one more surprising experience: a Taoist "mass for the dead". This happened in Wuhan during a visit to the Changchun monastery 長春觀 , belonging to the Quanzhen order but of Longhua lineage. The temple had just recently been opened, only the front hall had been restored so far; behind this hall there was a temporary gate, closed to the public. After some negotiations I was allowed to climb the steps to the hall on top of the hill, where we could go and see the abbot. Before he appeared, I was attracted by the noise of some small cymbals and bells; to my surprise I found six old monks sitting around a table in the center of the shrine hall; they were reciting texts from old-looking handwritten manuals. On their left side, near the wall, were five or six persons standing and following the proceedings. Outside the gate, on a table, offerings of fruit were laid out, and together with those, there was a little paper-and-bamboo house in many colours. That explained everything: the priests were performing a ritual for a recently deceased person, and the family was in attendance. Since the priests looked all very old, in their sixties or seventies, I assume it must have meant a recent revival of a practice forbidden for a long time: younger monks have probably not yet learned the liturgy. Also, the fact that it took place in a secluded area, almost in secret, points to the fact that there still is hesitation in the temples about the acceptability of some religious rituals.

What I found most amazing was that Taoist priests performed this ritual. In Taiwan I have several times witnessed Buddhist nuns and occasionally monks perform these prayers for the dead. But the Taiwan Taoists have different types of ceremonies. The one I saw in the Wuhan temple looked very similar to the Buddhist type.

Several Buddhist monasteries, if not all, have daily, or at least bi-monthly recitation rituals. In the Jade Buddha Temple 玉 佛 寺 of Shanghai, I saw and heard a large group of monks, mostly young men, recite the evening prayers in the Mahavira hall 大 雄 殿 . They recited a text of the names of the Buddha, and my guide explained that this takes place in the mornings as well. A rather numerous group of lay visitors attended the recitation. In other temples, lay devotees are also welcome to join the community prayers.

But my greatest surprise came on the Qingming 清 明 Festival, April 5, at the Shanghai Longhua monastery 龍 華 寺 . Here another facet of the old rituals had returned to life: the *pudu* 普 度 festival. A whole week of special recitations for the dead and for the living had been organized. On April 10 the final ceremonies would take place, but unhappily I could not attend. Even on the day of my visit, the temple was in great turmoil; hundreds of visitors and devotees came and went. In the Mahavira hall a large number of monks recited sutras for the dead: they were dressed in dark brown robes, covered with bright red tunics that hung over one shoulder. In smaller rooms along the main courtyard, groups of elder monks sat in armchairs, waiting their turn to enter the main hall for recitation: one group would recite the *Lotus sutra* , another the *Lankavatara sutra* , and a third one would chant the *Huayen sutra* .

In two larger rooms, each on one side of the courtyard, tables were temporarily set up as altars: they contained incense burners and offerings to the Buddhas. On the wall were a great number of yellow and red cards with names of devotees written on each one. The left side room was for the dead, the cards were yellow; the right side room had red cards, with the names of the living devotees, and prayers would be said for a long life of happiness. Registration for any of these two took place at a desk; it was a source of revenue for the temple.

The highlight of the festival was likely to be an oldstyle *pudu* celebration, to liberate all the orphan souls (or hungry ghosts). At the back of one temple hall, hidden by a cloth screen from the view of the visitors, several craftsmen were engaged in manufacturing the traditional paraphernalia; a huge boat of Guan Yin to ferry the souls across the ocean to the Buddha's paradise, had already been completed. Now other figures were being made, as well as horses: the materials were light bamboo sticks and multicoloured paper, similar to those seen in Taiwan. But it was altogether a surprise for me to see this happening in China.

Inside the temple some elder ladies were folding "gold" paper into ingots; they were teaching younger women how to do it. All these objects as well as the "money" would be burnt on the last day of the festival.

Taoist and Buddhist Monasteries

Since the publication of the *Journal of Chinese Religions* in 1983, some new materials are available and new events have taken place.

In Beijing the most important event was the reopening of the Baiyun Temple in the western district of the city. There was an official opening in March 1984, and a colourful brochure has been printed. I was able to meet some of the Taoist priests and members of the Taoist Association of China. The journal mentioned by Professor Jan Yün-hua: *Daoxiehuikan* 道協會刊 is still not available for purchase by foreign scholars.

The temple was the setting for last year's general meeting of the TAC: on December 4, 1984, Taoist administrators and abbots from seventeen provinces and/or metropolitan cities attended the third plenary congress of the TAC. A newspaper report said the number of Taoist priests now permanently residing in temples or monasteries has reached several thousands. In the provinces and cities, thirteen local chapters of the TAC have been established, while the administration of a number of important national Taoist temples has been returned to the Taoist clergy. In his opening address, president Li Yuhang explained the objectives of this one week meeting: to examine methods for nurturing and training young Taoist talent; to discuss ways to promote Taoist research and scholarship; and to exchange management experiences among the personnel of various temples.

This Baiyun temple continues to organize training sessions for young recruits from all over the country, and among the activities of their own personnel they make a serious effort to train young monks in the performance of the rituals. Even at this time, many of the priests are not yet fully capable of participating in the ceremonies.

Another prestigious temple of old, now closed to religious life and to visitors, is the Temple of the Eastern Peak (*Dongyue miao* 東 嶽 廟).[1] Although I had sincerely hoped that a visit would be arranged, it was fruitless. The temple is occupied, I was told, by a police school unit and was off limits for all visitors. I had to satisfy my curiosity by a look at the old gateway, now on one side of the road, and the main entrance, at the other side. Of the buildings within the complex, not much was visible. I heard that the famous presentations of the hells and their tortures situated in an adjoining temple had all been destroyed, even before the "Ten Years' Chaos", since that was clearly an example of feudal superstition. I wonder whether this temple of the Eastern Peak will ever be "liberated" again.

I do hope so, for at one time, in pre-liberation China, the temple of the Eastern Peak used to enjoy a greater popularity than the Baiyun monastery; especially on New Year's Day huge crowds came here to worship.[2]

The Beijing Temple of Confucius is visited with mixed feelings; the ancient architectural style of the buildings and the impressive old trees in the courtyards underline the previous glory of the temple; also, a set of stone tablets engraved with the names of all the successful candidates in the imperial exams of the Yuan, Ming and Qing dynasties are a unique historical record; finally in another courtyard, one finds the thirteen ancient classics engraved in stone tablets. These are all reminders of a splendid past. However, today the temple looks rather gloomy: the main hall contains sets of ancient musical instruments: bells and stone chimes. But the hall is dusty and poorly maintained. To crown the depressing atmosphere, one sees near the back wall the dust-covered statues (in plaster!) of Confucius' seventy-two disciples. One wishes that the Red Guards had taken exception to those rather ugly creatures instead of smashing valuable art pieces.

Alvin Cohen also saw the temple and writes, "I visited the Confucius Temple in Beijing. It is strictly a museum and tourist attraction. It is ill kept and everything inside is covered with dust. There is not even a pretense of cult activity."[3]

Of course, in a country where Confucius and Confucianism

have known frightening ups and downs in political evaluation, one can hardly expect that his temples would recover quickly from these upheavals.

On the slopes of the Western Hills, west of Beijing, are two old and beautiful Buddhist monasteries. The Tanzhe monastery was first built during the Jin dynasty (AD 265-316); it is the oldest temple in the Beijing area; even today it breathes an atmosphere of quietness and solitary meditation. Of course, it is also because we are the only visitors today, and no monks or nuns live here any longer. Some of the huge trees are said to have grown from seeds brought from India. In a small hall up the steep hill is a piece of floorstone with two foot impressions: legend holds that these were made by a Mongol princess, a daughter of Kublai Khan, and a devout Buddhist, who stood and worshipped for many years in front of the Buddha images.

In another small pavilion there hangs a large stone-carved fish: its colour is greyish-black and very weird: it is believed that the stone is a meteorite, fallen from the sky.

Another attraction is a pavilion which is said to have been often visited by Emperor Kangxi and later by Dowager Cixi. The floor of the pavilion is designed as a network of small canals: if one looks at it from the south side, one sees it as a dragon head; if one looks from the north side, it is like a tiger's head (*nan long bei hu* 南 龍 北 虎). The canal is connected with a water spring at the back. Emperor and empress came here to play betting games: they would release goldfish in the small canals, and also let cups of wine drift on the water: the loser of the goldfish race would have to empty a wine cup.

Some of the pavilions are still locked up: they are empty now. The ones reopened since 1981 have new, but rather ugly, Buddha images. The old statues, it is said, have all been smashed by Red Guards.

Down the hill, on a larger area adjoining the temple grounds, is an impressive "park of pagodas": it contains 72 stone tomb pagodas of different age, size and structure. Some are very old, from the Jin period, but others date from Yuan, Ming and Qing times. These miniature pagodas are actually tomb decorations; one of them was built on the grave of the above mentioned Mongolian princess, but mostly they are tombs of eminent Buddhist monks.

Not too distant from here is another Buddhist monastery, Jietai si 戒 台 寺 which literally means "Initiation Platform" Monastery. There are only three such temples left in China today. It was first built in the early T'ang period (AD 622). The temple grounds are beautiful and quiet and especially famous for five extraordinary pine trees, each several centuries old. Since it had just snowed a few days before our visit, the contrast between the snowcovered branches and the blue sky was fascinating.

We had a cold picnic lunch here in a very cold little side building. But walking up and down the steps I could gradually warm up my freezing feet. I can imagine the hardships which monks of past generations must have undergone here.

In Beijing, monastic life continues within some old Buddhist temples: the Fayuan si 法 源 寺 is the seat of the Buddhist seminary; in each of the two years of training there are about forty candidates. The Guangji si 廣 濟 寺 is usually closed to tourists, but I was allowed to visit the premises with a young monk as my guide. Since 1953 this temple has become the national headquarters of the Buddhist Association of China. They seem to have an excellent library with some old manuscripts; altogether a collection of more than 100,000 books. Other treasures include bronze statues from the Ming dynasty; Maitreya, Weito, and the eighteen arhats; and a set of the Chinese Tripitaka printed during the Ming dynasty.

While these two temples are centers of spiritual and intellectual training and study, another temple is much better known as a tourist attraction: the Yonghe temple 雍 和 宮 , located in the northeast. It was originally built as a palace for an imperial prince (1694), but in 1723 it was converted into a Buddhist temple for Tibetan lamas. The whole complex consists of five central halls with smaller side-buildings. In the last hall one sees an 18 meter high statue of the Buddha Maitreya, carved from one single sandalwood tree. The hall must have been built around the statue, or at least around the tree trunk, once it was erected.

In all the halls, there is a Buddhist monk on guard. At first unaware that taking photographs inside the halls is not permitted, I prepared to take a picture of a bodhisattva, and suddenly out the blue, a monk appeared with an angry look: "no pictures allowed!" A continuous flow of tourists visit the temple. Most of them are Chinese from the countryside on their very first visit to the capital. Many of them burn incense, which is available in each hall, and many also donate cash: the large transparent offering boxes in each hall are very well furnished with bank notes. I asked an elder monk-on-guard about the temple personnel: he said that there were now about eighty monks living here (lamas), half of them younger men in training. They each have to spend many hours a day doing surveillance and even the younger recruits are not exempt. In those cold buildings, that in itself is quite an ordeal. On the 1st and 15th day of each lunar month, the monks congregate and recite the scriptures in one of the main halls. Devotees are allowed to attend.

In early March I made a special excursion to the capital city of Hebei province, Shijiazhuang 石家莊 .[4] Outside the city in a neighbouring county about 18 km away, I visited the Longxing monastery 隆興寺 dating from the Sui dynasty. It must once have been a beautiful temple. The architecture is remarkable and some wonderful statues are still to be seen in the inner halls, but otherwise the grounds give a desolate impression: broken floorstones, dust everywhere, even covering most of the beautiful sculpture; and most of the halls are rather empty. One hall had collapsed, perhaps in the republican period, but the foundation stones are still visible. The authorities have started to restore the buildings; once restored and cleaned, it will become a great monument again, but like many other temples, it is as yet empty of religious life, a mere tourist attraction.

It was already mid-March when I started out for Sichuan. In and around Chengdu several famous temples have resumed an active program. Two ancient Buddhist monasteries are well known: Baoguang si 寶光寺 and Wenshu yuan 文殊院 .

The 'Precious Light' monastery (Baoguang si) is in the small neighbouring town of Xindu 新都 : a very large temple complex situated in a park of trees and bamboo. This temple has a long history and has been quoted as "one of the most famous ordination places in China ... ".[5] T'ang emperor Xizong 僖宗 found refuge here in 881 during the uprising of Huang Chao 黃巢 , and had a pagoda built and the temple renovated.

Arranged along a north-south axis, the monastery follows a common layout pattern, found in many other temples: the gate-shrine, the hall of the four heavenly kings (*devaraja*), the pagoda (its location is an exception to the ordinary layout), the hall of the Seven Buddhas, the great Buddha hall or *mahavira* hall, and at the back end, a two storey building to store the Tripitaka. East from the central axis one finds a large impressive building: the Lohan hall or Hall of the Five Hundred Arhats. It is specially designed in the form of the Chinese character *tian* 田 or 'field' and contains, right in the center, a thousand-armed Guan Yin statue. There also are statues of a Buddhist triad, of six bodhisattvas, fifty patriarchs and 500 Buddhist saints or arhats. The statues average 2 meters in height and are gilded and polychromed. An impressive collection, happily saved from the iconoclasts of the sixties.[6]

At one time this temple was very famous and counted over 2000 monks during the Song dynasty. One can still see the meditation hall and recitation hall, once used for the joint practices of Chan (Zen) meditation and Pure Land devotionalism.

The temple buzzes with activity: worshippers come and go in large numbers. They burn incense, especially in front of the pagoda and in the major Buddha hall. Many devotees kneel down, and bow or kowtow in front of the images. Some remain kneeling in silent prayer. The outer screen wall with a large character *fu* 福 painted on it attracts numberless visitors. They estimate the distance from where they stand and the character on the wall; then, with closed eyes they approach. Wherever they stop, they stretch out their hand: if they can touch the character *fu* , their good luck is ensured. Other practices on the border-line between religion and magic are also coming back. In the same temple, in one of the inner court-yards, there is a large tank-like brass water container, filled to the brim. Visitors drop coins onto the surface: if the coin floats, it is an omen of good fortune. (I tried and my coin floated, but the Chinese coins are extremely light.) In other temples, I believe in Shanghai, devotees touched the large brass incense burner with both hands and then rubbed their own bodies to ensure good luck.

These and similar activities are spontaneous expressions of the people's rich imagination and their need for comfort and encouragement, although they are rather on the periphery of true religious faith.

Outside the temple in the market place there is the ubiquitous crowd of shoppers and watchers. Dozens of small stalls selling devotional articles are set up along the street side: they will be described in part two.

Another highlight of my stay in Chengdu was the visit to the Taoist Qingyang gong 青 羊 宮 . The abbot and his secretary, both over 70 years old, were very friendly. During the 'Ten Years Chaos' they had been forced to do manual labour in the adjoining park, which once belonged to the temple, but had been separated from it in 1958 and had become a cultural park. It is only since 1983 that the monks have returned to full-time monastic life in the temple. In 1985, there were only eight monks and three young students; there also were eight nuns and three young novices. The main training center for Taoists in Sichuan is on Qingcheng mountain (青 城 山).

The monastery has a long history. It was first established in the Later Han dynasty. During the T'ang it was visited by emperor Xuanzong. One of the present halls dates from the Ming period but has not yet been restored. When I prepared to take a photograph of it, the abbot asked me in a friendly way not to do it.

At the very back of the temple grounds, there is a hall where the original woodblocks of the *Daozang jiyao* 道 藏 輯 要 ("Supplement to the Taoist Canon") are kept. A new printing is under way, but only 100 sets are in the making. I looked at some of the printed sheets and was impressed: the characters are rather large and very clear. The woodblocks date from the Daoguang 道 光 period of the Qing dynasty (1821-50); they cover at least two walls of a large storage room.[7]

Today is a Sunday and worshippers are quite numerous. They burn incense and light candles. They even burn "paper money" (certainly a recent innovation!), but only one variety of paper money is available: very plain, light yellow paper, quite appalling compared to the rich variety of paper money for sale in Taiwan and Hong Kong. One woman was kowtowing in front of the statue of Guan Yin, while a nun accompanied her movements with beating a bell. Several other nuns were engaged in looking after the needs of worshippers, or otherwise busy selling articles of devotion. Near the entrance there was a side building, named "vegetarian food hall", which also doubled as a tea-parlour. But I saw quite a few pieces of chicken and other meat hanging up.

One of the priests told me that there is an average of one thousand visitors a day. On New Year's day over 10,000 showed up. It seems that religion is Sichuan is not dead. But there is a side-effect: the monks and nuns are too much involved in the "running of the business". How much time is there left for personal spiritual cultivation? Yet, the temple needs the revenue for living expenses, and upkeep or restoration of the buildings.

The following morning we start out for a trip to the famous Taoist mountain Qingchengshan. Arriving at the foot of the mountain, we encounter a large group of old women pilgrims (I heard there were 107), members of a pilgrimage association, who travel and climb all the sacred mountains in China. They had just recently visited Omeishan. For their age, in their sixties and seventies, they are still very mobile and move fast. Since the antireligious pressure is off, these old ladies must feel very excited and fulfilled in being able to travel like this, having great fun and gathering merits at the same time.

The temple complex is situated in the middle of beautiful mountain scenery. There is an astounding tree with wide branches and foliage, said to have been planted by Zhang Daoling 張 道 陵, the founder of religious Taoism in the 3rd century A. D. According to tradition, Heavenly Master Zhang resided on this mountain in one of the sacred caves for spiritual cultivation.[8] There still is a special shrine here dedicated to him where his statue is enthroned. But the central hall of the monastery is the Hall of the Three Pure Ones (三 清 殿), where most pilgrims go and kneel to worship.

During an interview with the abbot, we find out that about 80 Taoists live in the monastic compound; fifty of them are young students, male and female, who have joined the order during the last two or three years. Before that time, only elder monks lived here; they are now in their seventies or eighties, some are even over ninety. The younger adepts come mostly from the province. According to one informant, this temple is one of the only three Taoist 'seminaries' in the country. Its affiliation is with the Longmen sect, subsect of the Quanzhen order. At a very early time the monastery had been a Buddhist center, but a T'ang emperor had donated it to the Taoists.

Life in this monastery seems to be very busy: although isolated in the mountains, it is within reach of many curious tourists and worshippers. Every day tour buses are parked at the mountain foot. As a result, old and young Taoists alike are engaged in duties related to receiving the guests, some of whom stay overnight. Besides staffing a registration office, they also run a little shop, a kitchen and a large dining hall, where Taoists function as waiters and waitresses, others are in charge of the guest quarters, others still supervise the several shrines spread around the mountain. Some even stay at the entrance gate at the foot of the mountain to sell and check entrance tickets; and, oh kindness! some strong young men take over the luggage bags of some "respected guests", bringing them up and down again. Even without carrying anything but a camera, my progress was slow!

Being so busily engaged in temporal concerns, it seems that these young adepts do not have very much time for instruction, spiritual practices and self discipline. There is hardly even a library. Upon my request, the abbot took me to a second floor room, on top of the Sanqing hall. This was their library, but it does not appear to be in use. There are several cases containing a silk-bound edition of the *Daozang*, published by the Commercial Press of Shanghai in the pre-war period and, further, a set of the *Daozang jiyao*, and some other books relating to literature and history. But this can hardly be called a library. Besides, the bookcases are locked and there are no desks in the room. Apparently nobody ever comes here.

For study, the young novices have their own textbooks, but I did not have a chance to see them, except for one short treatise of Taoist history. The abbot said that there is time for spiritual practice in the evening. Prayers are recited in the main hall every morning and evening, but only by one person. Only on the 1st and 15th day of the lunar month, is there a joint recitation.

My guide, Mr Li, said that most of the younger disciples have had a high school education, but are not too bright in academic pursuits. Personally, I was not overly impressed with the training of novices here, although my short stay does not allow me to draw definite conclusions.

Back in Chengdu I visited a very old Buddhist temple, the Wenshu yuan 文殊院, the headquarters of the Buddhist Association of Sichuan. A huge crowd was visiting the premises. Many old ladies prostrated in front of the Holy images, and burnt incense. Another crowd was sitting along the walls and reciting their rosaries "Omitofo", or chatting when tired of praying. This was the first time I had seen the return of the rosaries. The Park adjoining the temple is a grand meeting place: some old ladies walk around reciting their rosaries, others, especially old gentlemen, bring their bird cages along, hang them in a tree and let the birds compete in singing while the old men sit, chat and smoke (*sit, chat, ananda:* being, knowing, bliss).

Outside the temple, all along the street leading to the temple gates, small stalls have been put up for small business related to temple worship: several stalls sell rosaries, incense and firecrackers, also good luck pendants and scrolls of spring festival couplets. There is an old shrine to the earth-spirit built into the wall; some engraved characters remain, but the statue has "walked out". Yet, several people are worshipping, prostrating in front of the small shrine and offering incense. Only a few meters away, there is a furnace for burning paper money. People came here perhaps to thank the local guardian for their newly acquired economic well-being?[9]

A few years ago, a campaign was launched in China against

"spiritual" pollution, and on the positive side, to create a "spiritual" culture. At that time, I was puzzled by it and wondered whether "spiritual" really meant something religious, as the English term seemed to imply. As I realized later, the campaign had nothing to do with religion. It was rather negative: a reaction against material cultural elements brought into China by the "materialistic" West (especially the USA!). As a result dance halls were closed down, and pornographic magazines, like *Playboy,* and films were confiscated. It was a short-lived campaign, however, and soon China was back to normal. I witnessed one remainder of this episode: in a small restaurant in Xindu, Sichuan, where we stopped one day for lunch (March 16, 85), I saw a poster stuck on the wall filled with slogans to promote a "spiritual" culture. That was the only positive side of the campaign that I witnessed. It can be called a campaign for "beautification (*mei* 美)[10] and consisted of four parts:

(i) "Beautification of Behaviour" 行為美
    "Beautification of Environment" 環境美
(ii) "Beautification of the Mind" 心靈美
    "Beautification of Language" 語言美
(iii) "Stress on Hygiene" 講衛生
    "Stress on Discipline" 講秩序
    "Stress on Morality" 講道德
(iv) "Stress on Decorum" 講文明
    "Stress on Good Manners" 講禮貌

Somehow, this poster reminded me of the old Confucian values being reinstated.

In the early morning of March 22, I left by train for Chongqing (war capital of the nationalist regime). It was a great experience, witnessing the colourful country side, but I have to return to my major theme. In this large city, I was told, only *one* single temple had been reopened. (Hard to believe, but how can you argue against your official guide, if you don't have any other evidence?) This temple is the Buddhist *Lohan si* 羅漢寺 and I was very pleased to visit its precincts. It had already been a famous Buddhist monastery during T'ang times, but when a lohan hall was added to it during Qing times, it was renamed *Lohan si* . Unfortunately, during the Ten Years' Chaos, all the 500 statues of the Buddhist saints were smashed. With some foresight, some other statues, a large Buddha and two arhats had been hidden and were now back in place.

The lohan hall was under reconstruction. The hall itself was already finished, but the life-size statues of 500 Buddhist saints were being remade. I saw workers mixing clay with straw and sand and carrying the raw material inside in bamboo baskets. In the hall other workers were shaping the rough body of each image, while artists from the Sichuan Fine Arts Institute were engaged in doing the precise moulding work. All they had as models were photographs of the surviving set of arhats taken in a Suzhou temple. I must admit that the new sculptures were of high artistic quality (according to one of the artists they were better than the old set, which had been smashed).

In the major Buddha hall of the temple a group of 3-4 girls and a woman came to worship (only those who actually want to worship are allowed inside). One of the young girls, about 12, was laughing all the time while she bowed in front of the Buddha. When I asked why she laughed, she answered that she was very nervous since it was the first time she had come to worship the Buddha. Perhaps she was more nervous because a foreigner was standing there and watching?

On the first and 15th day of each lunar month, there is a special recitation service, to which members-devotees are invited. Pillows used at that time were now stacked high up in the corners. I was told that several hundreds of devotees participate.

The temple counts 30 to 40 members now; several of them are younger adepts, accepted during the last few years. Some of them have been sent to Beijing for study and training.

There is a hall once used as a "meditation hall" (*chan tang* 禪堂), but presently transformed into partial residence. Along two sides small partitions or cubicles have been made for the monks to live in. There are plans to reinstate a formal meditation hall in the future.

From March 24 to 26 I am floating on the Yangzi. Although the sky is overcast, the scenery is just immensely splendid. But there is no space for poetry here! Arriving in Wuhan on March 26, early next morning we are to leave for Wudang Shan 武當山, Hubei province. That comes as a surprise. Although I had requested a visit to this famous Taoist monastery in my correspondence, I had never received any confirmation but was very pleased when I heard the news. We shall travel in a private car: a driver, my guide Mr. Yang and myself. Later I heard about the distance we travelled: almost 600 km one way to the mountain. It took us about 5 days to complete the round trip (partially due to poor planning!). There is no space to describe the trip, not even the highlights. Some details will return in a later section (door gods, etc.). On the morning of March 29 we set out for the mountain: a convoy of two jeeps to take me there. Bodyguards, or people also curious to visit the mountain?

The monastery has a long history. Most of the buildings were constructed during the Ming dynasty. The major deity worshipped here is Zhenwu Dadi 眞武大帝 . or the god of the Pole Star (in Taiwan preferably called Xuantian Shangdi 玄天上帝) and patron deity of several martial arts groups. It is said that Zhang Sanfeng 張三豐 , the founder of the Taijichuan form, lived in this temple.

Restoration of the major halls is in full swing. The central hall still looks dilapidated, in contrast to the photographs I had seen earlier. But it takes time to restore so many cultural centres. Fortunately, most of the temples spread around on this mountain had been saved from red guard vandalism. The local population had just blocked their way; instead they went to a temple nearby on the lower hills, Taizi po 太子坡 , and destroyed its statuary dating from the Ming period.

We had lunch with the local Taoists in their guest dining room; only vegetarian dishes were served, mostly made of *doufu* (bean curd) but of excellent quality. Our hosts were two Taoist priests, the president and vice-president of the local branch of the Taoist Association. I was told that about 80 Taoists are now living on the mountain; sixty of them are younger men and women, who came here from one to three years ago. In one of the rooms their daily schedule was posted, and I took care to copy it.

5:00 AM	rise
6:00-6:30	morning prayers
6:30-7:00	breakfast
7:00-11:00	work
11:30-12:00	lunch
12:00-2:30 PM	rest
3:00-6:00	work
6:30-7:00	evening prayers
7:00-7:30	supper
8:00-9:30	study
10:00	lights out

Besides this time schedule, four times a week there is practice in recitation: Sunday, Monday, Wednesday and Friday, 8:00-9:30. I do not recall whether it is a.m. or p.m., but I presume it is the latter, coinciding exactly with the study period. That again was a disappointment to me; so much emphasis on work (and tourism?), and so little time devoted to the real purpose of monastic life: spiritual cultivation. Is it that the monasteries lack qualified leaders and teachers?

In the afternoon we visited other shrines spread around in the hills; I did not have the courage to climb to the "golden pavilion" on the highest peak, although up there spiritual practice is taken more seriously. The mountain is open to tourism; at the foot of the mountain there is a large hostel for tourists and many touring cars are stationed on the parking lot. No wonder that this uninterrupted contact with the secular world dilutes the power of spiritual life in the temple.

On our return trip, probably 20-30 kilometers away from the mountain top we stopped by the ancient archway leading up to the monastery. It used to be the beginning of the *shendao* 神道 : here pilgrims would start their spiritual ascent to the top in silent meditation or prayer. My escorts took their leave (in two jeeps!) and we were on our way back to Wuhan. Not far from the town Sanglang there is a famous shrine to Zhugeliang 諸葛亮 of the period of the Three Kingdoms. It is situated in an attractive park with pavilions, visited by crowds of tourists. Nearby we found an old five-towered pagoda, similar to the one in Beijing, but apparently not known to the tourist industry. Yet, there are still some beautiful Buddha statues placed all over in niches. The central pagoda is built in dagoba style, and dates perhaps from the Yuan period.

In Wuhan I visited two temples: the Buddhist Guiyuansi 歸元寺 and the Taoist Changchun guan 長春觀 . In the Buddhist temple there is a hall of 500 arhats. I was told there were about 80 inmates here, mostly male, and quite a few young faces among them. The devotees coming to worship were not as numerous as in Chengdu.

Outside the temple grounds, as in Chengdu, there is a street full of little souvenir stands, selling devotional articles. What is most fascinating is the return of protective amulets and of small porcelain Guan Yin statuettes (see below).

The Taoist Changchun monastery, already mentioned previously, is still worth some extra attention. It is being restored: the entrance gate and the major shrine are already completed and look bright and colourful. It was reopened only in 1984. The major shrine has some unique features, which I did not see anywhere else in China: on the central altar is a large image of Laojun (or Lao zi), and altars on the left and right sides with images of Zhuangzi (Nanhua zhenren 南 華 眞 人) and a disciple of Laozi unknown to me. The two side walls have been painted recently, in August of 1984, with two large frescoes each; on the right side: Laozi riding his ox, travelling through the mountain pass; and Laozi teaching his followers. On the left wall there is a scene of Confucius kneeling in front of the Old Master, while three of his disciples stand respectfully in the doorway. (The 4th painting I was unable to identify.) The large frescoes are something totally new to me.

This brings me full circle back to Shanghai. Monastic life in China is slowly gaining new momentum. Young people are attracted to the religious life, not in large numbers, but perhaps sufficient to ensure the continuity of the traditions. It remains to be seen whether this revival is solid enough to cause a complete renaissance of monastic life. My most serious worry is that the younger generation of recruits are not seriously trained in the spiritual disciplines of Buddhism and Taoism, and will lose themselves in externals, driven by economic and relevance considerations. There appears to be a commercialization of religious life, a hollowing out of true religious experience. That is perhaps a greater threat than political pressures or even persecution.

Return of the Door Spirits

I have already mentioned the revival of magico-religious practices such as in the Baoguang si in Xindu, Sichuan. In the countryside and in smaller towns, life is different from that in major centres. It does not merely apply to religious customs; for example during my two months' stay in Beijing, I had not seen one single dog or cat. In Chengdu I saw two cats (one of them was tied up) and later, travelling by train from Chengdu to Chongqing, I could see many dogs running around small farmhouses.

My first hint that rural life is different from city life came to me during an excursion to the Western hills near Beijing. It was around the Spring festival, and from a distance, I could see small houses decorated with the traditional colourful spring couplets. On our way to the Ming tombs and the Great Wall, I had the same experience, although I did not see any door gods. In Beijing however, and later in Shanghai or Wuhan, not a single such case: no couplets, no door gods, not even any decorations on the outside gates. Firecrackers, on the other hand, were very abundant in Beijing. Already days before the Lunar Festival, I could hear from my hotel room their explosions all around. On the eve of the festival, it was a crescendo which suddenly exploded into an orgy of noise at midnight. From my hotel window, I could see a fairly large section of the city: the noise and sight were unbelievable. It lasted about 20 minutes and then started to taper off. In the local markets or small stands spread around the city, I had noticed quite a variety of firecrackers for sale, and eager customers competing to buy them. Yet, I think this is a recent development: with increased economic well-being, the Spring Festival regains more and more of its ancient splendour.

I do not know how or whether city dwellers decorate the inside of their houses for the occasion. It was hardly possible to visit people at their own houses. But in the countryside, many of the restrictions have been cast away. People cannot live forever in an atmosphere of ideological purity, of puritanical constraint and joyless sacrifice. If there is a little economic improvement, one wants to enjoy it. And the Chinese people are artists in doing just that.

I had two great opportunities to catch some glimpses of country life: first my trip from Chengdu to Xindu and Qingchengshan and later a five-day return journey by car from Wuhan to Wudang mountain. On both occasions, we passed through a large number of smaller towns and villages and I made special efforts to stop occasionally and photograph the returning door gods. My guide and driver used to laugh at me: they are city dwellers and probably look down on this folklore of the country-side. They were obliging but I had to insist, otherwise I would have totally missed the opportunity to have a closer look at this new emergence of door spirits.

Menshen 門神

It is not necessary here to study the origins of this ancient custom. According to several authors the tradition goes back to prehistoric times, and a mythical story connects the origin with the Yellow Emperor.[11] A new dimension was added in T'ang times, when emperor Taizong had images of two fierce-looking generals painted and attached to the doors of his palace room. These two historical generals, named Weichi Gong 尉遲恭 and Qin Qiong 秦瓊, became prototypes of many later pairs of door spirits, some of whom were likewise brave generals, while others derive from legendary stories.

The variety of door spirits which I saw in China in 1985 was really amazing and fascinating. The specimens which I was able to collect derive from two kinds of sources: those pasted on the people's doors and those I purchased either in stores or museums.

Among the former group is a great variey of posters, photographed along the road in small villages of Sichuan and Hubei. On our return trip from Wudang mountain to Wuhan, we stopped in a new settlement village: all the newly built brick houses had door spirits pasted on the door wings. In that one street, I counted at least ten different varieties. Here are some samples.

Fig. 1 (Sichuan): Two military commanders in full armour. The one on the left side has a dark face, and brandishes what looks like knotted bamboo cudgels; the one on the right side, fair faced, brandishes two swords.

Fig. 2-a (Sichuan): is a variation of **fig. 1**; the paintings, especially the military costumes are more elaborate. But white face and black face are similar; so are the two types of arms. These generals are also identified: the black faced general is Jing De 敬 德 ; the fair faced one is Qin Qiong 秦 瓊 . **Fig. 2-b** is a modern woodblock print, a variation of the same *men shen* , purchased in the provincial museum of Hubei in Wuhan.

Fig. 2-c is another modern variety.

Fig. 3 (Sichuan): another pair of identified generals; on the left side: Gao Chong 高 寵 ; on the right side: Yang Zaixing 楊 再 興 . They look more fierce than those in **figs. 1 and 2;** each holds a lance in one hand, and grasps a sword suspended at his back. This is the first case of complete symmetry or of mirror images: the two generals are the same in all the details, except here and there in the colouration. I found many other examples of mirror images, expecially in block prints.

Fig. 4 (Sichuan): a set of two generals on horseback. This sample I found in a Sichuan store, with some luck, for since the Spring Festival was already over most stores we tried had run out of all their posters. This poster, not yet divided into 2, measures 75 x 52 cm. The general on the left is named Zhang Xianzhong 張 獻 忠 ; the one the right is Li Zicheng 李自成 the former is holding a spear in his right hand, whereas general Li holds a sword in his left hand.

Since I bought this poster in the store, brand new, I was able to read in the margin some interesting details about its production. The artist is named Zhang Zhineng 張 志 能 . The printing had been done in Chongqing, Sichuan in 1984 on two dates: 10,000 copies in March; 146,000 more copies in May. The sale price is 16 cents (in Chinese currency). The large number of copies at the 2nd

177

printing indicates the success and quick sell-out of the first edition (compare with **fig. 14**: in July 1984 the first edition ran to 13,600 copies).

Fig. 5 (Hubei): two generals on horseback. The general on the left is certainly Guan Gong 關 公 . One recognizes his stereotyped face and the gesture of his hand patting his beautiful long beard. The other general has a long and full white beard. Both have a bodyguard holding the horse's reins in one hand, and a heavy halberd in the other hand. One can easily recognize Guan Gong's stereotyped bodyguard. The two pictures are very symmetrical, the horses and the background decorations are mirror images.

Fig. 6 (Hubei): two more generals, on foot. The one on the right, once again, is Guan Gong, holding his halberd. His beautiful long beard and whiskers are a solid clue to his identity; but more so is the character *Guan* seen near his right knee. On the corresponding place of the left picture figure the character *Zhang* 張 , giving away the identity of the general: Zhang Fei 張 飛 , armed with a horn-shaped halberd.

Fig. 7 (Tianjin; but purchased in a Hong Kong art store in April, 1985). It is a set of long scrolls with the paintings themselves measuring 54 x 108 cm. Both sides are almost mirror images except for the arms. The figure is Zhong Kui 鍾 馗 , the famous legendary devil catcher. The printings are made of woodblock prints, reproduced in the Yang Liuqing studio of Tianjin. No date is indicated but one can guess its recent creation. On the very top is a seal script, reading *quxie zumo* 驅 邪 逐 魔 : exorcize evil and chase devils.

Fig. 8 (Taiwan). Here is a pair to contrast with the mainland samples. I found them in a Taichung market in Dec. 1978; the pictures measure 26 1/2 x 39 cm.
The drawings are much simpler than any of the above figures, but whereas their artistic value is limited, their symbolism is interesting: the two personalities are not military commanders, but literary officials, which is unusual for door guardians. The symbolism expressed is one no longer popular on the mainland: high rank and emoluments. Indeed in lieu of names, the left picture has *chin lu* 晉 祿 , "to bestow emoluments". The right picture shows *chia kuan* 加 冠 , "to add a crown". In their hands the two mandarins are holding a *ju-i* , an official scepter, and a "deer", pro-

nounced *lu* , and symbolizing the *lu* : "emoluments"; or a "crown", pronounced *kuan* : an official. Here the traditional symbolism of door guards has been replaced by a more "bourgeois" ideal of aspiring for a high career in life.

The following figures are samples of changes that have been taking place in the traditional themes of the door spirits. It would be most interesting to have a more complete selection of samples.

Several changes can be seen to be happening: male generals are occasionally replaced by females. The religious connotation of door spirits is being neutralized or eliminated by new themes: either of a Marxist nature or of a more secular symbolism.

Fig. 9 (Hubei): a pair of amazons on horseback galloping and in combat with lance and bow. I saw this set on several occasions, and found it very impressive. In modern China, it shows the equality of the sexes: even the exclusively male role of door guardians is now being shared by women.

Fig. 10 (Hubei). This is a pair of more graceful women, apparently in military apparel. On the banner flapping in the background are two characters: on the left picture: *mu* 穆 ; on the right side: *liang* 梁 . In the foreground are two pieces of furniture which I cannot identify but which must have symbolic value.

Fig. 11 (Sichuan). An unusual set of new door guards: half religious, half folkloristic: an image of Sun Wukong (the monkey king) on the left side; he jumps over the clouds carrying his heavy steel rod. On the right side is the boy god Third Prince, or Prince Nato: he rides on his flaming wheels, while in his hands he carries a lance and his magic ring to capture demons.

Although these two personalities derive from mythology, they have become so much part of popular culture through opera, etc. that their religious origin has been pushed to the background. In a socialist society, this theme is perhaps more plausible. However, there are other attempts to remove the religious themes from the Spring celebration.

Fig. 12 (Hubei). Here the old heroes have been replaced by youngsters of modern China: boys and girls brandishing swords and other weapons, in what appears to be a martial arts performance. In the background are dragons and phoenixes as minor images; the rest of the picture is strongly parallel.

Fig. 13 (Hubei). This is a perfect sample of the new spirit that has emerged in socialist China. Each side shows three girls and three boys; some are musicians (certainly), others are dancers and singers (probably). In the background are communist symbols and an explosion of firecrackers. I assume that posters like these are purchased by party members and cadres, who are not expected to share in the so-called popular superstitions. This new variety of "door spirits" is certainly more acceptable from the viewpoint of socialist orthodoxy.

Fig. 14 (Sichuan). This is another poster purchased in a Sichuan store. It is an attractive and very colourful set showing five young children, boys and girls, in multinational dancing costumes, each holding a lit lantern in their hands. On the lanterns are characters signifying "Socialism is Wonderful!" (left) and "The Communist Party is Dearest!".

The artist is identified as Liu Yuling 劉玉玲 (see other details under **fig. 4**).

Fig. 15 (Hubei). This is perhaps the most surprising new motif of the group of samples. Does it express the family's desire for offspring? Or is it an alternative to get away from undesirable religious themes? In any case, it is another example of the secular tendency creeping into the celebration of the Spring Festival.

Fig. 16 (Sichuan). In a small but crowded market village near Chengdu, we stopped to photograph some door god posters.. At one house, the woman was unusually cooperative. "Wait", she said, "You can have this one." Before I could protest, she ripped off the poster and gave it to me. I did not clearly see the other half, pasted inside the front room. (As a compensation, the woman asked me to take her photograph and mail it to her. Being the center of the crowd's attention she looked proud and radiant!)

This single poster represents another martial hero in full battle dress, standing; in his right hand he grasps the sword at his back, in his right hand he holds a halberd.

On the poster's margin I read that the artist is Hou Shiwu 侯世武. The first printing was done in June 1984, by the Sichuan People's Printing Company; one set costs $0.16. Although there is no indication of the door god's identity, his popularity may be guessed from the large number of copies printed: 324,300! (Compare with **figs. 4 and 14**). His features indicate that this is a variation of **figure 3**, most likely Yang Zaixing.

Return of Magic Charms and Other Customs

Whereas Al Cohen's article deals with the rich variety of silver amulets and charms used by the Chinese people in pre-Cultural Revolution times, I would like to show the emergence of new types appearing on the popular market. The antique silver charms have all but disappeared in China: they can still be purchased (at inflated prices) in the Friendship Stores and hotel souvenir shops or in the antique stores of Hong Kong, but I am certain that enormous quantities have left China.

Now that religious freedom has been restored, the old custom of protecting young children is once again a lively concern. It is even more urgent today: because of the one-child policy, parents who are lucky enough to produce a baby boy cannot take any chances and one of the best ways to protect babies against evil fortune is to have them wear good luck amulets.

I photographed a married couple carrying their young baby between them. He wore the traditional silver "lock" (see Al Cohen, **figs. 3, 4 and 5)**; it was an older one of good quality, the only one of its type I saw being used in China.

The new types surfacing on the market, especially near the popular temples, are cheap and ugly imitations of the old prototypes. I only purchased five samples, but I estimate that two years later a much greater variety will be available. The ones I purchased came from Chengdu or Wuhan; another one I found in Nanjing, and one I received from a friend in Beijing.

Fig. 17 (from a small souvenir stand, Ming tomb in Nanjing) is the only clear reproduction of the old silver lock; on the front side are the characters 百 家 鎮 ; on the back side is a small image of a deity riding a *jilin*. The size: 4 x 5 cm. Made of very light silvertone aluminum.

Figs. 18 and 19 are almost identical: the son-bestowing goddess seated on a *jilin* (unicorn). In her right hand she holds a flower, in her left hand the traditional wind instrument, called *sheng* 笙 (bamboo flute). The scales of the jilin are clearly visible.

Both charms are made of gilded plastic and are finished on both sides, in contrast with the old silver ones. Except for a minimal difference in size (6 x 7 1/2 cm vs. 6 1/2 x 8 cm), the major difference lies in the decoration attached at the bottom: **fig. 18** has 4 tiny copper (gilded) bells: **fig. 19** has 4 red silk tassels.

Further compare these two charms with Cohen's **figs. 7-11:**

among his samples none of the figures holds a *sheng* ; in my collection of old amulets, there is one in which the goddess holds a tiny *sheng* . Perhaps a musical instrument is more orthodox than the more traditional *ju-i* ?

Fig. 20 (and **fig. 21**) are both flat aluminum plates, copper tone, with red background. **Fig. 20** shows a deity riding a *jilin*. It is hard to distinguish what the deity carries in his/her hands. The white characters are *qilin songzi* 麒 鱗 送 子 . At the top are 2 bats, symbols of luck. At the bottom are attached 4 tiny copper bells. **Fig. 21** is very similar in shape to **fig. 20**, also with a red background but the decorative patterns are more complex: in the centre is a *yin-yang* symbol, above it 4 characters: *jixiang ruyi* 吉 祥 如 意 . On either side of the central symbol figure are a dragon (right) and a phoenix (left), and finally, toward the outside edge, all around, are the 12 animals of the Chinese Year cycle in the traditional order. Only 2 small tiny bells are attached but a necklace of (plastic) pearls goes with it. Size 5 1/2 × 7 1/2.cm.

I have seen other variations of **figs. 20 and 21** in the market place; sometimes with an image of Milofo. I saw one young nicely dressed child wearing one amulet. But the photograph is not sharp enough to distinguish the details of the picture.

All in all, compared to the rich variety of amulet types used in earlier times (see Cohen's article), the ones reappearing today are still very limited: so far only 3 basic types have come to my attention. As time goes on, probably new ones will resurface.

As various religious and magico-religious practices are reviving after a long period of hibernation, it is only a matter of time before many other old traditions will regain acceptance (see, for example, **fig. 22**). This especially affects the rural areas, where possibly many traditions never fully disappeared. As the economic well-being of the people rises, it is only to be expected that surplus money will find its way into temple funds and into the hands of religious practitioners and diviners. In Chengdu near the Wenshu yuan, I saw the first palm-reading stall: a simple mat spread on the street was all the paraphernalia needed; yet this was not secret any longer, but in full sight of all passers-by.

What about old rituals surrounding births, weddings and funerals? Here again, a whole field is open for investigation. From my own brief glimpses into rural life, I can only guess that old traditions are being revived. The following are just such brief glimpses, too meager to permit generalization, yet I am confident they are sufficient to indicate a much stronger presence.

In a village on our way to Wudang mountain we passed a small procession walking on the roadside: one sona player and several others who were beating gongs; other people were just walking along. I guessed it to be a wedding procession; this turned out to be correct. The bride, walking ahead, held a handkerchief on her mouth (to wipe the traditional tears?). Both my guide and the driver were surprised to see this. Weddings nowadays are very simple, and this little extra festivity was unusual for them.

There was more evidence of ancestral cult practices. Although I regret not having had a chance to visit a cemetery at the Qingming festival (my Shanghai hosts did not support the arranging of such an excursion), on several occasions I caught short glimpses of ancestral cult-related activities. On our drive through a Hubei village, we passed what turned out to be a funeral truck. Many people were seated in the truck and a good many paper flower wreaths were stacked on it. The wreaths were different from those I used to see in Taiwan; they had a much lighter frame, and besides the white background, multicoloured flowers were stuck into them.

In another village I saw a single man walking at the roadside, carrying several colourful objects made of bamboo and paper, such as a bridal chair, a horse and a boat. My guide explained them as offerings to be deposited on an ancestral grave: the ancestors may be in need of vehicles to travel to the Western Paradise.

Finally, in many rural areas, usually a short distance away from the village, I noticed groups of tumuli or small mounds, variously shaped, in each province (Sichuan, Hubei, and around Nanjing). In Hubei they tended to be sharp pointed; around Nanjing more rounded at the top: sometimes they were covered with some kind of a container, probably an earthenware plate, with a bamboo twig planted in it. I have not seen those anywhere else, and do not understand their significance. Since this was shortly after Qingming, I presume they are a sort of decoration or offerings to the ancestors. On very rare occasions I saw some lonely stone graves near the railway track, on a hillside.

Although according to official reports, burials in modern China are always performed by cremation, this appears to be theory rather than practice. In the countryside the oldfashioned burial still seems to take place. (Too far away from crematoriums? or too expensive?) In the larger cities cremation is the rule.

This is the end of my story. Although my experiences are not systematic and only sketchy, what emerges is yet a clear indication that popular religious practices are not defunct. The next few years will probably allow us to obtain more complete information. We may be in for great surprises!

NOTES

1 Anne Swann Goodrich, *The Peking Temple of the Eastern Peak: The Tung-yueh Miao in Peking and its Lore* . Nagoya (Japan): Monumenta Serica, 1964. See also A. Swann Goodrich, *Chinese Hells. The Peking Temple of Eighteen Hells* . Monumenta Serica, 1981.

2 H. Y. Lowe, *The Adventures of Wu. The Life Cycle of a Peking Man* (2 vols. in 1). Princeton Univ. Press, 1983 (1st ed: Peking, 1940-41).

3 A. Cohen, p. 8 of unpublished manuscript, 1985.

4 Shijiazhuang and Saskatoon, my hometown in Canada, were twinned in 1984 and that prompted my visit there. Within the city there is a large cemetery of the revolutionary heroes, which also contains the tomb and statue of Dr. Norman Bethune. This medical doctor has taken the fancy of the Chinese people; he has inspired numerous art works and possibly legends. In older times, he might have become the object of a religious cult.

5 J. Prip-Moller, *Chinese Buddhist Monasteries* (Copenhagen, G.E.C. Gad, 1937, and Hong Kong, Hong Kong University Press 1967), p. 312. The book contains many photographs and drawings of the temple

6 In J. Needham's *Science & Civilization in China* , vol. 5 part 5 (p.36) one sees the photo-reproduction of one lohan or arhat, holding the yin-yang symbol in his hands. The photograph was taken in 1972.

7 Not too long ago, in 1986, I was informed that the set has been published.

8 There are a large number of such sacred caves or *dadongtian* 大 洞 天 in China. See the article by T. Hahn, appendix. It is doubtful whether Master Zhang stayed on this mountain; it is more likely that he retired for spiritual practices on another Sichuan mountain:

9 This is the second earth spirit shrine that I encountered in China. The very first one was along the ascending steps of Mount Qingcheng by the roadside: also a very small shrine with a Tudigong statue in the niche, covered with glass. I wonder how many of these shrines have survived in rural China.

10 Ironically, the character *mei* is also used to romanize "America" or *mei guo* ("beautiful country"). I'm sure, however, that the campaign did not mean "americanization"!

11 Studies about the door gods are quite numerous, although I am not aware of a single detailed monograph. Here are a few references:
Alfred Koehn, "Harbingers of Happiness: the Door Gods of China", *Monumenta Nipponica* , 10 (1954), 81-106.
E.T.C. Werner, *A Dictionary of Chinese Mythology.* New York: The Julian Press, 1969 *(Men-shen* , pp 311-2).
C.B. Day, *Chinese Peasant Cults* (2nd ed: Taipei: Ch'eng Wen Publ. Co. 1969). Pictures of Peking and Canton door gods: opposite page 48;

of Chekiang and Hunan door gods: opposite page 90.

C. A. S. Williams, *Outlines of Chinese Symbolism and Art Motives* , Shanghai, 2nd revised ed., 1932. Drawings of *Men-shen* are on pages 129 (civil) and 130 (military).

Kuo Li-cheng 郭 立 誠 , *Min-su hsieh-chu* 民 俗 擷 趣 (Taipei: Ch'u-pan-chia wen-hua shih-yeh, 1977), pp 55-66.

About Chung K'uei, see Jonathan Chamberlain, *Chinese Gods* (Hong Kong: Long Island Publishers, 1983), pages 156-160; also Chiu Kuan-liang, "Dance of Chung Kuei", *Echo* , 6, no. 7 (1976), 17-24.

Fig. 1

Fig. 2-a

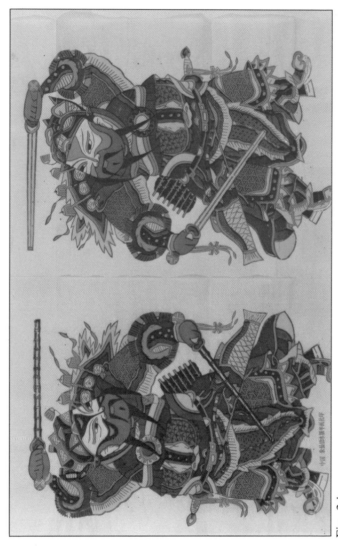

中国冬皇帝头圆号弥尼甲

Fig. 2-b

Fig. 2-c

189

Fig. 3

Fig. 4

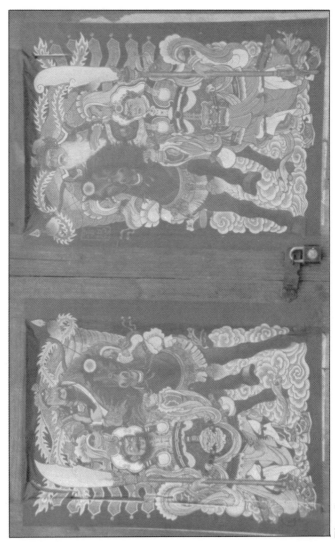

Fig. 5

192

Fig. 6

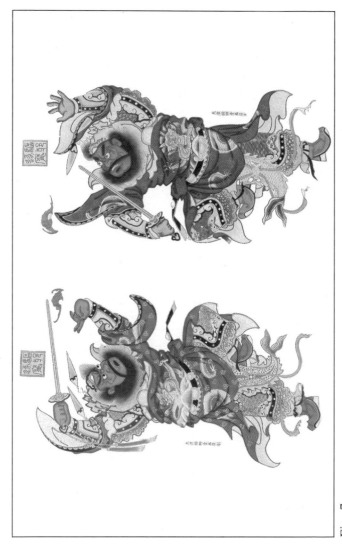

Fig. 7

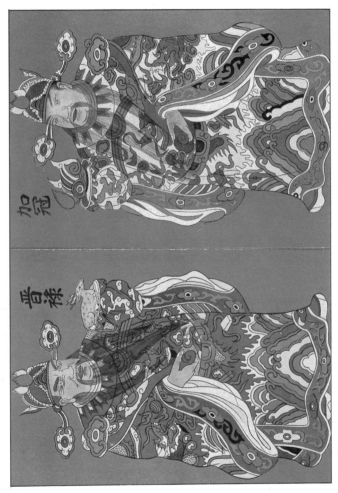

加冠

晋禄

Fig. 8

195

Fig . 9

196

Fig. 10

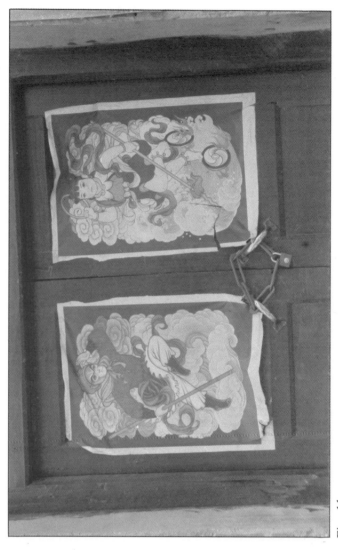

Fig. 11

198

Fig. 12

199

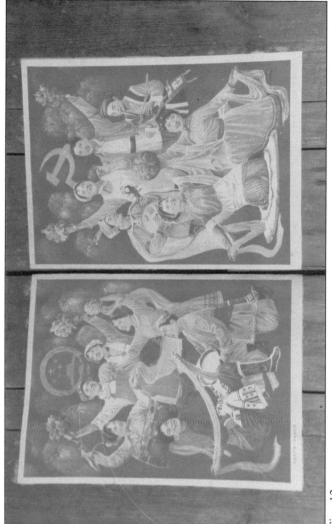

Fig. 13

200

Fig. 14

201

Fig. 15

202

Fig. 16

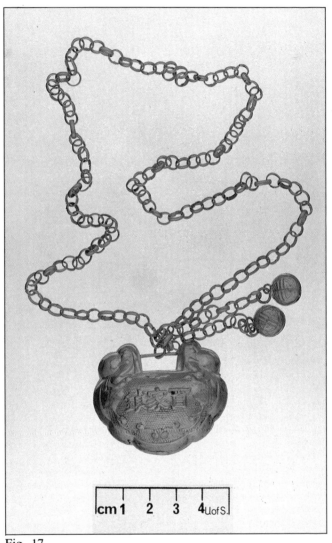

Fig. 17

Fig. 18

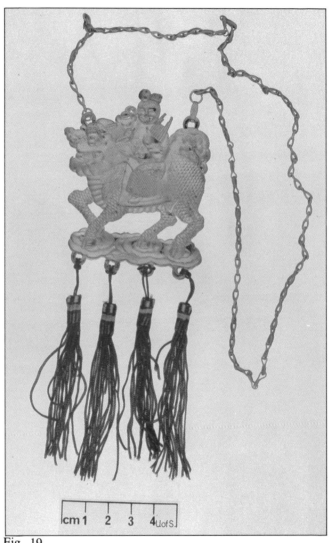

Fig. 19

Fig. 20

Fig. 21

Fig. 22

209

A REVIVAL OF CONFUCIAN CEREMONIES IN CHINA

By Paula Swart and Barry Till

If anyone thought that Confucius was no longer respected in modern China of the 1980's, they would be sorely mistaken. As recently as September 1986, a very significant event took place in the town of Qufu in Shandong province, where Confucius lived and taught in the 6th and 5th centuries B.C. Grand ceremonies to celebrate the anniversary of Confucius' birthday were held here for the first time in mainland China since liberation in 1949.

During the Cultural Revolution (1966-1976), Confucius and his ideas were severely criticized and the town of Qufu, which contains a large amount of Confucian monuments and cultural treasures, became a major target in the anti-Confucius movement. Truckloads of Red Guards poured into the city to denounce the Sage and vandalize anything associated with him. Their acts included the desecration of his tomb, in Chinese eyes the ultimate expression of contempt. However, the last decade has seen the transformation of Qufu from a dusty provincial town into one of the leading tourist attractions of China. Many new buildings now line the paved streets and a magnificent new hotel called the Queli Hotel (Queli being the name of Confucius' birthplace) was constructed, attesting to a changed attitude towards the town where Confucius was born. Since the late seventies, Qufu has opened its doors not only to foreign visitors, but also to large numbers of Chinese tourists. No doubt, the reason for the costly restoration of the town was mainly to attract the tourist dollar, but there also appear to be increasing numbers of Chinese streaming into the city to show their growing awareness and pride in their cultural heritage. Some of the most important sites to be visited are the Confucian Temple, the Kong (Confucius)[1] family mansion and the cemetery containing the tombs of Confucius and many of his descendants.

Confucius, born in 551 B.C., was not generally recognized as a great sage during his lifetime, but later generations claimed him as such, and the philosophy of Confucius came to dominate Chinese thought for the next two thousand years. Confucianism is a system of social-political ethics and correct interpersonal relationships rather than a religion; its aim is to lay down the rules for the ideal moral society. It represents high intellectual ideals, but there is no supernatural sanction to be dreaded by one who fails to live up to them:

> Tzu-lu asked about serving the spiritual beings. Confucius said, "If we are not yet able to serve man, how can we serve spiritual beings?" "I venture to ask about death." Confucius said, "If we do not yet know about life, how can we know about death?"[2]

Qufu, his birthplace, came to be revered as a sacred site of Confucianism, attracting scholars and pilgrims from all over Asia. In the past, his direct descendants were given official appointments and honorary titles, accommodated in relatively luxurious residences in Qufu and treated with utmost respect. In pre-1949 China, grand ceremonies were held at Qufu on the birthday of Confucius. Today his birthday is still celebrated at Confucian temples in Taiwan, Korea, Japan and the Ryukyus. Its recent celebration for the first time in decades at the Confucian temple in Qufu marks a turning point in modern Chinese history. Previously, the Communists had accused Confucius of impeding progress by trying to preserve the past, criticized his teachings as feudalistic and autocratic, and sought to break up the core of Confucian teaching by breaking up family loyalty.[3] Now the Chinese seem to be re-assessing Confucius' position in history.

According to one tradition, the Confucian temple in Qufu was originally constructed in 478 B.C. by Duke Ai of the state of Lu, a year after Confucius' death.[4] Over the centuries it has been rebuilt and enlarged, so that most present buildings of the large complex date from the Ming and Qing Dynasties. At one time the temple occupied one fifth of Qufu's total area and owned several hectares of land which were used for the sole purpose of raising the large number of pigs, sheep and cattle required for the seasonal sacrifices. Instead of burning cheap paper effigies, the worshippers performed the full sacrificial rites laid down in the books of old. The temple, one of the architectural marvels in China, looks like an imperial palace with magnificent yellow-tiled buildings within a surrounding red wall and with nine interior gates opening to individual courtyards set off by stretches of ancient pines and cypresses.

The Dacheng Hall (Hall of Great Perfection), the principal shrine for paying homage to Confucius and the main structure in the Confucian temple, formed the central stage for the recent ceremonies, which were performed with great pageantry to the delight of the spectators: see **Figs. 1-7**.

The cult of Confucius combines aspects of both the worship of nature deities and the ancestor cult. During the Han dynasty in 136 B.C. Emperor Wu made the Confucian classics the basic discipline for the professional training of government officials. The special status of the ancient philosopher dates from that time and whenever a strong central government ruled China, Confucius was held in high official esteem. Posthumous titles were heaped on Confucius. He was named Duke, Prince, Venerable Sage of former Times and Sacred Teacher of Antiquity. No wonder that, with so many honours bestowed on his person, the ceremonies held in his memory became increasingly more elaborate and were ruled by many regulations. In due time, Confucius was raised to the rank of a deity and awarded the same sacrifices as the sun and the moon. This process of deification found its final culmination in 1906 when the last Manchu emperor elevated Confucius to a position on a par with Heaven and Earth, the highest objects of worship. Sacrifices to him were held on a number of occasions. In addition to the ceremonies to celebrate the anniversaries of the birth and death of the Sage, other important ones, known as the four *ding* sacrifices, were held on the fourth day of the first month of each of the four seasons. With the increase of the prestige of Confucius, ceremonies came to be held on two levels: on a family level in Qufu by the living descendants, and on a national level. Local scholars and officials conducted the ceremonies in the various Confucian temples which were erected in all major cities, and in the capital, the Emperor himself offered the sacrifices.

At the recent memorial services in Qufu, solemn music of drums and ancient bells once again floated through the spacious courtyards of the temple as the long procession of ceremonial attendants, dressed in silk gowns, approached the large marble terrace in front of the Dacheng Hall. The participants, from the master of ceremonies, dancers, musicians, to the sacrificial attendants were all dressed in magnificent costumes of different colours according to their status. Ancient sacrificial vessels including bronze wine pitchers, wine cups, meat containers, lamps, etc. were laid out on tables. The sacrificial offerings on this occasion consisted of the *tailao* sacrifice (an ox, a sheep and a pig), salt, cat's blood, rice, fish and water chestnuts. All items and vessels were arranged under the magnificent double yellow-tiled roof of the Dacheng Hall, with its richly ornamented beams and rafters, and finely carved stone pillars of interlaced dragons and tracery.

In order to perform the almost forgotten ancient dances and music that accompany the great ceremonies, extensive research obviously had to be done. Old books like *The Music of the Sage's Mansion, The Rites of the Sage's Mansion* and *The Annuals of Queli* were consulted, and people, who had actually witnessed the sacrificial rituals before 1949, had been requested to help perfect the performance of the Qufu Opera Troupe, which had been selected to perform the ceremony.[5]

Arranged on the side of the terrace, the "antique" orchestra played the traditional instruments, some of which were reserved for the cult of Confucius alone and were rarely seen elsewhere, like the mouth organs (*sheng*) and pan-pipes. The sets of bronze bells and musical stones, drums, wind and string instruments produced grave and dignified music appropriate for the memory of Confucius. In the centre, the dancers struck their poses, and moved their wands, tipped with long pheasant feathers, in unison with the music.

At the booming sound of the great drum, the prescribed *Three-time Sacrifices* started amid melodious ancient music known as the *Six-movement Music*. To receive the spirit, the first movement called *Manifesting Harmony* is played, followed by the *Proclaim Harmony Movement* which marks the initial offering rites. While the music plays, the master of ceremonies kneels and kowtows in front of the Apricot Terrace, which is said to be the place where Confucius lectured to his disciples. Here the master of ceremonies offers incense and wine and is then led by the sacrificial attendants into the Dacheng hall, where he kneels and kowtows again and presents vessels in front of a statue of the Sage. Silk, cotton and wine are then offered during the recitation of prayers. To mark the completion of the offering and to conclude the first sacrifice, the *Describing Harmony Movement* is played. The second and third Sacrifices were identical to the first one, the only difference being that no prayers were recited. This solemn and interesting ceremony came to an end when the music to bid farewell to the spirits, entitled *Beautiful Peace of Moral Virtue Movement* was played, while the master of ceremonies again kowtowed nine times. Next he took offerings of silk and cotton to be burnt and returned to his position in front of the Apricot Terrace.

The recent ceremonies coincided with an important conference in Qufu, where scholars from all over the world had gathered to discuss and reassess Confucian thought. The symposium was no doubt instigated by the Association for Research on Confucius, which was established in 1985.

Despite the great fanfare attached to the recent Confucian ceremonies of the Qufu Opera Troupe, the performances are just performances and nothing more. To have an authentic ceremony at Qufu, the most direct descendant of Confucius must conduct the rituals. The most direct descendant, who also happens to have the world's oldest family tree, is the 77th descendant of Confucius, Kong Decheng, who was granted the official title 'Ritual Master of the Supreme Sage and Teacher' in 1945. He fled with his family to Taiwan following the Communist takeover in 1949. Only time will tell if he or one of his descendants will return to legitimize the ceremonies in Qufu.

NOTES

[1] *Kong* is Confucius' family name; Confucius being the latinized form of the Chinese name *Kong fuzi* meaning "Grand Master Kong". For an account of the ducal family in the 20th century, see Kong Demao and Kelan, *The House of Confucius* , (London. Hodder and Stoughton, 1988).

[2] Wing-tsit Chan, Trans. *A Source Book In Chinese Philosophy* (Princeton University Press, 1963), p.36. The quotation is from *Analects* 11:11.

[3] Kam Louie. *Critique of Confucius in Contemporary China.* (Chinese University of Hong Kong & New York: St. Martin's Press, 1980.)

[4] This tradition is historically unreliable. See John K. Shryock. *The Origin and Development of the State Cult of Confucius.* (New York: Paragon Books Reprint, 1966; 1st ed.: 1932), pp.93-97.

[5] It is amusing and ironical to see that the ritual was performed by opera artists. In traditional times, only officials of government, even emperors, were entitled to perform those elaborate sacrifices. Today, because of their Marxist and therefore atheistic commitments, no government official would want or dare to participate (in contrast with Taiwan, where even today city mayors and high government officials celebrate the rituals on Confucius' birthday, September 28). The second best choice are opera artists: in the eyes of the Marxist government, such a sacrifice is probably just 'play-acting' anyway.(Note by the editor)

Fig. 1 Gate to the Kong family cemetery, which contains Confucius' tomb.

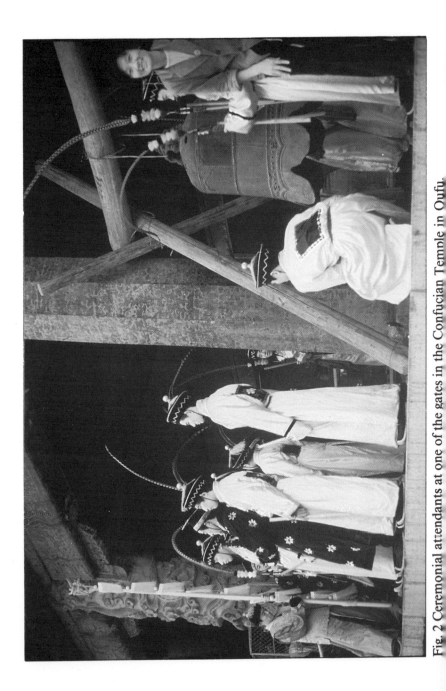

Fig. 2 Ceremonial attendants at one of the gates in the Confucian Temple in Oufu.

Fig. 3 Ceremonies in front of the Dacheng Hall (Hall of Great Perfection), the most important sanctuary in the Confucian Temple.

Fig. 4 Ceremonial dancers with pheasant feather tipped wands

Fig. 5 The same.

Fig. 6 Master of Ceremonies crossing courtyard in the Confucian
    Temple.

Fig. 7 Procession of ceremonial attendants to the Dacheng Hall (Hall of Great Perfection) in the Confucian Temple in Qufu, September 1986.

221

THE CATHOLIC CHURCH IN CHINA

JOSEPH J. SPAE

Part I: Elements of the Situation

Much has happened since the death of Mao Zedong (1893-1976) and much has been written about "reconciliation" between the PRC and the Catholic Church. For sheer theological reasons, China remains Christianity's monumental challenge of the century. One billion Chinese face one billion Christians. Both groups, Christians believe, are jointly committed to partnership in the service of the Chinese people. This need, and its corresponding mission, is acutely felt by the Catholic Church. The quantity and quality of those involved indicates the magnitude of the task ahead.

The Official Policy on Religion and the Catholic Church

The Party organ in charge of religious affairs is the United Front Work Department (UFWD). The United Front controls and staffs the government's Religious Affairs Bureau (RAB). These organs work in close cooperation with the Ministry of Public Security, the national police system. The RAB controls five "People's Religious Organizations": The Protestant Three-Self Patriotic Movement (TSPM), the Chinese Catholic Patriotic Association (CCPA), the Islamic, Buddhist and Taoist Associations. These five Associations have essentially the same political task: to lead their followers in support of Party and state goals, such as the Four Modernizations; to cut short any "anti-revolutionary" movements.

The CCPA was established under duress in 1953. With the avowed aim of separating a government organization from a religious organization, two supplementary organizations were added: The Catholic National Administrative Commission and the Catholic Bishops College. In fact, the distinction between the three organs is theoretical. Their officials and the higher direction which they receive are largely the same. No decision can be taken without government consent.

The spirit in which government control must be exercised is, theoretically at least, that of the constitutional "freedom of religion." In practice, however, it is made explicit in an abundance of statements, such as that of the 1982 Document 19 which stipulates that "all religious affairs must be conducted only under and through the Patriotic Association," and "religious infiltration, especially by the Vatican and Protestant missions must be resisted." China is wrenching its new version of a Marxist economy into the 20th century. In 1986, there were signs that this could also mean greater freedom of religion. But the changes are subtle and brittle, and often contradicted by reprisals against Catholic leaders such as Bishop Joseph Fan Xueyan and his vicar-general, Father Huo Binzhang, in 1983 for "colluding with anti-Chinese foreign forces (the Vatican) to jeopardize the security of the motherland." The penalty: 10 years in prison. The bishop had previously been jailed from 1958 to 1979 for his religious beliefs. There is also this other side of the coin: Party Secretary Hu Yaobang declared at a news conference in Rome on June 21, 1986, that "China is willing to discuss relations with the Vatican after we have overcome some difficulties." Officially, these difficulties are two: the diplomatic ties between Taiwan and the Vatican, and the independence of the Chinese Church from the Pope. Both problems, it is well known, are actively studied by the Vatican and are no doubt open to diplomatic negotiations.

Freedom of Religion in Action

Chinese papers like to announce that "Freedom of Religion Wins Respect" (*Ta Kung Pao*, Nov. 17, 1985). We are told that there are "tens of millions of religious believers in China today." This "freedom" is attributed to the Central Committee and its new regulations of March, 1982, the already mentioned Document 19. The new policy stipulates that "Non-believers are forbidden to preach atheism in and around temples, monasteries and churches." Its results are for all to see. "After complete renovation, 40,000 temples, monasteries and mosques have opened to the public. Most were badly damaged or destroyed during the Cultural Revolution." A carping mind will remark that these repairs were done for the sake of tourism and not for the benefit of religion. This is true; but the same places also provide an outlet for folk devotions hitherto forbidden as "superstition".

Here follows a short roundup of some facts related to religious freedom in practice. China's leaders advocate freedom *from* religion. Communists are liberated people; hence Party members dare not adhere to any religion. Marxism believes in the gradual disappearance of religion. Interestingly, Chinese leaders have recently admitted that "religion" is part and parcel of life in "liberated" countries such as France, England and the USA. There is a budding pride, expressed in 1986 by tourist guides, in "a

223

revival of religion." The guides are unaware of the 1984 and 1985 Amnesty International reports stating that "it is impossible to estimate the number of people imprisoned for their beliefs, but published estimates by officials of several prisons put the number of 'political' prisoners at 3%." In China, religion belongs to the domain of politics. Thus the best-known prisoner is the legitimate Catholic bishop of Shanghai, Gong Pinmei. He was arrested on Sept. 8, 1958 and condemned to life imprisonment in 1960 on charges of leading "a counter-revolutionary clique under the cloak of religion." After much foreign prodding, Gong was released on parole on July 3, 1985, exchanging prison for what looks like house arrest. Nevertheless, as we shall point out below, there are signs of improvement, also for the Catholic Church. Thus repeated assurances have been that the situation of the Catholic Church in Hong Kong will not change after 1997. Such assurances have met with considerable scepticism and July 1986 was a month of prayer for the Colony. Meanwhile, Ren Wuzhi, director of RAB, published a much remarked article on "China's policy of Religion" in the October 1985 issue of the *You-sheng, Voice of Friendship* , in which he reminds his readers that "the government supports the efforts of churches rejecting interference in our country's religious affairs by foreign religious persons who attempt to control China's churches once again." Mr. Ren referred to the worldwide Catholic Church and to the Vatican.

The CCPA, Mouthpiece of the Government

One of the most tantalizing aspects of the Catholic Church in China is the nature and behavior of the CCPA. Mention the word, and you are forced to declare colors. In the foreign press, no topic has led to so many contradictory statements as that of the CCPA. For the government and those few Catholics willing to follow its directives, the CCPA, although an "association" and not a "Church", is the only fold to which Catholics do or should belong. It is recalled that, in 1953, the government set up the CCPA, forcing priests and laity to sign a statement pledging allegiance to the PRC and rejecting loyalty to Rome. Refusal often meant return to prison. From 1955 on, the government ordered the illicit consecration of bishops. Since that time some 50 or more priests were consecrated bishops. It is reported that some of them have been asked to swear on oath not to have any dealings with the Pope. The names of all the bishops and their sees have never been published. One bishop recently told this author: "I don't know how many bishops there are in China: we have no contact with one another." The CCPA publishes at irregular intervals *Catholic*

Church in China, a government-controlled gyroscope for religious news. Few bishops take it seriously, and not all receive it. Who belongs to the CCPA? There is no adequate answer to this question, due in part to the secrecy with which the government surrounds the non-official Catholic Church in China. After many visits, the last one being in May 1986, I have come to the conclusion that the overwhelming majority of Catholics refuse to have anything to do with CCPA but strongly retain their allegiance to Rome. It is true that many of them accept the sacraments from the hands of "patriotic" priests, provided they are not married. But, on any occasion, these faithful Catholics will vote with their feet. Such occasions were the visit of Cardinal Sin to Beijing in October 1985; that of Bishop Wu of Hong Kong in January 1986, when visiting his mother; and an announcement that Bishop Gong Pinmei would say a public Mass in Shanghai. On each of these occasions, thousands of Catholics turned out to pay homage to ecclesiastics close to the Pope or imprisoned for his sake. In the cases of Bishop Gong and Wu, the faithful or "the people" were driven away at the request of the CCPA, appalled at the size of the crowds who came to greet them.

Bishop Fu Tiesha, Spokeman for the CCPA
No one illustrates better the official attitude of the CCPA than the bishop of Beijing, Michael Fu Tieshan. In numerous interviews and statements, he has never left a stone unturned to attack the Pope and praise the government. The Pope has said that he has tried "on several occasions to make contact with Bishop Fu, but all in vain." The flavor of the official CCPA leadership — and no doubt also of the government's thinking — can be tested in some of his significant utterances, such as the following:

> We believe in God. At the same time, we are citizens of the PRC. We support the Communist Party although we do not believe in atheism. — It is my duty to help the government implement the policy of religious freedom under the leadership of the Communist Party until all believers have acquired the spirit of patriotism, and follow the policies of the state. — We are not a schismatic Church but an independent Church. — The Vatican is interfering with China's internal affairs through contacts with Taiwan Catholics and the secret appointment of Chinese bishops. These actions by the Vatican, in disregard of the right of Chinese Catholics to self-government, have obstructed the exchanges between Chinese Catholics and the Vatican and have created confusion among Chinese Catholics. As an independent and autono-

mous religious organisation, the Chinese Catholic Patriotic Association is willing to conduct friendly exchanges with Catholics worldwide on the basis of equality and friendship. The Association has already established contacts with religious believers in many parts of the world since the early 1980s.

The last quotation is taken from an interview of Bishop Fu with the official New China News Agency, April 12, 1986. To which the agency adds:

Bishop Fu has been elected as standing committee. member of the National Committee of the Chinese People's Political Consultative Conference on April 11. In the past few years, he has met more than 30,000 Catholics from other countries and visited Canada (in 1981), Belgium (1985) and the United States (1986). He has been active in expanding Sino-foreign Catholic exchanges.

Bishop Jin as Mediator

During the last two years, the Patriotic Auxiliary Bishop of Shanghai, Aloysius Jin Luxian, has practically stolen the limelight from Bishop Fu Tieshan. He has travelled to Hong Kong and Macau, to Manila and Germany, where he spent 10 days, in April and May 1986. All agree that Jin is a most capable man who has spent long years in prison, working in the service of the government as a translator. He speaks several European languages and is a self-taught reader of Russian. He is affable and articulate. Perhaps he is the best spokesman for China's religious policy, not so much by defending it but by putting the burden of the situation squarely on the Vatican. During a 90 minute conference at St. Augustin near Bonn on April 19, Jin repeated all the old arguments: Rome's intent to dominate the Chinese Church; its colonialist and aggressive attitude towards China; its refusal to go along with Communism; its slowness in appointing Chinese bishops, etc. Never did he say a word about the bloody persecution under which the Church has suffered and continues to suffer; about the complete subservience of the CCPA to the government or about the abnormal canonical situation in which the Chinese bishops find themselves. Jin hints at the possibility of his serving as a mediator between Beijing and Rome, and it has been reported — without any proof — that Rome would not be adverse to using his services.

Jin is rector of the Sheshan seminary near Shanghai. The seminary has recently been enlarged, thanks to Jin's worldwide efforts to collect funds. The new seminary costs US$705,000 of which the

Shanghai diocese paid half. In a May 13, 1985 interview, Jin clearly outlined his doctrinal position: "We truly have religious freedom and it is permanent. We are Catholics: we want communion, as brothers not as subjects. Our relationship with the Vatican is a complex problem. We want to be autonomous but united with the whole Church. Seminaries do not have to join the CCPA. I myself am not a member. We want them to be celibate. The government helps us in religious life. It gives us a free hand." Jin's statement raised more than a few questions. But then, so does the very nature of the CCPA itself.

Impressions of the CCPA

It is imperative to keep in mind the basic distinction between the CCPA and the Catholic Church in China, not only because the CCPA correctly declares that it is not a Church, but above all, because, according to the convergent testimonies of many recent visitors, agreement among Catholics with the CCPA is by now a rare phenomenon. As Bishop Fu Tieshan put it: "Many Catholics despise the CCPA." In fact, this may be an understatement. Yet, in all honesty, it must be admitted that not enough is known about the Church in China to come to acceptable conclusions. How many "patriotic bishops" feel that they should heed the CCPA directives? How many of them have secretly contacted the Pope to express their unswerving loyalty? If the government's pressure were removed, how many of them would refuse to enter in communion with the Vatican? No one knows, but from the repeated utterances of Bishop Fu and Bishop Tu Shihua, rector of the National seminary at Beijing, much is happening beyond the control of the government, and there is a thriving "underground Church." The government is not unaware of the fact that the Catholic Church in China largely repudiates the package deal which it has imposed on them. Nobody believes its communiques on religion as they appear in the official press. The new official mood wisely decides to leave well enough alone and to deflect foreign attention from the lack of religious freedom in the country. The Catholic Church profits from this benign neglect. It is an open secret that a number of bishops have recently been consecrated with Rome's approval, and that hundreds of young men are being prepared for the priesthood by individual priests all over the country.

In all this, it seems, the government is taking a leisurely U-turn, the way it does in matters economic and political. In foreign circles writing about China, there is a constant and insidious tendency to treat the Catholic Church in the PRC as if it were on the eve of becoming a type of Anglican Church. Nothing could be farther from the truth. The blood of martyrs does not become the seed of schismatics.

There is no better illustration of the above than a recent event brought to the attention of the world by Amnesty International. Briefly: "China's forcible closure of an 'illegal' Catholic seminary loyal to the Vatican, during which novice nuns were allegedly sexually assaulted, has triggered a worldwide campaign, requesting that letters of protest be addressed to Premier Zhao Ziyang." (*South China Morning Post*, Sept. 23 and October 8, 1986). There was no mention of this ugly fact in the Chinese press. (For full details, see *China Update*, Winter, 1986, pp.51-52.)

The CCPA's Achievements

What has just been said, it is emphatically stressed, does not imply any judgement on the subjective dispositions of bishops and others who sincerely believe in the role of the CCPA. One sympathizes with well-meaning people, cut off for a long time from objective news and often after long periods in jail, who gave their name to the CCPA and allowed themselves to be consecrated bishops. You hear them say: "This was the only way to save the Church from worse." And they may well be right, at least if one takes into account their perplexity in the face of martyrdom, and their honest desire to work for their faith. Nor is there any reason to question some material achievements of the CCPA. It claims that some 400 churches have now been opened. (In 1938, there were 2,187 churches and 13,000 chapels.) There are now some 200 seminarians in seven training centers. (In 1948, there were 924 seminarians.) Many priests have been set free from work camps, although often without restoration of their civil rights. Recently, priests faithful to Rome have been allowed to say Mass in the patriotic churches, i.e., the only churches open to the public. On December 24, 1985, the Beitang, Beijing's venerable cathedral, was reopened for worship to the acclaim of thousands of Catholics. The reopening of this church, soon to be followed by a fourth Catholic Church in the capital, is in line with a government policy which restores to the care of the CCPA many Catholic churches, partly or completely destroyed. The faithful have been invited to solicit foreign funds to rebuild their churches. The release of Bishop Gong Pinmei has been mentioned. The role which the CCPA played in this is uncertain. But the acting bishop of Shanghai, his illegitimate successor, has often expressed disapproval of Gong's conduct. Now the venerable prisoner is in his care, a charade, which, it was reported, moved Gong to ask permission to return to prison. Impartial observers are convinced that the government is wisely preparing the CCPA's sinking ship. This could be done through a new *modus vivendi* with the Vatican, discussed later. Meanwhile, the government took note of Bishop Fu's words: "There could be 3 million Catholics in China who have kept a secret alliance to the Vatican." (*South China Morning Post* , Nov. 7, 1984). This means 80% of all Chinese Catholics. Reconciliation with the Vatican, therefore, has become a useful need.

How Will the Church Survive? Can it Survive?

Catholic doctrine insists on the sacramental life, a life which largely depends on the presence of priests and bishops. A church without priests and bishops, under normal circumstances, could not long survive. Hence the need for seminaries. In this matter, it would seem, the wishes of the CCPA coincide with a greater openness on the government's side, and seven new seminaries have recently been established, parallel to similar institutions among Protestants, Muslims, Buddhists and Taoists. We are well informed about the Catholic training institutes. Visitors have reported on their curriculum — the classic one — and on their students: they are alert, eager to learn, but also, according to Bishop Jin, undisciplined.

They recently made an excellent impression on this author. The management of half a dozen or more Catholic seminaries is entrusted to the CCPA. The students are often housed in restricted quarters and lack competent professors and books. It has been reported that some of them are not *bona fide* applicants but government agents, a fact vigorously denied by Bishop Jin. Are they likely to become good priests? One hopes so, but, obviously not at the expense of Catholic orthodoxy. That such a fear is not unthinkable stand out from the fact that, in most if not all, seminaries, the government (or the CCPA) has an agent whose job it is to control every detail of seminary life. Without his permission no one enters, and no one can talk to the students or the staff. More disquieting is the attitude of at least one well-known seminary rector, the already mentioned Bishop Tu Shihua who said: "Each bishop, after he has been selected by the clergy and the laity in the respective local church, immediately receives the power to govern his church through episcopal consecration... The bishop of a local church neither receives his powers by the appointment or ratification of the pope, nor exercises them as his delegate. The local church, headed by the bishop, possesses the right of independent self-rule and self-management... No other bishop is allowed to interfere, to restrict, still less to deprive a bishop of the ordinary, immediate, proper power awarded by God. On the contrary, it must be respected." We shall return to this difficult question which seems greatly to occupy the CCPA — and the government.

The Church Survives: the Faithful

Recent visitors have brought back abundant details of Catholic life outside the major cities. Increasingly, reports tell of the fervor of the faithful in attending Mass and receiving the sacraments. They are even convinced that there is a significant increase in the number of Catholics. It is not rare to see them travel a hundred kilometers and more to attend the Mass of a visiting foreign priest. There are no reports on their relationships with the CCPA, because these are simply non-existent. We also know about their life of prayer and mutual support. One expression of this support is seen in the fact that, in predominantly Catholic villages, they help one another in the observance of the one-child quota by arranging the adoption of children by otherwise childless families. They bow to the government and accept official work, but without putting their faith at risk. In many Christian villages, particularly in the North, officials are Catholic and leave complete freedom to the community. It is also reported that many ministries, usually reserved to priests, have been taken over by the laity, such as the dispensing of some sacraments. There have been reports of wondrous facts, even miracles. Be this as it may, the greatest miracle is the sheer survival of these simple people under extreme duress. As one old man told me: "The most they can do is send us back to jail!"

As Chinese, Catholics have retained the essential qualities of their noble nation: moderation and inclination to compromise rather than violence and revolt; respect for authority and the natural order of things; strength of will hidden beneath a courteous gentleness; a gay and brave resignation coupled with endless and ingenious patience; a boundless self-respect submerged in a lasting sense of humor, which, like poor Ah Q, Luxun's immortal character, must justify itself to itself to save its face. They quote an age-old saying: *"Shen chu, gui mo"* or "gods appear, ghosts vanish." History is unpredictable. But this much already is certain: in the book of life, theirs will be a glorious chapter.

Part II: The Vatican and the Church in China

Towards Reconciliation: an Independent View

The Vatican receives abundant advice on how to reach reconciliation with the PRC, but rarely from Chinese writers as qualified as Father Louis Wei. Wei is a former Chinese diplomat once stationed in Rome, a well-known historian, author of *Le Saint*

Siège et la Chine (Paris: Allain, 1967), and *La Politique Missionaire de la France en Chine* (Paris: Nouvelles Editions Latines, 1960). In September 1984, Wei went to Rome and presented a lengthy document which, perhaps for the first time, stated the conditions on which an understanding between Beijing and Rome might be possible. This document has been published in its entirety in *China Update*, 15, Spring 1986. Here follows a summary translated from the French. The author stresses that he speaks for himself. Wei begins his exposé with the statement that reconciliation is possible but on one basic condition: the Vatican should break off diplomatic relations with Taiwan. He makes eight points:

1. *The ambassador to the Vatican should leave*. While the Vatican Nunciature in Taipei is practically vacant since 1971 — at this moment, there is only a Charge d'Affaires — Taiwan has maintained its Ambassador at the Vatican. This high official should be withdrawn according to diplomatic custom. Such would be a small but important step in the right direction.

2. *Normalization of diplomatic relations between the Vatican and the PRC*. Only then could the next step be set: the Vatican should dispatch a high official to Beijing for negotiations. Such a move could be prepared at the United Nations between the representatives of both countries there.

3. *Canonical recognition of all patriotic bishops by the Vatican*. This would imply their confirmation to the sees they now occupy. What has been done in other places, namely, the appointment of bishops by the local people, could also obtain in China.

4. *The Catholic Church in China should no longer be dependent on a Roman office, such as the Propagation of Faith*. Forty years ago, the Vatican established the Chinese hierarchy, an eloquent sign that the Church there had attained maturity. The time has come for her to be disenfranchised from an office that symbolizes the "missionary period." Catholics in China want to be on an even rank with those of England and France.

5. *Conditional amnesty for bishops, priests and faithful still in Chinese prisons*. Wei cites as example Bishop Gong Pinmei, since released (see above), and Archbishop Tang Yiming, exiled in Hong Kong. He calls them "the victims of the Vatican foreign relations since 1949." This would be an admirable starting point toward reconciliation, much like the Concordat concluded in 1801 between Pius VII and

231

Napoleon. As this Concordat brought peace and unity to the French Church, a similar gesture in China could obtain similar benefits.

6. *The practice of religious liberty in the PRC* . The Catholic Church in China is on an equal footing with that of the other religions as guaranteed by the Constitution. Hence, she need not ask for special privileges, such as the protection by an outside power. As for the return of her former works of social servive and education, this is a matter to be treated by the Chinese Episcopal Conference and the government.

7. *The state of the CCPA* . There is only one Church in China. There is no private or underground Church. The CCPA is not a "Church". It is a civil and not a religious institution. It merely serves as a liaison organ between the government and a religious entity.

8. *The matter of Taiwan* . Taiwan is an integral part of Chinese territory. Normalization of Sino-Vatican relations would not imply the abandonment of the Church in Taiwan where the Vatican could appoint a representative without diplomatic credentials. The Taiwan embassy to the Holy See would be closed. This would be a provisional arrangement until Taiwan becomes reunited with China.

The Pope and the Church in China

The pronouncements of and interventions by Pope John Paul II on behalf of the Church in China are too many for enumeration here. On the occasion of the election of Bishop Fu Tieshan to the see of Beijing, the Vatican press bureau let it be known on August 3, 1979 that "the Holy Father did not authorize, let alone approve of, the election." A few days later, on August 19, this provoked an angry retort from Tang Ludao, general secretary of the CCPA: "We denounce the Vatican's blatant interference in the internal affairs of the churches in other countries. . . the power of nominating bishops directly comes from God. The voice of the people is the voice of God." Similar invective language has often been repeated since then. On the Pope's side, there have been frequent references to "an improvement in the situation, justifying hope for the future," the earliest one appearing in the *Osservatore Romano* , Eng. ed., August 27, 1979. Bishop Fu angrily reacted: "It is for the Vatican to recognize the independence of the Chinese Catholic Church. Until this happens, there is no room for dialogue with the Vatican on religious matters." (*Le Monde* , August 29, 1979) Since that date, it would seem, Fu has not changed his mind.

Meanwhile the Pope kept repeating what was in his innermost heart: "As Christmas draws near, I send my greetings and my good wishes to the sons of the Catholic Church in China, as to all the members of that great nation, renewing the hope that there may be positive developments which will mark for our brothers and sisters of the Chinese continent the possibility of enjoying full religious freedom." (OR, Aug.27, 1979.) At the Ricci celebration in Rome, October 25, 1982, the Pope referred "to the common heritage of the Church and China which appears as a solemn and symbolic reference point for a constructive dialogue directed toward the future since, as I have said in Manila on February 18, 1981, it is the future that we have to look to." (OR, Nov. 12, 1982.) In the same vein as his Manila speech, the Pope pleaded from Seoul, Korea, with the CCPA "to live the faith in full communion with the universal Church, to the joy and enrichment of all." Since the Manila speech, it is no secret that the Pope desires some day to visit China. He is not one easily to despair.

Meanwhile many keep vigil with the Pope at the Vatican. "The Vatican acknowledged yesterday that, despite repeated efforts, it had not been able to establish any official contact with government or Catholic Church authorities in China." To these words, Cardinal Casaroli, the Pope's Secretary of State, added, "An extremely bold gesture might be needed to establish regular contacts with China." The cardinal did not elaborate. (AP, Vatican City, Nov. 19, 1984.) Cardinal Casaroli's remarks obviously were an indirect reference to the results of Cardinal Sin's trip to China.

We have mentioned the two apparently chief obstacles to reconciliation, one of them being Vatican-Taiwan diplomatic relations. Casaroli is on record as having said that "this matter could conveniently be solved." The second obstacle remains. As AP puts it, it is "the refusal by the CCPA to accept papal authority." But is this a fact? Cardinal Sin and other visitors to China have pointed out that the refusal of communion with the Vatican voiced by the CCPA is really not that of Chinese churchmen but the voice of their master, the Communist Party. It is the general conviction of well-informed observers that, if the Chinese government stopped making a mockery of religious freedom and cleared their name with the civilized nations of the world, all they would have to do is open free communications between the Chinese Church, patriotic and/or other, so that Catholics in China can begin serious conversations with Rome. Why should the government fear to let the chips fall where they may, when it has a fine opportunity to gain respect throughout the word and lay to rest all misgivings related to the freedom of religion in their country?

What does it have to fear from a tiny minority?

We quoted above an important remark by Cardinal Casaroli. The cardinal made it at an impromptu interview with reporters on November 19, 1984. He added: "China has so far rebuffed papal efforts for dialogue. We hope that regular contact will become possible." Reacting to this statement, a Beijing spokesman repeated "my government's unvaried position." Wang Yaobing, speaking for the RAB, stated: "As long as the Vatican maintains ties with Taiwan, and as long as it fails to change its discriminatory attitude toward patriotic Catholics, there will be no contacts between China and the Vatican."

Cardinal Casaroli was further asked to comment on how the Vatican can hope to establish relations with mainland China without breaking off relations with Taiwan. The Secretary of State gave this diplomatic answer: "That is an interesting question that merits a well-thought-out answer. I'll think about it!" Archbishop Achille Silvestrini, in charge of the Church's Bureau of Extraordinary Affairs, let it be known that "I have great confidence that things will change." He saw three reasons for optimism: 1. greater openness in China to the rest of the world; 2. greater appreciation for the papacy's position in international affairs; 3. reevaluation of the role of missionaries in China. "But," he added, "China is an exceptional and difficult case. Persecution of Catholics by the government and a tendency toward schism by some Chinese Catholics are still major stumbling blocks to better relations. The Pope is aware of the Chinese situation. He has always said that the Church sees China as a great nation and a great civilization." Returning to the Pope, the *Osservatore Romano*, Nov. 25, 1985, carried his speech, entitled, "May the Church in mainland China join in full communion." Tang Ludao did not fail to voice his criticism; "Relations between the Chinese Catholic Church and China have been severed for decades... This is the responsibility of the Holy See. For decades, the Vatican has disregarded the sovereignty of the Chinese Church." These were the offending words of the Pope:

> The Church (of China), so dear to me, is continually in my mind and I daily beseech the Spirit that the day soon may come when, after the obstacles of various kinds have been removed, there will come the desired moment of communion fully lived, fully expressed and enjoyed. In the meantime, there is entrusted to us the fruitful mission of praying for those communities, that their faith in the Redeemer of humanity may be lively and deeply experienced in the communion of the one, holy, catholic and apostolic Church,

which has in Peter and his successors the "permanent and visible source and foundation both of the bishops and the whole company of the faithful." (*Lumen Gentium ,* 23)

The Pope had said nothing new. He spoke the same words in Manila and Seoul. Tang Ludao said nothing new either. He merely repeated what he had said on August 19, 1979 (see above).

In March 1986, the Vatican organised a secret three-day summit meeting on the relations between Communist China and the Catholic Church. The meeting, attended by 15 experts, was called by Cardinal Jozef Tomko, Prefect of the Congregation for the Evangelisation of Peoples *(Propaganda Fide)*. It was rumored that guidelines were about to be published on contact with the CCPA and other Catholics. The meeting, according to Bishop Sanchez, secretary of the Congregation, "discussed the delicate problem of the Chinese government's attempts to control the Catholic Church" *(Sunday Examiner ,* April 25, 1986). This last statement is important although its implications are often forgotten, even by Catholic authors. To put it bluntly: should a reconciliation between the Chinese Government and the Vatican occur, it will not be dictated or inspired by the CCPA but by Beijing's pragmatic interests. Whatever the government's decision, it will be meekly followed by the CCPA. This hard fact determines the parameters of mutual dialogue, not only from the political, but, above all, from the theological point of view.

Obstacles to Reconciliation

From a Roman Catholic point of view, it is necessary to put the obstacles to reconciliation in their true *theological* light. Anything less would be irrelevant and counter-productive. Let us review the main obstacles, taking for granted the loyalty to the faith and to the Pope of the overwhelming majority of the Chinese Catholics, priests and bishops. It must be remarked at this point that Catholics in China are usually said to number from about 3 million to 6 million. As about 1400 priests and 2000 religious sisters have been sent to work as private citizens in the former communes, or are in jail or labor camps, the task of teaching the faith to children and catechumens is borne by parents, friends and "barefoot catechists", most of them women. The loss of clergy and the attitude of the CCPA which accepts without questioning the government's directives and moral standards on marriage, divorce, abortion and contraception, throw a number of delicate problems in the path of pastoral workers. Hence the new type of "evangelists" is known to and appreciated by the masses.

235

1. *The first obstacle*, at least from government side, is said to be Taiwan and the fact that the Holy See keeps up diplomatic relations with that "part of China". This obstacle is more fictitious than real. The Vatican has discussed all future eventualities with the bishops of Taiwan; their faithful would accept whichever solution their pastors propose. As for the Beijing side, if the matter is, as they say, of a diplomatic nature, it can be solved through diplomatic negotiations. Mr. Hu Yaobang's recent visit to Rome and his reluctance to meet Vatican representatives seems to indicate that no serious dialogue has begun. China, while blaming the Vatican for a diplomatic impasse refuses to engage in dialogue with the Vatican in a diplomatic way. In this matter of Vatican relations with Taiwan, as seen by the PRC, impartial observers spot serious contradictions. They advert to the fact that, while Taiwan receives all sorts of help from the USA, China nevertheless entertains good relations with that country. Also, Beijing often repeats that the nuncio, Archbishop Riberi, left China for Taiwan. The truth is that the Communists threw him out. He did not choose Taiwan. Why not ask the Vatican representative to return to Beijing?

2. *The second obstacle* is raised by the Dec. 4, 1982 Constitution. Article 36 refers to religion. It is one of the 24 articles defining the "Fundamental Rights and Duties of citizens." This is the text:

> Citizens of the People's Republic of China enjoy freedom of religious belief. No state organ, public organization or individual may compel citizens to believe in, or not to believe in, any religion. The state protects normal religious activities. No one may make use of religion to engage in activities that disrupt public order, impair the health of citizens or interfere with the educational system of the state. Religious bodies and religious affairs are not subject to any foreign domination.

I have meticulously analysed all available comments on this last phrase, also comparing them with China's former Constitutions. The intent of this phrase has been clearly expressed by Anglican Bishop K.H. Ting, president of the Protestant Three-Self Movement, in these words:

> The last statement says that religion is not to be controlled or dominated by a foreign country. I personally supported it. That is, supported the inclusion of this sentence in this arti-

cle... When our committee discussed and decided to put in this clause, we were thinking entirely of the interference of certain Roman Catholic groups in the domestic scene of China.

The bishop of Beijing, Fu Tieshan, also came forth with his comments on Article 36. He believed that it would assure "a healthy development of all religion." He accused "some foreign churches" of trying to control the churches in China, adding: "The provision in the new Constitution which stipulates no foreign Churches may dominate Chinese ones is in agreement with Chinese Catholics who run their Churches independently." The publication of Fu's comment was followed by a flood of similar remarks. Most analysts deplored the vagueness of the terms used here. Others quite plainly read in them that "if a Catholic is not a member of the CCPA, he is being unconstitutional." Or in other words: "If you are pro-party and pro-government, you can enjoy freedom of religion." In China (and elsewhere) policy is one thing, implementation is another. It must be said, however, that in the four or five years which have passed since the Constitution was voted, with few exceptions, it has not been applied lopsidedly against Catholic believers. This is entirely due to the goodwill of the government, rather than to that of the CCPA. We shall discuss the problem of "independence from foreign domination" hereafter. Meanwhile be it said that the whole problem of freedom of religion lies in the respect of basic human rights. By no other standard should the new Constitution be judged; with no other standard can the Catholic Church — or the Chinese people — be satisfied.

3. *The third obstacle* is the constitutional clause: "Religious bodies and religious affairs are not subject to any foreign domination." Legal minds have complained about the imprecision of this clause. K.H. Ting may be right as to its origin and intent, but he is not an authoritative exegete of the Constitution. The "independence" of the Chinese Church from the Vatican has meanwhile been stressed much more often by the CCPA than by the government. Recent official attitudes seem to leave the door wide open for future accommodations. It is necessary, therefore to draw a sharp distinction between the "political" implications of the clause and its "theological" content. As to the political aspect, the Vatican has repeated time and again, with the agreement of the world Church, particularly since Vatican II: papal authority excludes any form of

"domination"; it is essentially a service dedicated to the unity of the Church. To ignore this fact could only betray gross ignorance or ill will. And in the case of China, we need not take either for granted. To clear up this point, let us now consider the theological aspects of the matter.

One could understand that the Chinese government thinks of the papal government as similar to itself, i.e., behaving as the emperors of yore, holding the life of citizens in its hands, and arrogating to its few leaders at the top all authority over the nation. People's Congresses and "the will of the masses" notwithstanding, such are the undisputed facts. It would be a travesty of the truth to project this pattern of rule and domination upon the leadership of the Church, and any CCPA bishop knows that much. Since Vatican II, the practice and formulation of papal authority has undergone considerable change. When Patriotic Bishop Jin Luxian says that "the Churches, the Patriotic Church and Rome, are equal, equal like the persons of the Holy Trinity," he is speaking theological nonsense, as we now hope to prove.

According to Vatican II and its authoritative commentators, the universal Church is the community of the local Churches. By local "Church" is meant every religious community that has a social organization such as a parish, a diocese, or a larger cultural area, say, China. About such churches, Vatican II asserts: "The individual bishops are the visible sources and foundation of unity in their own particular Churches, which are constituted after the model of the universal church; it is these and formed out of them that the one and unique Catholic Church exists" (*Lumen Gentium*, §23). The above doctrine serves as a substantial correction to the view of the local Church as an administrative division under the sway of the universal Church. The truth is that the one Church exists in the local Churches, and that the one Church consists of the local Churches. The universal representation of this fact — held sacred since the beginning of the Church — is the community of the bishops together with the bishop of the Church in Rome. In this sense the Pope can be called the visible principle and foundation of the unity of the bishops and the multitude of the faithful, and the bishops and the visible principle and foundation of the unity of their local Churches (*Lumen Gentium*, §13).

As for the relationship between the bishop of Rome and the other bishops, here too Vatican II has restored the authentic tradition: the bishop does not receive his power to perform sacred actions, such as the ordaining of priests and bishops, from the Pope's "fullness of power" upon being given his canonical mission by the Pope. This is perhaps the old vision in which the Chinese bishops have been educated. It developed only from the later Middle Ages when the appointment of bishops was made between the "canonical mission" on the one hand, and the sacramental ordination on the other. Post-Vatican II theology asserts that episcopal ordination transmits the fullness of the episcopal ministry as a united office sanctifying, teaching and governing: an office that naturally is transmitted and accepted in fellowship with the other Churches and their bishops. (*Lumen Gentium* §21; new *Codes of Canon Law*, canon 375 §2.) Bishops are not deputies of the bishop in Rome. They possess their own authority that is not suppressed by the pope's universal authority but on the contrary confirmed, strengthened and defended by it (*Lumen Gentium* §21). By virtue of their office, bishops act on their own responsibility, without prejudice to the competence of the Pope to reserve certain matters to himself when these matters really affect the general interest and leave the bishop's normal competence intact. Such is the matter of Church unity and orthodoxy.

Both the Pope and bishops act in and from the community of their apostolic college, the seat of unity within the one Church. Rome has always been the center of that unity. Hence, the bishops of Rome, since ancient days, have exercised authority over all the Churches and their bishops because of their special responsibility for this unity. The whole matter has been competently examined in the international theological periodical *Concilium*, June 1986.

It is now possible to apply this doctrine to the Church in China, and lay to rest a number of misgivings and misconceptions as they have appeared in discourses of Chinese bishops mentioned above. Anyone conversant with Church affairs — or, for that matter, with government affairs, be it in China or elsewhere — knows that administrative changes are very slow. There may have been reasons to grumble against centralisation of the Church, or against the power of an "Italian" Curia. But the years after the Council have muted this grumble. The implementation of the doctrine on the true nature of the local Church and the universal primacy of the Pope, as seen by the Council, has made more progress than most Chinese bishops — and also the Chinese government — can imagine. Some immediate conclusions stand out: if necessary, the Pope, to promote unity, could allow the Chinese bishops to be chosen by their clergy and faithful. If necessary, the Pope could even grant to the government a veto right, as long as its interference is not plainly the result of its traditional hatred of religion — and on this point, the situation seems to be slowly changing. For the rest, the Chinese bishops would be as individuals and as members of the Episcopal Conference in charge of their own affairs.

As for the constitutional clause on religious bodies subject to "foreign domination" it makes little sense when applied to the Catholic Church. The principle of unity and orthodoxy is not predicated on "domination" but on "communion." This communion, no doubt, is both an internal and external affair. Insofar as it is internal, the government could not object to it; insofar as it is external, as the Constitution pleads in favor of friendly relations with foreign Churches, its acceptance by Church government has made China become more and more aware of late that it puts itself outside the pale of nations by refusing normal relations with the Vatican. What is the advantage of behaving like a second Russia or Albania?

In the case of Catholics, no apology is required for this lengthy theological exposé. To disregard it would to be to vitiate from the beginning any efforts at reconciliation between the Church and China. To harp, on either side, on the shortcomings of the past when so much is at stake, would be a betrayal of the future.

Conditions for Reconciliation on the Catholic Side

If our previous survey of the situation is correct, the Catholic Church in China, by and large, is anxious to reach a public state of communion with the world Church and with the Vatican. It is a misrepresentation of this situation to talk about "two Churches opposing one another." It must be repeated time and again that such an opposition, sometimes expressed by the CCPA, would melt away as soon as a spring of government tolerance appears. Most observers agree that at least some 95% of the faithful are anxious to leave the official ghetto in which they are kept. It is a public secret — frequently pilloried by representatives of the CCPA — that not a few of their bishops, officially on their side, are known for their allegiance to the Pope. The CCPA has not arisen out of the ashes of a xenophobic nationalism. It is the result of extreme pressures by the government.

These pressures continue: so far no Roman prelate or other bishop has had free access to the Chinese bishops. Why this quarantine? Could it be that they cannot be trusted because, in their heart, they are loyal to the Pope? But, in the light of what we have said, what is wrong with such loyalty?

For all that, a deep and respectful sensibility to Chinese culture, past and present, is now required of the Church. The point has been made eloquently by the Pope in his address of November 25, 1985, from which we have already quoted: "The Church looks with attentive interest to all aspects of a delicate moral sensibility, already recognized as values by traditional Chinese humanism (*OR*, Nov. 25, 1985). This sensibility has many facets.

1. A sympathetic understanding of a Church, no longer "a mission Church", but a totally indigenous Church, similar to the Church in France and other countries. The acculturation which this presupposes should not be led by antiquarian enthusiasts trying to beaver away to construct a local Chinese Church displaying the classical features of the missionary past, but rather intent to express its national identity in contemporary terms, using an idiom that develops alongside the idiom of the times. Such an acculturation must be left to the Chinese themselves. Specifically, it must be left to the Church leadership how best to educate the faithful in a patriotic collaboration with the modernization of their country, as well as in a pastoral and liturgical renewal based on Vatican II.

2. A thorough study of the past and present of the Catholic Church in China in an effort to "learn from facts." It must no longer be taken for granted in 1986 that the Beijing government clearly intends to separate the CCPA membership from the overwhelming majority of the faithful. There are good reasons to believe that the contrary is true. Beijing stands to gain by the internal unity of the Chinese Church, and by its acceptance by the world Church. Here as in other sectors, their pragmatism will prevail. The constitutional clause on "foreign domination" is not a theological but an ideological clause. As we indicated, it presents no real difficulties to an understanding between Beijing and Rome.

3. A further study of the evolving fate of Chinese communism and Chinese atheism. We have entered upon "A New Age of Capitalism" (*Time*, July 28, 1986). Marx has been publicly debunked, and his theories said to be antiquated. Former ideological distinctions have been thrown overboard. The leadership, as this author experienced only a few months ago, seems anxious to obtain Catholic collaboration (*China Update*, Autumn 1986). Traditional fears of a communist regime are less warranted. Instead, frank collaboration is called for by the world Church, through cultural and scientific exchange, through contact with Chinese officials and bishops, through joint educational efforts, such as scholarships, etc.

4. This spirit of understanding and flexibility involves a persistent effort which could facilitate diplomatic contact between Beijing and the Holy See. It is, indeed, at this level that the matter of Taiwan must be broached, as well as the official standing of the CCPA. In this effort, the Churches of Asia must all share a common concern for China and its Church by their example as neigh-

boring Churches, free from all "domination of a foreign power." In turn, the unwavering attitude of millions of Chinese Catholics should be their inspiration.

5. Goodwill and diplomacy call for mutual trust and openness. Such openness was expressed recently by the retired Archbishop of Taipei, Lo Kuang: "Our stand is that we must keep the unity of the Church, union with the universal Church, union with the Pope. That is an article of faith. . . For the Holy Father, the Holy See, it is very important to recognize Chinese Catholics. We will also recommend that the Vatican recognize them" (*Asia Focus*, Sept. 13, 1985). Lo Kuang does not distinguish between "patriotic" and "loyal" Catholics. We have already pointed out the reasons for this stand. It deserves imitation.

6. Taking the Church in China as a whole, which means without undue attention to the divisions under which it labors, and leaving aside the fact that the government severely restricts its activities, it must be the common goal of Catholics to invent a new point of encounter with China and her Church. No turning back to the past; no yearning for the unfortunate privileges of the French Protectorate. Rather, there must be esteem and honor for a great people with its common sense, with its exacting morality trying to put community above self, with its patience and forbearance under suffering and oppression. On the Chinese as well as on the foreign side, mutual mythologies must be dissipated.

Recently, Mr. Hu Yaobang did not meet the Pope, he did not talk to any Vatican dignitary. One wonders whether, on such occasions, diplomatic niceties could not give way to diplomatic daring. The stakes are too high; to a large extent, the future of the world depends on it. Could there not have been, as Cardinal Casaroli suggested, "an extraordinary gesture"?[1]

NOTES

[1] Background information and literature related to this essay may be found in *China Update*, a quarterly newsletter published by the author. *CUP*, Dennenlaan 8, B-3031 Oud-Heverlee, Belgium.

THE PROTESTANT CHURCH IN CHINA

Joe Dunn

Introduction

Less than 10 years ago, the existence of Christianity in China was in grave doubt. Then, in 1976, with the smashing of the "Gang of Four" signalling the end of the Cultural Revolution, a sense of deep relief spread throughout the country. Social, cultural and political activities sprang up like mushrooms after a long, chilly winter. The Chinese people began to shake off the trauma of the ultra-leftist terror and took a new beginning on the road of social-ist modernization and prosperity. The Christians too came out to worship after a suspension of activities for nearly 20 years.

When the first Protestant church was reopened for the public on April 8, 1979 in Ningbo, Zhejiang province, many people still did not believe that the government's religious policy would be implemented. At best, they took a wait-and-see attitude, doubting that an open church would be truly free to preach the gospel. However, the news that the church there was open again spread fast and far. Despite some sceptics, many flocked to the doors of the church. Not a few of them came from a hundred kilometres away. It usually took them days to come to church on foot. Rev. Cai Wenhao, president of the Zhejiang Christian Council, re-called, when the church was later re-opened in Hangzhou, that he had to preach every time more than one hour. He had no intention of preaching such a long sermon. But every time when he looked at the thirsty audience, he simply could not stop. "Did I preach too long?" he tried to ask. "No," came the resolute answer, "we've been waiting several days to hear the word of God".

One interesting question is: Where do these Christians come from? As we know, during the Cultural Revolution (1966-1976), there was no visible sign of organized Christian activities in China. All churches were closed down and the clergy were sent to factory jobs or to farm work. It was only after 1979 that they gradually came back to the Church. What had happened in the meantime? The fact was that even in the most difficult years, during a time of increasing political and social pressure against religious belief, some Christians still believed that the Lord would not fail them or forsake them. They tried to meet whenever and wherever it was possible. Sometimes they might just kneel down behind closed doors without saying a word. Sometimes they whispered a few broken words. Sometimes they met in the streets as passers-by and greeted each other by nodding their heads or making a special gesture. When the atmosphere was more relaxed, they met by twos or threes. They recited a few Bible verses from memory or read from a Bible which they otherwise kept in a safe place. There were lots of prayers. When the atmosphere further relaxed after 1979, they invited friends to join in their worship. They began to sing aloud and bore witness to the unfathomable love of the Lord. More and more, they went around to tell others the good news of Jesus. And a gathering of a few individuals thus swelled into a community of a few dozen and sometimes over one hundred believers. They met mainly in houses, but also in courtyards and on sunning grounds. Though some of these home gatherings were led by ordained clergy or individuals who had been church leaders in the past, many more were organized and carried out on the initiative of lay men and women.

These are people who are, generally speaking, second or third generation Christians themselves. Quite a number of them are illiterate or not well educated. Their knowledge of the Bible is superficial. Yet their devotion and enthusiasm are contagious. Their young children are so moved by the words and deeds of their parents that they voluntarily commit themselves to the Lord Jesus and help their parents out. These young people too are not well educated, but at least they are able to read the Bible and teach hymn-singing and do some of the organizational work of the small Christian community.

When the home gatherings began to grow, most of them were isolated from one another. But as they grew in number, a more trained leadership was needed. As a result, they began to merge. Some developed more quickly and saw themselves as part of a much wider fellowship and established regular patterns of sharing and communication. Thus, from the seed of the Word, then a flock of people and finally some kind of organization, the home gatherings gradually developed into meeting points which are more or less organized and formal, but have emerged in a spontaneous and natural way. Then finally Church properties were re-claimed and re-opened for public worship services.

The growth of home gatherings both in cities and rural areas bore witness to Jesus Christ over against the pretentions of ultra-leftism in China. In this way they helped to sustain and renew the faith of the Church. Rev. Cai Wenhao frankly admitted: "I saw during the Cultural Revolution that my own faith was weaker than that of many lay Christians. It is they who have now become the motive force of our church. We would not have survived without the laity during the Cultural Revolution. We were weak, but they continued to do evangelistic work. We were pushed forward by our congregations."

Indeed, the religious phenomenon in China is such that it is not the shepherd who goes out to seek the flock, but the flock that is going around to seek the shepherd. Everywhere in China, you hear a similar description of their situation over and over again: "Plentiful is the harvest, but few are the laborers." They flock to the doors of the Church and the shepherd has to take care of them.

To cope with the situation, the Church in China has set up two organizations. One is the committee of the *Three-Self Patriotic Movement* (TSPM): self-government self-support, self-propagation. The other is the *China Christian Council* (CCC). The TSPM was formed in 1954, but did not function during the Cultural Revolution; it was revived in 1979. Its objective is to unite all Christians in China so as to take part in the reconstruction of the country. The defence of the Three-Self principle is certainly a Chinese response to Western domination. Yet it is also more than that. The Chinese Church wants to be independent, not only of foreign churches, but also of pressures coming from within the country itself. The underlying motive is to be free, not just financially and organizationally, but also psychologically and spiritually. Further, the Chinese Christians must find a common language with their own people. They can do so, provided (in the words of Bishop K.H. Ting), the Church "rids itself of its colonial nature, ceases to be a replica of foreign Christianity, does not antagonise or dissociate or alienate itself from the cause of the Chinese people, but joins them in that cause, plants its roots in Chinese culture, forms a Chinese self and becomes a Chinese entity."

The China Christian Council, established in October 1980, is envisaged as an organization to serve the pastoral and ecclesiastical needs of Chinese Christian communities, including the printing of the Bible, the development of new Christian publications and the training of new leaders. The relationship between the Council and the TSPM has been described as that between "two hands and one body." Their respective constitutions emphasize the co-operation and division of labor between the two organizations. Leadership at the national, provincial and local levels tends to overlap, and local councils and TSPM committees will often work together on common concerns. For example, in negotiation for the return of a church building, access to printing facilities or the organization of lay training classes, both organizations would naturally be involved. This is an expression of their essential unity. The distinction between the two is that the Council is specifically designed to meet the ecclesiastical needs of the churches, whereas the TSPM is more or less a political arm of the Chinese Church. Both assume a comprehensive understanding of faith according to the principle of mutual respect in matters of belief.

China now has an estimated four million Protestant Christians. The Church has re-opened more than 4,000 churches and tens of thousands of meeting points, has established eleven seminaries and many short-term Bible training classes and has printed 1.8 million copies of the Bible in 4 languages, namely Han (mandarin), Korean, Lisu, and Miao. The CCC publishes *Tian Feng* (a monthly church journal) and a *Theological Review*, a quarterly *Sermon Collections* and a quarterly *Syllabus* for lay communities which has 40,000 subscribers, a new Hymnal and a common Catechism, booklets for the training of volunteer church workers and other devotional books. All of these have been accomplished with limited human and financial resources.

United Front and the TSPM

Yet China, under the leadership of the Chinese Communist Party (CCP), is an atheist state. The fact that Christians in China are not only tolerated, but are today encouraged to play their active role in Chinese social and political life is because of this united front policy. Before the CCP came to power, the united front was a strategy to promote national and anti-colonial movements on the one hand, and to work towards the ultimate victory of the CCP on the other. After 1949, the task shifted more to getting on with economic construction and the consolidation of national unity. Since the building up of the country can never be the work of one party, but of all the people, the effort to unite all who can be united for the common struggle is of utmost importance and necessity. Bishop K.H. Ting, President of the CCC, spoke at the formation of the Council in 1980: "Like all other citizens, those who believe in Jesus Christ ardently desire a strong and prosperous motherland and look forward to the early realization of the Four Modernizations; it is only natural that Christians are part of the united front. There is freedom to maintain any religious faith and outlook or any worldview, under the principle of mutual respect. It is in this way that unity and stability, the formation of the united front and the realization of the Four Modernizations are made possible."

The ideal of the united front is the principle of "seeking the common ground while reserving differences." With regard to religious policy, for example, the common ground is that of patriotism, socialist reconstruction or modernization, while the differences are the differences in ideological or religious belief and world view. In order to make the united front work, the principle of mutual respect must be upheld. But in a country like China where the Party assumes sole leadership, where there is a lack of democratic tradition, and where the implementation of the law still leaves much to be desired, it is sometimes difficult to silence critics who think the united front is only an expedient.

Expedient or not, this is the context in which Chinese Christians have found themselves since 1949. Either they take a passive stance towards the new government and thus leave no room for the individual to work deliberately in and for society, or they support the Communist revolution and wrestle with their understanding of God and God's world in the new political reality.

Admittedly, many Christians in the early 1950's seemed quite hesitant to take the second option. However, those who affirmed the revolution did so out of their Christian conviction. For they were able to discover a sense of hopefulness in the changes which the revolution was bringing to Chinese society.

Take Y.C. Wu for example. He was a leading figure in the Chinese Christian TSPM from its inception until the Cultural Revolution. As early as the 1940's when the Chinese Communists were still being hard pressed by the Nationalists, he had moved from the social gospel towards a limited synthesis of communism and Christianity. When he initiated the "Christian Manifesto" in 1950, a political statement which lent full support to the people's government and put the Church within the united front under the leadership of the CCP, he surely cannot be condemned as a mere opportunist.

On the other hand, we must understand that his lifelong concern had been the relevance of Christianity to social questions. In those days China was constantly plagued with crises. Many a Christian was anxious to seek a way to help and if possible, even a Christian way. But no adequate answer was given. To find a viable position for the Church to stand on in the mighty stream of time, Wu placed great emphasis on God's immanence in history and nature. He cautioned that undue attention given to the divine transcendence would make one pessimistic. Since Jesus did not tell us to create another world, why did we not immerse ourselves into a broader social movement and respond to God's promise there? He therefore did not argue for an independent Christian program, but for Christian participation in a united front of the Christian people. He proposed that Christians should accept the leadership of the new government in opposing imperialism, feudalism and bureaucratic capitalism, and urged Christians to take part in the efforts to build a new China.

Under the driving force of Wu and other like-minded Chinese Christians, the TSPM was born. The idea of the Three-Self was not first formulated by the Chinese Christians themselves. In the last century, mission board secretaries like Henry Venn (Church Mission Society) and Rufus Anderson (American Board of Commissioners for Foreign Missions) were already talking in terms of the Three-Self. In the 1920's, certain Chinese Protestant leaders, encouraged by missionaries in the field, adopted the Three-Self as a goal for the Church in China. Yet because China in those years was too much in disarray, and because most denominations depended heavily on foreign help, both financially and personnel-wise, the talk about Three-Self could only be superficial.

By the 1950's, China was again united. No matter what problems and faults the country has, it is a sovereign nation. It is also the context in which Christians in China have their share of joy and sorrow. Leaders of the TSPM like to use the word "patriotic" to characterize the way they feel towards their country. Being patriotic in the 50's meant that the Church had to cast off the foreign and colonial character of Christianity and to express its loyalty to the people's government.

This patriotic understanding of severing the relationship with foreign mission boards and throwing themselves into nation building was exactly what the government would expect Christians to do. In fact, as the Communists understand the concept of united front, no basic differences are permitted on the political level. What differences exist there, must be overcome and transformed. Reserved differences are allowed only on the ideological level. This is the pre-condition for the cooperation between Christians and Communists in China.

The development of the TSPM, however, was closely bound up with the vicissitudes of the Chinese social and political climate. As the CCP turned increasingly towards ultra-leftism in the 50's, the TSPM, being a public organization, fell an easy prey to the many political campaigns. This was reflected in an overzealousness in the pursuit of the patriotic goals of the TSPM: an extreme politicization of the Christian message, an inadequate understanding of the concrete realities of Chinese Christianity, and a poor grasp of the united front approach.

It was not until the Third Plenum of the 11th Party Congress of the CCP (Dec. 1978) that the Party was able to seriously re-evaluate its past history. It decided to strengthen socialist democracy and the legal system. It called for emancipating the mind and seeking truth from facts. The work of the united front was resuscitated and given a new mandate. With the new open-door policy, China has indeed changed tremendously.

In the field of the academic study of religion, there are indications of a breakthrough. A different approach to religious study seems to be emerging. Some Marxist scholars are ready to investigate several sources (traditional Marxists usually insist that the sole source of religion is the suffering of the people brought about by their exploitation in a class society) as possible origins for the continuing existence of religious belief in China. They are even ready to accept the positive contributions of religion as a motivation for ethical behavior and as support for socialist values. This more open-minded Marxist view of religion identifies five (rather than two, namely reactionary and deceptive) characteristics of religion: the mass, ethnic, international, protracted and complex natures. They would put the emphasis on religious freedom and toleration as the basis for a united front which will promote unity, socialism and patriotism among believers and non-believers.

In the meantime, the reactivated TSPM must have thought a lot about its own excesses and abuses made in the movement's formative years. The impatience, inexperience and lack of discernment of some of the leaders proved to be costly and dangerous. Yet, the faith in the grace and steadfast love of their Lord gives them hope and courage to face the future. Instead of falling back, they move ahead with a renewed understanding of their selfhood as a Chinese Church and of their unity based on a common faith in Jesus Christ. They are thankful for the fact that Christians from various denominational backgrounds have come much closer together than in any previous period. Indeed, the two main schools of theological thought — the fundamentalists or the evangelicals, and the liberal theologians or the social gospelers — are learning to speak out, to pray, to praise, to publish and to run seminaries together. They came together because they both want to move closer to the centre which is Christ. By showing mutual respect, they can now enjoy the fruit of mutual enrichment.

As K.H. Ting recently remarked, when the TSPM was organized, its task was primarily negative. It had to fight against feudalism, imperialism and bureaucratic capitalism. At that time, patriotism, support for the leadership of the CCP and a sense of identification with the Chinese people were very precious for the Chinese Christians as the context for their unity within the common ground. It is not that patriotism is no longer important, but because the TSPM is at the same time a Christian movement, its task must have a more positive content, i.e., they must try their best to better govern, better support the body of Christ and better propagate for the sake of Christ.

The optimism of the TSPM is not without substantial backing. As Chinese Christian leaders affirmed in 1980: "God's rod and staff were never very far from us as we moved through the valley of the shadow of death." In fact, even in the darkest time when nearly all visible and institutional expressions of religious life seemed to have terminated, there was always the existence of a persevering faith. The seed fell on good soil. It blossomed as soon as the sky cleared up. After the Cultural Revolution, the home workshop gatherings began to grow rapidly as self-governing, self-supporting and self-propagating Christian communities. The message is therefore loud and clear: whether in churches open for public worship or in home gatherings, the Chinese Church is, by the grace of God, alive and growing. It has moved towards Three-Self, so that Christianity has at last become a Chinese religion. The TSPM, now with its back free, can better manifest its transcendent loyalty in the Chinese context.

"House Churches" vs. "Three-Self Church"?

Despite the fact that Chinese Christianity is more unified now than ever before, some overseas Christian groups like to argue that churches in China are divided into "house churches" and "Three-Self churches." They claim that the "house churches", having been tested by fire, have become the mainstay of the Chinese Church. Other reports stress the rapid growth of such churches: indeed, some maintain that in the span of a couple of years the number of Christians in these churches has reached several dozen million. They often maintain that these are persecuted churches and that members suffer in faith through deep waters and scorching fire, i.e., the extreme misery. Finally, there are statements that they and the "official church" are diametrically opposed to each other with the "house churches" put forward as the genuine Christians and the "Three-Self" as a tool of the government whose mission is to eliminate religion from Chinese society altogther.

Some clarification of terms is at this point in order. What, after all, are the so-called "house churches"? How are they related to the Three-Self or "Official Church"?

First, we must say that there is no such thing as a "Three-Self Church" or an "official" Church in China today. These labels and classifications have been created and applied to China by Christian groups in Hong Kong and overseas. In China, as we know, there is a Christian Three-Self Patriotic Movement Committee (TSPM). It is called the Three-Self Committee, but has never been known as the "Three-Self Church", because it is not a church. Somewhat more like a church and parallel to the Three-Self organization is the China Christian Council which was formed in 1980. At present, the CCC takes care of all kinds of Church affairs. At the county or village level where Christian Councils may have not yet been formed, the Three-Self committees or Three-Self subgroups are responsible for administering Church affairs.

With regard to "house churches," this terminology actually is not being used in China. There, Christian groups who meet outside a church building are, generally speaking, called "meeting points" or "gospel points." Only a few of these groups understand what a visitor means when talking about the "house churches." That proves only that they are informed about the name. It does not mean that they themselves like to use that name to describe themselves. The term "home gathering" or "home worship" on the other hand, is often used in China to refer to informal Christian meetings which take place in homes sometimes. The name itself is not what really matters. Of more interest is why there are such meeting points in the first place.

Prior to Liberation, i.e., before the founding of the People's Republic in 1949, China had become a fertile ground for indigenous churches and sects. Most of these groups, such as the True Jesus Church (founded in 1917), the Jesus Family (1921), and the "Little Flock"(1922), were headed by charismatic leaders who maintained strict standards of social control guided by theologies which were as idiosyncratic as they were fundamentalist. As a rule, they were hostile or indifferent to society and state. Indeed, they gained their identity by assuming a position over against the "world." But "worldly" were not only secular authorities; all other church organizations and teachings which they did not accept as biblical were also labelled as "worldly". As a result, they set themselves apart not only from the mainstream denominations, but even from one another.

The indigenous sects, each in their own ways, were quite successful in drawing a great number of followers at a time when China was facing innumerable crises, including a resistance war against Japan and a civil war between the Communists and the Nationalists. Other factors may also account for their rapid growth. For example, most of these groups understood themselves to have recovered true Biblical teachings based upon apostolic principles which were untainted by modern perversions of the gospel. At the same time, they also represented an incipient na-

251

tionalism in Chinese Christianity. Rejecting tutelage, foreign structures and modern values, many of these sects believed that only they could serve the cause of Christ in China.

Take for example, the True Jesus Church, founded in 1917. They believed that their coming into existence helped end the First World War. They claimed to be the only true Church in the whole world. How do they know it? Because "there is salvation in no one else" (Acts 4:12), i.e., only in the true Jesus. And "all who came before me (true Jesus) are thieves and robbers" (John 10:8). No doubt, there are churches which came earlier or later than the True Jesus Church. But they are said to have made a prey of the Christians by philosophy and empty deceit against the spirit of Christ (Col. 2:8), or they have turned to a different gospel, trying to please men (Gal. 1:6, 10). More importantly, the True Jesus Church was revealed in the Bible. That was a great mystery, they claimed. How? Look up I Cor.11:3 "I want you to understand that the head of every man is Christ." Their reason is the following: Christ is the head of every man, so put Christ on top of every man. But Christ is Jesus. He is also the True God. And every man is the Church. There you have the

True	God
Jesus	Christ
Church	every man

True Jesus Church revealed in the Word of God! Further, I John 2:8 testifies, "I am writing you a new commandment, which is true in him and in you."

Only Jesus is true and we, the Church, are under Jesus, so we are the "True Jesus Church."

| True in him | True Jesus |
| True in you | Church |

As evidence of their possessing the Holy Spirit, they speak in tongues, leap and dance before the Lord, mourn and rejoice in the Holy Spirit. And they are to be baptized in living water with the face downward. The True Jesus might be bad exegesis. But one must remember that they were and still are one of the most appealing groups in rural China.

Thus, on the eve of Liberation, in addition to the heavy dependence of most Chinese churches on foreign help which made it hard for them to cooperate with each other, there had emerged an independent tradition which fostered the family-type worship and which, from the very beginning, distrusted the mainline denominations and even went so far as to describe the latter as evil. Their arguments against the big churches, however, were heavily charged with emotion. Politics did not seem to be a major factor.

After 1950, the CCP was in complete control in China and promoted a very strong political program right from the start. Further, it never wavered in its confession of atheism. Christians of all stripes, colors and shades were certainly frightened by this new situation. Face to face with the Communist reality, the mistrust among groups deepened. Worldly anxiety was fought out under the cloak of theological debate. In order to preserve their identity and to survive, sacred fences were built. For example, the doctrine of "justification by faith" was stretched to assert that Christians and non-Christians were two different kinds of people and thus should not work together. The doctrine of total depravity of the human being was used to justify the escape from worldly affairs and concerns. And since faith and morality did not complement, but rather excluded, each other, any human endeavor was futile. Only God and His words, nothing else, could save.

Already, in the early '50s, unceasing struggle began to surface, with increasing intensity and radicalization, both within society and within the Church. Most indigenous groups, insisting on the dichotomy of "believers" and "non-believers" as outlined in II Cor. 6:14, were hesitant to accept either the Three-Self or the political leadership of the CCP. To the extent that they rejected integration, they were viewed as threatening social harmony and were suppressed by the authorities. The TSPM, operating within the framework of a united front led by the CCP, took the side of the government to argue against the separatist sects. As people were humiliated and hurt, both physically and spiritually, resentments grew. Not a few Christians vented their spleen against the Three-Self Committee. They seemed to be more convinced than ever to have nothing to do with the big churches. When it became evident that only the "normal" churches were considered legitimate by the government, those who could not follow chose rather to meet in private homes.

On the other hand, the rural Christians, who represented the majority of Chinese believers at the time, experienced the Revolution somewhat differently. They professed a simple faith, one both practical and enthusiastic. During the time of the land reform movement in the early '50s, churches in many places were closed by the authorities. The faithful could then only meet in their homes. After the completion of the land reform, from 1955-56 onward, a number of churches were re-opened and peasants returned to worship as before. Yet, after the great Leap Forward movements in 1957-58, as leftism prevailed over all China, church activities everywhere dwindled away; they came to a complete halt during the Cultural Revolution from 1966-76.

Finally, after a hiatus of nearly twenty years, the churches were reopened. The news spread very fast. Christians, who had met only in their homes in groups of two or three, became more and more open themselves. Before, they did not even sing for fear of being discovered: in Wenzhou an old pastor said that in the darkest days of the Cultural Revolution, they often met behind closed doors all day long, simply kneeling down in silent prayer. Now they extend invitations to their neighbours and friends to come to sing and pray aloud. Just as in the open churches, their homes are packed to bursting. Growth in numbers of such home gatherings seems to have gained momentum after 1980. But since 1983, there seems to be a slow-down.

As it was in the past, so it is also now: the rural Church in China is numerically stronger, but structurally, financially and intellectually weaker than its urban counterpart. So far, over 90% of them come to church because they believe in and have experienced the healing power of Jesus. What they hope and pray for is a peaceful, secure existence for themselves and their children. They do not demand much of life. In fact, from 1979-82, most of these rural Christian groups in China had yet to hear the name "Three-Self Movement." Far from having gained an opinion of its principles, they had not decided how they saw themselves relating to it. On the contrary, the growth of these groups made it necessary for them to voluntarily seek the help of the Three-Self committees which had also been consolidating their work and becoming better known on both the national and intermediate levels. Matters such as the reclaiming and restoration or opening of former church buildings, Bibles, pastoral work, teaching materials, these were all subjects about which the Three-Self committees were knowledgeable and experienced. They were willing to help local communities in these matters. A house group leader in Huaxian, Guangdong Province, was asked three years ago why he had decided to join the Three-Self Movement. He answered, "What have we to lose? Now I can preach Jesus openly and boldly." Indeed, he mounted a loudspeaker in front of the chapel located at the centre of the town's major marketplace. Every Sunday he preaches the gospel, not only to those who drop in, but also to those who come from the surrounding villages for business.

Beyond any doubt, the overwhelming majority of meeting points has joined the local Three-Self Committees. They join for a variety of practical reasons, but to say today that many groups are opposed to the Three-Self Movement is pure conjecture.

That does not mean, of course, that there are no Christian groups in China which still oppose the Three-Self Movement.

There are some who do so out of conviction. These include certain people of the Little Flock background, some followers of Wang Mingdao, a noted preacher living in Shanghai after many years' imprisonment because of his uncooperative attitude toward the government, and a few intellectuals formerly associated with the Inter-Varsity Fellowship and who still have contact with their colleagues overseas. The latter are also the ones who might greet people from abroad in the street in English and play the role of Nicodemus, i.e. visit you in your hotel room at night. They certainly suffered a lot in the past: the question is whether China's present religious problems can be solved if there is no courage to forget, forgive, and move forward.

Certain Little Flock groups in China, directly influenced by the head person of the international organization in Los Angeles, Li Changshou, have gained the nick-name of "Yellers," because they yell when they pray. These groups are perhaps the most vocal in their opposition to the Three-Self Movement. They equate people working in the movement with the "great harlots" as recorded in the book of Revelation. They receive orders and financial help, as well as literature, from abroad. A couple of years ago, they organized people in China to disrupt the meetings of non-Yellers and they engaged in acts of vandalism and coercion against other Christians in Dongyang and Yiwu in Zhejiang Province. At the end of 1983, the "Yellers Sect" was declared illegal by the Public Security Bureau throughout China. Yet here and there one still hears of their activities going on, now mainly in secret.

Ultra-leftism has wrought the biggest damage to the country. The TSPM, too, has not been able to rid itself of "leftist" sectarianism since its founding.

There are people in the Three-Self Movement who are so callous that they hurt the feelings of their fellow Christians and violate their own principle of operating on the basis of "mutual respect" for those who come from differing denominational backgrounds and theological points of view.

Or they are so much after fame and power that the development of the Church no longer excites them. This has, in some cases, caused some groups of Christians who joined the Three-Self Movement to withdraw from it some time later. Negligence, insensitivity to minority views and practices will likely lead to disaffection by some Christians and thus hinder the post-denominational goal of Church unity.

Finally, the wisdom of certain cadres in the government's Religious Affairs Bureau leaves much to be desired. In certain places, they simply do not allow Christians to run their own affairs.

Everything is "arranged" for the Christians. They even write up so-called "patriotic covenants" for the Church, covenants which are neither patriotic nor Christian since the Christians are asked to accept them without question. A simple illustration will show how a good thing can be turned into a disaster by the foolishness of certain government officials.

Four years ago, a certain Christian group north of Kunming in Yunnan Province was accused of having a strong anti-government stance. A team from the Institute of Social Sciences in Shanghai was sent out to investigate. They found out that since 1906 missionaries of the China Inland Mission, Seventh-day Adventists and the Christian Independent Church had been active among the tribal peoples in that area. Little Stone Bridge is entirely populated by the Miao people. All the Christians, 60-70 strong, were hard-working. Before the Cultural Revolution they were several times awarded the title of "Advanced Brigade." Yet, during the Cultural Revolution, they were repeatedly harrassed. In 1969, forced to choose between Jesus and Chairman Mao, they chose Jesus. They were then told in no uncertain terms that "the land belongs to Mao; you cannot till it; the cattle also belong to Mao; you cannot put them out to pasture. Each piece of wood and grass belongs to Mao." The brigade leader and the treasurer were dragged off to the commune headquarters, tortured and imprisoned. The others could do nothing to retaliate. They tried their best to find a way to survive in a situation where their lives were already quite miserable. They turned to individual farming. Then, in 1973, there was a campaign to criticize "the tendency to return to individual farming". Surely, without leadership, these people did not know where to turn. Under heavy criticism, they decided to give up individual farming. Their lives became more miserable than before. Their relatives in near-by villages tried to smuggle food to them by night. That too was stopped by some local cadres. What hope was there if not that seen in the earnest desire for the immediate coming of Jesus Christ?

Even after the arrest of the "Gang of Four" in 1976, the situation at Little Stone Bridge did not change much. In March 1978, the local authorities planned another struggle against the "anti-revolutionaries existing under the cloak of religion." Rifles and machine-guns were set up around the hilltop where the Christians lived. Two hundred militiamen were deployed. They shouted to those living on the hilltop, demanding that all people immediately "surrender", i.e. give up their faith in Christ. No one answered. Finally the militiamen charged the hilltop and bound every male, hands and feet together just like cattle, carrying them to the prison in town. The leader was sentenced to seven years' imprisonment. After a year, he was released. The cadres apologized to the people, putting the blame on the "Gang of Four." They answered: "We don't blame you at all. All this truly indicates what is prophesied in the Bible. Christians must suffer, the gate to heaven is narrow. In the past, we did not believe enough; now our faith is strengthened."

Leftist practices, however, die hard. Soon they organized other Christians to turn against their brothers and sisters at Little Stone Bridge and to criticize them for being unpatriotic. The Christians at Little Stone Bridge responded by saying that "Yes, we cannot be 'so patriotic' as others who just drift with the tide." They, therefore, insist that they are the "little stream" Church which goes through the narrow gate, while those who claim to be patriotic, they name the "big stream" Church. The "little stream" Church is now spreading its influence among other tribal people. They do not want to be united with the "big stream" Church. This is a clear example of how foolish and short-sighted people in authority can mess things up, whether in the Church or in society.

In reality, Christians at Little Stone Bridge are against neither the government nor the Three-Self principle of the Church. What they cannot understand is the fact that certain people in the name of the government *and* of the Church have treated them in such a foolish and cruel way. In fact, when representatives of the national Three-Self Committee from Nanjing and Shanghai visited them lately, they were most warmly received by local Christians. A more conciliatory attitude has also developed at Little Stone Bridge over the past two years.

In a country as large as China, there are certainly individual Christians or groups who disagree with the TSPM and members of certain meeting points tend to be individualistic, other-worldly and divisive. But the vast majority seem to be thankful that they can again worship openly, in their homes and in re-opened churches. The TSPM and the China Christian Council are grateful for the spontaneity and lay participation of the home gatherings. Such Christians were the motive force of Chinese Christianity and the major sources of Christian renewal and evangelical witness during the most difficult years. The home gatherings, on the other hand, are thankful for the organization and theological leadership which only the TSPM and the China Christian Council can provide; otherwise they would remain isolated, divisive and strait-jacketed. An apt description is that the emergence of the home gatherings is the main source for renewal of the Chinese Church *from below* , while the re-organization of the TSPM and the China

Christian Council are the primary expressions of the renewal of the Chinese Church *from above* . Both types of renewal are necessary. They are complementary to each other. Neither is perfect, yet together they are working out something better than they had before.

Christians in China want to walk on a path none have walked before, a path appropriate for themselves. They must become a Chinese Church with its own characteristic features. Having said that, they categorically deny that self-isolation is implied in their Three-Self principle. In fact, in recent years they have gone out to visit many countries and invited Christians from all over the world to visit them. What they do say, though, is that without self-reliance there can be no talk about a contribution to the universal Church. The Church in China, like churches elsewhere, is imperfect. The Christians there are also imperfect. Yet they are willing to come together, with the help of the Holy Spirit, to walk the path of faith. For this reason, despite their many shortcomings and mistakes, the Church in China is very much alive and growing, and this is very important.

FOLK RELIGION IN HONG KONG AND THE NEW TERRITORIES TODAY

David Faure

My Hong Kong research has, up to now, focussed primarily on the history of the New Territories, and I have come into touch with folk religion through observation at village festivals and interviews with elderly villagers. As background preparation for my New Territories project, I was also engaged in copying down historical inscriptions found all over Hong Kong, many of which were found in temples and ancestral halls. Out of that project, I learnt a little of the growth of temples in the urban setting. Neither research project, however, has prepared me to answer the question that must be crucial for this paper. That is, what is it in folk religion that people in Hong Kong and the New Territories today have continued to believe in? And why? In a day and age when one should expect life and culture to be very much more secular than in times past, the question is crucial, for it must not be assumed that folk religion continues in the city as one might have found it in the villages fifty or more years ago.[1]

Let me begin, then, with the simple statement that religious practices must have changed in the process of urbanization. In the not-so-distant past, the religious expression of individual experience and of family and community relationships formed an integral whole within the village world view. That was because a dominant social theory had gained ground that subordinated the individual to family welfare, and presented the household — rather than the family — as the building block of the community. In present day Hong Kong, the traditional ideology has given way to social standards that allow more room for individual exertion beyond the family, as the individual finds himself being more involved in settings over which the family has little control. In the new environment, the housewife still worships on behalf of her family, but her children, working away from home, seek religious protection from deities other than the ones she venerates. Hong Kong's housing estates offer no communal earth gods, and the temples that thrive in the midst of economic prosperity draw their clientele from far beyond the neighbourhood. Other than in the funeral, the modern Hong Kong family has no occasion to worship together, even as the household earth god lingers on and is offered sacrifices daily.[2]

Yet, folk religion holds its own, and it does so, as always, because the deities are efficacious and because evil influences are about that must be guarded against.[3] Indications that the belief in deities and their opposite numbers is thriving are not very hard to find, there being few taboos against the expression of devotion. Modern-looking businesses display the eight trigrams, in the middle of which a mirror deflects evil influences to the other side of the road. The Wong Tai Sin 黃 大 仙 Temple and the Che Kung 車 公 Temple are thronged by crowds at Chinese New Year seeking protection from the resident deities and counsel from the fortune-tellers. The General's Stone (*Jiangjun shi* 將 軍 石) on Bowen Road above the commercial-cum-residential district of Wanchai 灣 仔 collected its following in the 1970s.[4] When my wife was expecting, we were told that regular worship at the Roman Catholic St. Anthony's Church would ensure that we had a son. (We didn't follow the advice, and my wife gave birth to a daughter.) The exception proves the rule, for where folk religion has clearly given way is in the decline of exorcism as a cure for illness. As the village priests have told me, now that medicine is advanced, there is little need to resort to this cure.

In quoting these examples, I use the expression "folk religion" as a synonym for *paai-shan* (*baishen* 拜 神). The term is widely used and understood. To *paai-shan* is to perform the act of devotion to the deities, through the offering of incense, paper goods, food and occasional service. As Julian Pas reminds us, it is not Buddhist or Taoist or Confucian.[5] It is the product of another world view, one that sees the deity as a protective spirit, who offers protection to worshippers who enter into a devotional relationship with it. The act of *paai-shan* may be contrasted with that of *siu-i (shaoyi* 燒 衣), the burning of offerings to satisfy the *kuei (gui* 鬼), that is, a ghost, a spirit unattached to any fixed abode and which consequently cannot be regularly sacrificed to by a following.[6]

Because the act of *paai-shan* can be quite simple, and because knowledge of *shan* and *kuei* is widespread, acts of devotion may be totally personal. Such acts do not have to be more than an expression of personal interest to trade sacrificial goods for protection against ill fortune. This is not to say that expertise does not hold a place in folk religion. Some people are inevitably more knowledgeable than others in the way of *shan* and *kuei*. In the village, one hears of the "old women", non-professionals who are recognized for their knowledge of village traditions, or the *paai-shan-p'oh (baishenpo* 拜 神 婆 , that is, *paai-shan* women), who, for a fee, *paai-shan* on one's behalf.[7] A variation of the *paai-shan-*

p'oh may be observed in the old women at open air markets in Hong Kong today who "beat the small people" (*ta-siu-yan, daxiaoren*, 打 小 人) for paying clients, the act being the exercise of magic to thwart the designs of doers of mischief and causers of ill fortune.[8] These "old women" may be distinguished from two other categories of experts who are privileged in matters of worship. One of these consists of mediums, whose bodies serve as temporary vehicles for the spirits, and the other of priests, who exercise their powers by virtue of their knowledge of written texts passed on by their masters. Writing about the mediums, John T. Myers has distinguished between the Cantonese *man-sheng-p'oh* (*wenxipo* 問 覡 婆) and the *kei-t'ung (jitong* 乩 童) found among the Chaozhou, Fukienese and Hailufeng people. The *man-sheng-p'oh* practises at home, and is consulted by clients who want to contact the deceased; the *kei-t'ung* whom Myers studies operated from a temple that had a following, were descended upon not by the spirits of the dead, but the deities, and held court to solve problems that were presented to them by the people worshipping regularly at the temple.[9] The priests, known as the *naam-mo-lo (nanfulao* 喃 嘸 佬 , or *nanwulao* 男 巫 佬) are quite different from the mediums. They do not speak with the voices of their masters, but are trusted with ritual authority by them, which they gain through their mastery of the written texts, used in rituals. That the content of the texts is little understood by the lay believers is a matter of concern to no-one, neither priests nor lay worshippers, and this fact does not lessen the recognition that certain rituals, such as funerals, require their professional service.[10] Despite their different expertise, the "old women", the mediums and the priests are similar in that they are all regarded with some disdain by their clients. One result of this popular down-grading of these professions is that the practitioners do not solicit clients. Fortune-tellers, geomancers and herbalists advertise openly: the "old women", the mediums and the priests do not.[11]

Because a fundamental concept in *paai-shan* is that the deities may be come to terms with if they are attached to fixed abodes, worship is closely related to the establishment of shrines and images. These shrines and images may be divided into three classes: personal, domestic and territorial. The taxi-driver who installs an image of the Buddha in his taxi does it as an act of personal devotion. It is conceivable that if an image is particularly efficacious, a following may gather, although I know of no deity yet who has arisen from a taxicab. Another example of shrines that are set up out of personal devotion would be tablets or graves of past boxing masters venerated by their followers, for boxing societies are founded on the charisma of their founders who are regarded as the heads of fictitious households of which all followers are members.[12] Domestic altars are set up by the family, and on these are placed tablets of the ancestors as well as images of various deities, the most popular being that of Guanyin. In urban Hong Kong, because ancestral halls are of necessity impractical, the domestic altar is the focus of a sense of lineal organization. Territorial shrines traditionally consist of earth gods and temples. Earth gods are uncommon in the modern city, although they were set up during the nineteenth century when an immigrant Chinese population gathered in the streets immediately adjacent to the Western-dominated business-and-administrative district. The territorial temple may be divided into two types. Some have grown from temples which originally served villages that became integrated into the city. The Ch'e Kung Temple in Shatin district, even now, serves as the territorial temple for Tin Sam Village as well as the union of villages of the district, even though its clientele has expanded far beyond its original territory.[13] Others were founded recognizably as temples of the city. The Shaukeiwan 筲箕灣 Shing Wong 城隍 Temple, for instance, as its name implies, was built to be a city temple.[14]

I do not think there is any indication that personal devotion is on the decline in the urban environment, and because shrines conveniently become centres of communal organization, there has been a continuous development of personal shrines into communal temples.[15] The T'aam Kung (Tan gong 譚公) Temple in Shaukeiwan was founded by a single believer who had brought his image to Hong Kong from his native place some time in the nineteenth century.[16] The Wong Tai Sin Temple, the most popular temple in Hong Kong, began as a cult centred on this one master who became a deity.[17] The I Paak Kung (Erbai gong 二伯公) Temple was founded by a descendant of the deity in 1889, but he came to be known for his power to cure venereal disease and was worshipped in particular by prostitutes.[18] The Sui Tsing Paak (Sujin bai 綏靖伯) was invited to Hong Kong in the plague year of 1869, but the worship continued.[19] Some of these temples, undoubtedly, were related to ethnic groups that had settled in Hong Kong. Gregory E. Guldin notes in his study of the Fukienese community of North Point District that as immigration expanded in the 1970s shrines were set up by Buddhist monks and nuns among the immigrant population in rented apartments. The deities' statues and ritual objects installed in one of these apartment-temples, to use Guldin's term, were brought to Hong Kong from the original temple in Nanan 南安 county, Fujian. These temples were well supported, and they served also as convenient meeting places for groups of women in the Fujianese community.[20]

The Hong Kong Government has always accepted Chinese religion as a fact of life, realizing that the temples served as foci of organization in the Chinese communities. In 1851, as result of a petition by Chinese residents, it granted a plot of land to be used for a temple at which spirit tablets might be deposited. The Paak Shing (Baixing 百 姓) Temple that was built soon became a place where the dying as well as the dead might be left. In 1869, the discovery that the Paak Shing Temple was used as a "dying house" led to considerable concern by the Government, and under its encouragement, the Tung Wah 東 華 Hospital was founded to provide better medical care for the Chinese population.[21]

The Tung Wah Hospital has remained to this day one of the leading charitable organizations in Hong Kong. Its committee has always consisted of the most prominent and aspiring among the Chinese population, and although its judicial authority has since declined, for some decades from the late 1870s it acted as a "quasi-official body with a measure of jurisdiction over the Chinese."[22] It probably inherited this position from the committee of the Man Mo (Wenwu 文 武) Temple on the outskirts of the business district (hardly a block away from the Paak Shing Temple or the office of the Tung Wah Hospital when it was founded). Thanks to Carl Smith's research, we now have an outline history of temple building among the Chinese immigrant population in Hong Kong in the 1840s and 50s which helps to explain the central position of the Man Mo Temple in the community and its relationship to the Tung Wah committee.[23]

The first temple founded by the Chinese community that grew up near the Western business-cum-administrative district was the Shing Wong (Chenghuang 城 隍) Temple, for which money was raised in 1843. In 1847, the Man Mo Temple was built, at a cost that would have dwarfed the Shing Wong Temple, and it soon received a grant of land from the Hong Kong Government for the founding of a school. A description of the temple published in 1872-3 noted that at this temple, the founders " 'judged the people' in public assembly", and that in 1851, the "shopkeepers of Sheung Wan 上 環 . . . repaired the Man Mo Temple, elected a Committee, and therein decided all cases of any public interest."[24] In 1862, an iron bell was donated to the temple by forty-two shops listed as members of the governing committee for the year, in conjunction with believers of the Sz-waan 四環 districts, the principal Chinese settlements near the business-cum-administrative area.[25] The 1872-3 article also noted that the Sz-waan districts held a Yulan 盂 蘭 celebration in 1857, and the connection between the Sz-waan districts, the Yulan celebration, and the Man Mo Temple is further borne out by an incense burner donated to the temple by Sz-waan Yulan managers in 1872. Another burner donated in the same year records fifteen shops as forming the governing committee.[26] These scraps of information point to close ties between the Man Mo Temple and the Chinese communities near the business area.

Not being a religious organization as such, the Tung Wah Hospital was not centred on a temple. Its inauguration in 1870, nonetheless, consisted primarily of the installation of Shennong 神 農 as the patron deity of the organization. The installation of the deity was apparently the result of a compromise reached between a majority on the Tung Wah committee that supported instituting religious worship in the committee and a minority that opposed it.[27] Unfortunately, the details of the dispute are not known, but the issue raises the question how the Tung Wah might have been related to the Man Mo Temple. We know that the Yulan continued to be held by the Sz-waan under its own management committee, for another incense burner was donated by it to the Man Mo Temple in 1882. Thirty-three shops are listed on it as making up the committee of the year.[28] Another clue may be found in the preamble to the 1908 Man Mo Temple ordinance, which noted that the "hereditaments and premises and the affairs of the said Temple (that is, the Man Mo Temple) have for many years been managed by the directors of the Tung Wah Hospital."[29] A possible relationship between the two might have been that between the devotee and the territorial deity, a relationship that held between the Man Mo deities at the temple and the Po Leung Kuk 保 良 局 , another charitable institution, borne out by an incense burner donated in 1893.[30] There are other examples to illustrate the close relationship between the shrine or temple and the groups which in the pre-Second World War decades increasingly came to be recognized by both the Hong Kong Government and the Chinese communities as middlemen between the two. In this respect, folk religion performed its traditional role among its adherents. Yet, in one sense, its present role must have been quite different from its traditional role: British officials in Hong Kong did not profess to believe in the deities that were worshipped. Nowhere is the difference more marked than in education, for in the schools Christianity might be taught but certainly not folk religion. Furthermore, under the new scientism that was to be a dominant part of the twentieth century curriculum, folk religion was increasingly considered to be the embodiment of superstition.[31] However, there was no reason, really, on religious grounds, for the Hong Kong Government to take an interest in the affairs of the temples. Its concern, when it came, had more to do with the management of temple property and donations.

The more obvious interest shown by the Hong Kong Government was the possible use of temple donations for charity, a matter that can be illustrated by the negotiation with the Yaumatei 油麻地 T'in Hau (Tianhou 天后) Temple in 1914. The temple, built in 1876, was well attended by both land and boat people in Kowloon. It was managed by a committee formed annually among the local population. According to an account given in a Tung Wah publication, a suggestion was made by some local people to transfer the management of the temple to the Kwong Wah 廣華 Hospital, the Tung Wah's counterpart in Kowloon. A meeting was then held between the Kwong Wah committee and the temple committee under the chairmanship of the Secretary for Chinese Affairs. Naturally enough, the temple committee objected to the transfer, but agreement was finally wrung out of it by the chairman. Despite this agreement, the temple committee continued to manage the temple, and no money was handed over to the Kwong Wah until 1928, when the Chinese Temples Ordinance was enacted.[32]

The Chinese Temples ordinance of 1928 may be interpreted as an attempt to put all Chinese temples on the same basis of management as the Man Mo Temple and, in principle at least, the Yaumatei T'in Hau Temple. It required the registration of all Chinese temples, and provided for the formation of a Chinese Temples Committee made up of Chinese community leaders and the Secretary for Chinese Affairs. The committee was entrusted with the power to let by tender the right to manage a temple, to appoint or remove temple managers, to transfer surplus temple income to a General Chinese Charities fund, and to take possession of or to close a temple if necessary. Five temples that were privately owned by individual families were excepted from the ordinance, but other than these five, the Chinese Temples Committee had as much power over the temples as the Tung Wah had over the Man Mo Temple.[33]

This description of the ordinance, however, does not explain a clause within it that made it an offence to operate a Chinese temple that was not in a "complete and separate building which is used for the purpose of the temple and for no other purpose."[34] The Hon. Sir Chow Shou-son 周壽臣, who moved the first reading of the bill, explained quite clearly the implication of the clause:

> The first point which I wish to make is that the Bill will not interfere in any way whatsoever with genuine Chinese religion. On the other hand, it will tend to prevent religion being made a source of private gain, and it will, I hope, go far towards preventing the misuse of religion by adventurers who prey on the more ignorant members of the community. Some of the so-called temples against which this Bill is aimed are moved about from place to place when the dupes of each locality have been dealt with by the keepers. One object of this Bill therefore is to assist genuine Chinese religion by helping it to get rid of adventurers who use it for their own selfish ends.[35]

It should be pointed out that the Chinese Temples Ordinance was introduced by the leading Chinese community organizations of the time.[36] These remarks on the objectives of the ordinance are, consequently, interesting indications of educated Chinese opinion on the operations of temples. The difference that was found to be significant was not that between religion and superstition, but that between genuine religion and religious worship organized for the sake of private profit.[37] In this respect, a fine distinction may also be drawn between the professionals employed in religious rituals, such as the *naam-mo* priests, and the operators of temples and shrines. As the *naam-mos'* altars were private, they were assured that the ordinance did not bar them from their practice.[38] The exemption allowed several privately owned temples also suggests that the law was primarily aimed at temple managers whose religious commitment was doubted rather than all private temple operators.

It should also be noted that the Chinese Temples Ordinance was introduced at a time when some public pressure was exerted in Guangzhou against temple operators who operated for profit.[39] In Hong Kong, it may also be related to an amendment of the Summary Offences Ordinance enacted in 1933 that made fortune-telling outside the temple a summary offence.[40] An anti-superstition stance probably had something to do with the proposal for the ordinance, but in its enacted form, this aspect was played down.

The Chinese Temples Ordinance is an interesting example that illustrates the relationship of the Hong Kong Government and folk religion. However, it was never strictly enforced as its terms might imply. Most Chinese temples, to this day, come under these terms, but are not registered. Where the ordinance has been successful is in imposing some control on the finance of the better endowed temples and channelling their income into charity. The ordinance leaves worship at the temples quite alone.

NOTES

1 For studies on folk religion in present-day Hong Kong, in addition to papers by Marjorie Topley, James Hayes and Keith Stevens referred to below, see also Morris I. Berkowiz, Frederick P. Brandauer and John H. Reed, *Folk Religion in an Urban Setting, a Study of Hakka Villagers in Transition* (Hong Hong: Christian Study Centre on Chinese Religion and Culture, 1969), and Issei Tanaka 田 仲 一 成 , *Chugoku shufu no kenkyu* 中 國 宗 族 の 研 究 (Tokyo: Japan Society for the Promotion of Science, 1981).

2 The change from the traditional to the present world view of Hong Kong people is practically an unstudied subject. Useful observations on folk religion may nonetheless be found in Graham E. Johnson, "From rural committee to spirit medium cult: voluntary associations in the development of a Chinese town", *Contributions to Asian Studies*, 1 (1971), 123-43, and Lawrence K. Hong, "The association of religion and family structure: the case of the Hong Kong family", *Sociological Analysis*, 33, no.1 (1972), 50-7.

3 Marjorie Topley, "Chinese traditional ideas and the treatment of disease : two examples from Hong Kong", *Man*, n.s., 5 (1970), 421-37; Keith Stevens, "Under Altars", *Journal of the Hong Kong Branch of the Royal Asiatic Society*, 17 (1977), 85-100, illustrate this. For more examples, see the papers included in *Some Traditional Chinese Ideas and Conceptions in Hong Kong Social Life Today, Week-end Symposium, October 1966* (Hong Kong Branch of the Royal Asiatic Society, 1967).

4 Qiao Jian 喬 健 , "Xianggang diqu shiji chutan" 香 港 地 區 的 石 祭 初 探 , in Lin Tianwei 林 天 蔚 , ed. *Tifangshi ziliao yanjiu lunwenji* 地 方 史 資 料 研 究 論 文 集 (Hong Kong: Centre of Asian Studies, University of Hong Kong, 1985), p. 148. This Stone is not to be confused with the Yan Yuen Shek 姻 緣 石 , or Marriage Stone, with its much older shrine, located in the same area.

5 Julian F. Pas, "Religious life in present-day Taiwan: a preliminary report", *JHKBRAS* 19 (1979), 182-5.

6 David Faure, *The Structure of Chinese Rural Society: Lineage and Village in the Eastern New Territories, Hong Kong* (Hong Kong : Oxford University Press, 1986), 144.

7 David Faure, *op. cit.,* 226-7, n. 20.

8 Qiao Jian 喬 健 and Liang Chu-an 梁 礎 安 , "Xianggang diqu di 'daxiaoren' yishi", 香 港 地 區 的「 打 小 人 」儀 式 ", *Zhongyang yanjiuyuan minzuxue yanjiusuo jikan* 中 央 研 究 院 民 族 學 研 究 所 集 刊 , 54 (1984).

9 John T. Myers, "A Hong Kong spirit-medium temple", *JHKBRAS* 15 (1975), 16-27.

10 I must confess I have not studied the *naam-mo-los* who practice in the city. For the difference between the priestly texts and the villagers'

beliefs, see my paper "The priestly tradition and its place in the village culture of Hong Kong's New Territories", presented to the Conference on Taoist Music and Rituals, Chinese University of Hong Kong, 1985.

[11] The advertising section of the Hong Kong telephone directory includes in its table of contents astrology, physiognomy, palmistry, fortune-telling, date-choosing, geomancy and the compilation of the almanacs, but not *man-sheng* or *naam-mo* .

[12] For some examples, see Graham E. Johnson, *op. cit* ., 141-2.

[13] On the Ch'e Kung temple in its rural setting, see David Faure, *Structure* , 119-21.

[14] According to the 1974 tablet set up by the Chinese Temples Committee in this temple, it was originally a Fuk Tak (Fude 福 德) temple built in 1877. This agrees with the name struck on the lintel dated 1905. The *chenghuang* was installed after 1928, when the temple was taken over by the Committee.

[15] In addition to examples given below, see Keith Stevens, "Chinese monasteries, temples, shrines and altars in Hong Kong and Macau", *JHKBRAS* 20 (1980), 1-33, and James Hayes, "Old British Kowloon", in his *The Rural Communities of Hong Kong, Studies and Themes* (Hong Kong: Oxford University Press, 1983), 48-60, and "Secular non-gentry leadership of temple and shrine organizations in urban British Hong Kong, *JHKBRAS* 23 (1983),113-36.

[16] This story is given in a commemorative pamplet published on the occasion of the joint celebration of the Tin Hau and Taam Kung 譚 公 festivals in 1981.

[17] Marjorie Topley and James Hayes, "Notes on temples and shrines in Tai Ping Shan Street area", in *Some Traditional Chinese Ideas* , 128-9.

[18] Tablet at the temple now held by the Hong Kong Museum of History; see David Faure, Bernard H.K. Luk, Alice Ngai-ha Lun Ng, comps. *The Historical Inscriptions of Hong Kong* , 231-2.

[19] Faure, Luk and Ng, *Historical Inscriptions* , 593-4.

[20] Gregory E. Guldin, "Little Fujian (Fukien) Sub-neighborhood and community in North Point", *JHKBRAS* 17 (1977), 121-3.

[21] James Hayes, "Historical origins of the Paak Shing Temple (Kwong Fuk T'sz)", in *Some Traditional Chinese Ideas* , 139-41, and for background of the area in which the temple was located, Daffyd Emrys Evans, "Chinatown in Hong Kong: the beginnings of Taipingshan", *JHKBRAS* 10 (1970) 69-78. The spirit tablets referred to in this paragraph were those deposited for spirits of the dead. The practice of installing such tablets at temples is still common in Hong Kong where, for various reasons, they cannot be deposited on a domestic shrine. One reason, among many, would be that the deceased has not left behind descendants in Hong Kong.

[22] H.J. Lethbridge, "A Chinese Association in Hong Kong: the Tung Wah", in his *Hong Kong: Stability and Change, A Collection of Essays* (Hong Kong: Oxford University Press, 1978), 61.

[23] Carl T. Smith, "Notes on Chinese temples in Hong Kong", *JHKBRAS* , 13 (1973), 133-9.

24 Carl T. Smith, "Notes on Chinese temples", 134-5, quoting from *The China Review*, 1 (1872-3), 133.

25 Faure, Luk, and Ng, *Historical Inscriptions*, p. 691.

26 H.J. Lethbridge, "The District Watch Committee: the Chinese Executive Council of Hong Kong?", in his *Hong Kong: Stability and Change*, 105-6, and Faure, Luk and Ng, *Historical Inscriptions*, 694-5.

27 "Donghua yiyuan chuangyuan jiushinian zhi yange" 東 華 醫 院 創 院 九 十 年 之 沿 革 , 1-2, in *Donghua sanyuan fazhanshi* 東 華 三 院 發 展 史 (Hong Kong, 1961). On the origins of the Tung Wah Hospital, see also Carl T. Smith, *Chinese Christians: Elites, Middlemen, and the Church in Hong Kong* (Hong Kong: Oxford University Press, 1985), 109-10, 125-6.

28 Faure, Luk, and Ng, *Historical Inscriptions*, 701-2.

29 Ordinance No. 6 of 1908.

30 Faure, Luk, and Ng, *Historical Inscriptions*, 710, and on the Po Leung Kuk, see H.J. Lethbridge, "The evolution of a Chinese voluntary association in Hong Kong; The Po Leung Kuk", in his *Hong Kong: Stability and Change*, 71-103.

31 D.W.Y. Kwok, *Scientism in Chinese Thought, 1900-1950* (New Haven: Yale University Press, 1965).

32 "Guanghua yiyuan chuangyan yange 廣 華 醫 院 創 院 沿 革 ", 3-4, 10, in *Donghua sanyan fazhanshi*.

33 These five temples were the Koon Yam 觀 音 Temple and the Sui Tsing Paak 綏 靖 伯 Temple of Tai Ping Shan 太 平 山, the Lin Fa Kung 蓮 花 宮 and the Tin Hau Temple of Tai Hang 大 坑 and Causeway Bay, and the T'aam Kung Temple of Shaukeiwan. Paragraph 20 of the "Objects and reasons" for the ordinance (Supplement of *Hong Kong Government Gazette*, 23rd March, 1928) states: "These are all old established temples which were founded by families or individuals and not by the community or sections of the community, and their administration is unobjectionable, though the profits do not go to the community. As these temples are private property, and are run for private benefits, section 12 also provides that no public collection or appeal must be made for them except within the precincts of the particular temple for which the collection or appeal is being made."

34 Section 4, Chinese Temples Ordinance, 1928; I have quoted the line from paragraph 11 of the "objects and reasons" for the ordinance, given in supplement to *Hong Kong Government Gazette*, 23rd March, 1928, 212.

35 *Hong Kong Hansard,* session 1928.

36 These were the District Watch Committee, the Po Leung Kuk 保 良 局 Committee, and the Kwong Wah Hospital Committee. On these various organisations, see H.J. Lethbridge, *Hong Kong: Stability and Change*.

37 The ordinance seems to be a compromise between the more restrictive tendency towards folk religion and the more tolerant attitude which

would have included leaving Chinese customs alone. The former is quite apparent in the following statement: "The Hong Kong Government has announced that it would put into effect [its policy] of abolishing unhealthy temples. On this matter, I had many times made suggestions to the Government. These temples not only lead people into superstition, but they also hide dirty matters. Such things are heard all the time. If they are not strictly abolished, it is not possible to break superstition and maintain public morals." (Feng Pingshan 馮 平 山 , *Jishi ce* 記事冊 ,n.p., n.d., Taibei reprint, n.d.) 37. Feng was one of the most prominent Chinese community leaders of the time.

[38] *Huazi ribao* 華 字 日 報 , June 28 and 29, 1928.

[39] *Huazi ribao* , May 21, June 30, and August 12, 1929.

[40] Ordinance No. 6 of 1933, "objects and reasons" in *Hong Kong Hansard* , session 1933, March 16, 1933. This should be compared with paragraph 4 of "objects and reasons" for the Chinese Temples Ordinance, 1928 in the Supplement to *Hong Kong Government Gazette* : "There has been an alarming growth of pseudo-religious establishments in recent years. Many of the keepers are simply fortune tellers of an unrecognized and objectionable kind. Some of these temples occupy a single floor for a few months at a time until they have dealt with all the dupes of the district, when they move elsewhere."

THE JIAO FESTIVAL IN HONG KONG AND THE NEW TERRITORIES

Tanaka Issei

Introduction

This essay is a study of the religious *jiao* 醮 festival in the agricultural and fishing villages of Hong Kong and the New Territories. There are two types of traditional *jiao* festival in Hong Kong today. One is the *jiao* festival in Fukienese tradition handed down from generation to generation by the Hailufeng immigrants. These immigrants moved to Hong Kong from the Haifeng 海豐 and Lufeng 陸豐 counties in Huizhou 惠州 prefecture of Guangdong 廣 東 province. This type of *jiao* is celebrated in Changzhou 長 洲 (Cheung Chau) and other places where the Hailufeng immigrants are concentrated. Another is the *jiao* festival in Hakka-Cantonese tradition commonly handed down from generation to generation by both Cantonese and Hakka immigrants. The Cantonese immigrants moved to Hong Kong from various counties in Guangzhou prefecture of Guangdong province and the Hakka immigrants came from Mei 梅 county in Jiaying 嘉 應 prefecture, Guishan 歸 善 county in Huizhou prefecture and other parts of Guangdong province. This type of *jiao* is celebrated in the agricultural areas and coastal fishing villages in the New Territories.

These two types of *jiao* festival are similar in many ways but are also different in other ways. In this essay, I plan to compare and contrast the characteristics of the *jiao* festival in the Hakka-Cantonese tradition with the *jiao* festival in the Fukienese tradition. This essay will consist of the following:

A) Comparison of the lay-out of the festival grounds and the rituals of a three-day *jiao* in the Hailufeng (Fukienese) tradition celebrated by the Hailufeng groups and a three-day *jiao* in the Cantonese tradition celebrated by the Cantonese group.

B) Analysis of the characteristics of the *jiao* in the Cantonese tradition, discussed in (A), followed by an analysis of a five-day *jiao*.

C) Finally a look at the historical background of the characteristics of the *jiao* in the Cantonese tradition.

271

A. Comparison of the *Jiao* Festivals in Changzhou (Cheung Chau) and Longyuetou (Lung Yeuk Tau),

I. Festival Grounds

The Hailufeng people in Changzhou, Hong Kong have celebrated a three-day *jiao* in the open space in front of the Beidi temple 北 帝 廟 , located in the northern part of Changzhou island, once every year in the early part of the 4th lunar month since the mid-Qing period. This is called **Xuantian Dadi Taiping Qingjiao** 玄 天 大 帝 太 平 清 醮 ("Purificatory Rites for Peace under Beidi's Patronage"). It is believed that hungry ghosts who died unnatural deaths in and around the island bring about disease or disasters, and to restore peace a *jiao* is celebrated in which the principal deity Beidi and other high ranking gods are invited to the festival grounds so that through their power the hungry ghosts will be pacified. The *jiao* was traditionally celebrated only when there was a plague or epidemic. However, nowadays it is celebrated at the same time annually. The Changzhou *jiao* is taken here as a representative example of a three-day *jiao* in the Fukienese tradition which has spread to Guangdong.[1]

Next, the "Purificatory Rites for Peace" celebrated once in ten years at the Longyuetou 龍 躍 頭 village in the eastern part of the New Territories is taken as an example of the *jiao* in the Cantonese tradition. The Cantonese-Hakka Deng 鄧 lineage which emigrated from Dongguan 東 莞 county in the Ming period lives in this village. This festival is celebrated for three days once every ten years in the ancestral hall which is next to the temple of the principal deity Tianhou 天 后 , and its frontyard. It is celebrated by inviting several gods, so that the village can be blessed with peace and the hungry ghosts who cause disasters, accidents, etc. can be pacified.

When comparing the two *jiao* celebrations, let us first look at the lay-out of the festival grounds. **Figs. 1** and **2** at pages 297 and 298 show the respective lay-outs of the sites of the *jiao* in Changzhou and Longyuetou.[2]

Fig. 1: Layout of the site of the *jiao* in Changzhou

1. Bei Di Temple
2. The Altar for the Three Pure Ones
3. The Altar for Local Deities
4. Paper Image of the Mountain Deity
5. Paper Image of the Earth God

6. Paper Image of the King of Ghosts
7. Matshed Theatre
8. The Heroes' Hall

Fig. 2: Layout of the Site of the *jiao* in Longyuetou
1. Tianhou Temple
2. The Altar for the Three Pure Ones
3. The Altar for the Zhang Tianshi (Heavenly Master)
4. The Altar for Local Deities
5. The Paper Image of the King of Ghosts
6. Paper Image of the City God
7. The Kitchen for Vegetable Foods
8. Matshed Theatre

There are many similarities in the lay-out of the sites of the two *jiao*. In a festival ground, the areas of focus are the Altar for the Three Pure Ones where the highest ranking gods are housed and the Altar for Local Deities where the various gods from within the region are housed. Beside these two altars, there is a matshed where the deities of Hell are housed, and a matshed where the Dashiwang 大士王 is positioned. He is the King of Ghosts, an incarnation of the Goddess of Mercy, Guanyin, who gives food and clothing to the hungry ghosts and takes them back to the netherworld. Apart from these altars, there also is a nearby kitchen with stoves where vegetarian food is prepared during the festival period. Bamboo poles bearing the different standards are erected around the festival ground as indicators to guide and gather the hungry ghosts to the festival. The poles mark the boundaries of the festival area.[3] I shall compare the arrangements in the festival grounds at both *jiao* as follows:
  (1) The **Jiao Altar** (the Altar of the Three Pure Ones)
      At both *jiao*, the *jiao* Altar where the Three Pure Ones are worshipped is built next to the main temple. There is difference in the arrangement of this altar in the two *jiao*. In the Changzhou *jiao* more than 10 scrolls including those of the Three Pure Ones are hung in the center with the White Tiger God on the left and the Azure Dragon God on the right; these hangings face the main entrance.
  (2) *The Altar of Zhang Tianshi* (Heavenly Master)
      An Altar for the Heavenly Master where the scroll of the Founder of the Taoist religion, Zhang Daoling 張道陵 (Zhang Tianshi 張天師) is hung, is built near the *jiao* Altar. At Longyuetou, it is at the entrance of the ancestral

hall. However, this is not seen in Changzhou.

(3) *The Altar of the Local Deities*

An Altar for the Local Deities is built near the *jiao* Altar for the images and the tablets of all local gods. At Changzhou all images of major gods under and including Beidi are put in this altar. At Longyuetou, except for the image of Tianhou, only the tablets of the local gods are put there.

(4) *Matshed for the Deities of Hell*

In both *jiao*, some distance away from the *jiao* Altar, besides the City God, the big images of its subordinates, the Earth God and the Mountain God are also put there. At Longyuetou, however, only the City God is worshipped.

(5) *Matshed for the King of Ghosts*

On both festival grounds there is a corner for a matshed for Dashiwang. It is said that he is the reincarnation of Guanyin. The Dashiwang is supposed to provide food and clothing to the hungry ghosts and later see that they return to the netherworld. The power of Dashiwang to keep the hungry ghosts who cause trouble to the villagers under control, is symbolised by its gigantic image and fierce-looking face. The Dashiwang at Changzhou was a black standing figure looking ferocious and powerful while that at Longyuetou was a coloured seated figure looking rather gentle, like an official.

(6) *Standards or Banners*

In principle, in a three-day *jiao*, three bamboo poles (called **fan** 旛) are erected as signals for the wandering spirits. At Changzhou, this principle is not adhered to. The poles are erected at nine separate places with a total of eleven poles. The original three poles are erected in the Heroes' Hall, a place where the heroes who died in battle against other groups are deified, about 300 m. south of the festival area. This is Fukienese style. The other eight poles are erected around the festival area to indicate its boundaries. This is Cantonese style. Therefore, we can say that the method of erecting poles at Changzhou is a combination of Fukienese and Cantonese styles. At Longyuetou it is Cantonese style where three poles serve to indicate the enclosure of the festival area.

(7) *The Vegetarian Food Kitchen*

At Changzhou, some distance away from the festival grounds at the Heroes' Hall is a kitchen where the purified stove for preparing vegetarian food is placed. At the Changzhou *jiao*, one can see that there is an awareness of

the heroes who died for the group. At Longyuetou the purified stove is at one corner of the ancestral hall which is near
the *jiao* Altar.

(8) *Matshed Theatre*
During the festival period, as a votive offering for enter-
tainment of the visiting gods, a matshed theatre is set up,
usually directly facing the altars. However, at both
Changzhou and Longyuetou, owing to limitations of the
festival area, they are built in an unusual direction.

In sum, the basic lay-out of the festival area is the same at both
jiao . However, where the number and types of gods and colours
are concerned, the *jiao* in Changzhou is more colourful and gor-
geous than that in Longyuetou. In the Fukienese Changzhou *jiao*
festival area, popular idolatry is a strong feature, while at the Cantonese Longuetou the *jiao* lay-out of the festival area is closer to
what the literati prefer, ideological belief.

II. Rituals
In the following pages, I will compare the rituals of the two *jiao*.
For easy comparison let us take a look at the detailed programs
(plan) given in Table 1.

TABLE 1 Order for a Three-Day *jiao*

		Changzhou	Longyuetou
10 months before			Choosing the leaders of worship (*yuanshou jueding* 緣 首 決 定)
7 months before			The 1st Announce-ment to heaven of the villagers' names (*shang diyi biao* 上 第 一 表)
3 months and a half before			The 2nd Announce-ment to heaven of the villagers' names (*shang dier biao* 上 第 二 表)
2 months and a half before			Beginning to construct the ground for jiao Festival (*kaigong zhazuo* 開 工 扎 作)

1 month and a half before			Building the Jiao Altar (*kaida jiaopeng* 開搭醮棚)
4 days before			Building the Stove for vegetable food (*zuo zhaizao* 作齋灶)
The day before	Morning (*zao* 早)		Drawing Water (*qushui* 取水)
	Noon (*wu* 午)	Inviting the Local Deities (*yingshen* 迎神)	Inviting the Local Deities (*yingshen* 迎神)
			The 3rd Announcement to Heaven (*shang disan biao* 上第三表)
	Evening (*wan* 晚)	Setting up the Jiao Altar (*putan* 舖壇)	Setting up the Jiao Altar (*putan* 舖壇)
		Opening eyes ritual for Paper Images (*kaiguang* 開光)	
		Inviting the Deities (*qingshen* 請神)	Announcing to the Spirits (*fazou* 發奏)
The 1st Day	Morning (*zao* 早)	Announcing to the Spirits (*fuzou* 敷奏)	
		Raising the Standards (*yangfan* 揚旛) Morning Audience 1 (*zaochao* 早朝)	Raising the Standards (*yangfan* 揚旛) Morning Audience 1 (*zaochao* 早朝)
		Reading the Three Officials Canon (*san guanjing* 三官經)	
	Noon (*wu* 午)	Noon Audience 1 (*wuchao* 午朝)	Noon Audience 1 (*wuchao* 午朝)
		Reading the Upper Origin and Middle Origin Canon (*shangyuan-zhongyuan baochan* 上元中元寶懺)	Reading the Canon of Repentance (*lichan* 禮懺)

		Public Presentation to Heaven of the Memorial with the List of Villagers' Names (*fabang* 發榜)	
	Evening (*wan* 晚)	Evening Audience 1 (*wanchao* 晚朝)	Evening Audience 1 (*wanchao* 晚朝)
		Reading the Lower Origin Canon (*xiayuan baochan* 下元寶懺)	Reading the Canon of Repentance (*lichan* 禮懺)
		Night Prayer to the Three Officials (*wan can* 晚參)	Lighting the New Fire (*fendeng* 分燈) Military Training of Heavenly Soldiers (*dawu* 打武) Purifying the Sacred Tan (*jintan* 禁壇)
The 2nd Day	Morning (*zao* 早)	Morning Audience 2 (*zaochao* 早朝)	Morning Audience 2 (*zaochao* 早朝)
	Noon (*wu* 午)	Noon Audience 2 (*wuchao 2*)	Noon Audience 2 (*wuchao 2*)
		Noon Offerings (*wugong* 午供 or *wuchao* 午朝)	
			Public Presentation to Heaven of the Memorial with the List of Villagers' Names (*yingbang* 迎榜)
		Reading the North Pole Canon (*beidou jing* 北斗經)	
	Evening (*wan* 晚)	Evening Audience 2 (*wanchao* 晚朝)	Evening Audience 2 (*wanchao* 晚朝)
		Feeding the Hungry Spirits in Water (*jishuiyou* 祭水幽)	Feeding the Hungry Spirits on a Small Scale (*xiaoyou* 小幽)
		Night Prayer to the Jade Emperor (*wan chao* 晚朝)	Inviting the Jade Emperor (*yingsheng* 迎聖)

277

The 3rd Day	Morning (*zao* 早)	Reading the Star Canon (*xingchenchan* 星晨懺)	
		Morning Audience 3 (*zaochao* 早朝) Hauling Down the Standards (*xiefan* 謝旛)	Morning Audience 3 (*zaochao* 早朝)
		Reading the North Pole Canon (*beidou jing* 北斗經)	
	Noon (*wu* 午)		Noon Audience 3 (*wuchao* 午朝)
			Reading the Canon of Repentance (*lichan* 禮懺)
		Sending Off the Flower Boat (*qianchuan* 遣船)	
		Distributing the Pardon and Talismans to the Villagers (*banshe* 頒赦)	Carrying a Pardon from Heaven (*zousheshu* 走赦書)
		Procession of Local Deities for Patrol (*huijing* 會景)	
		Freeing the Living Creatures (*fangsheng* 放生)	Freeing the Living Creatures (*fangsheng* 放生)
	Evening (*wan* 晚)		Evening Audience 3 (*wanchao* 晚朝) Hauling Down the Standards (*xiefan* 謝旛)
			Reading the Canon of Repentance (*lichan* 禮懺 : end)
		Feeding the Hungry Spirits (*chaoyou* 超幽)	Feeding the Hungry Spirits on a Large Scale (*dayou* 大幽)

The Day After	Morning (zao 早)	Sending Off the Spirits (songshen 送神)	Sending Off the Spirits (songshen 送神)
	Noon (wu 午)	Procession of Local Deities for Return (huijing 會景)	Carrying Off the Paper Boat (laya pachuan 拉鴨扒船)

In the following pages I shall compare the major rituals of the two *jiao* item by item.

(I) Preparing the *Jiao*

(1) *Selecting the Leaders-of-worship (yuanshou 緣首)*

The *yuanshou* are chosen by the deity to represent the villagers to participate in the rituals according to the instructions given by the Taoist priests. This is a sacred role. At Changzhou, they used to select more than ten young men to be *yuanshou.* However, this custom is not observed today. For this reason the religious part of the *jiao* ritual has become the exclusive territory of the Taoist masters. The villagers are not involved. At Longyuetou ten months before the *jiao,* during the first lunar month, fifteen young men are selected by divination at the Tianhou Temple. In this way the joint relationship between the Taoist masters and the villagers is maintained.[4]

(2) *Announcement to the Spirits (fabiao 發表)*

Before the beginning of a *jiao* , a ritual called *fabiao* in which the day and the time of the *jiao* are announced to Tiandi (Emperor of the Heavens) is performed. In the same ritual, the names of the villagers who are asking for Tiandi's divine protection through the *jiao* festival are sent up to the heavens. This ritual is not performed at Changzhou but at Longyuetou it is repeated three times. The first was held in the first lunar month, the second in the eighth lunar month and the third on the day before the *jiao* . This meticulousness of the *fabiao* can be said to be a characteristic of the *jiao* in the Cantonese Tradition.

(3) *Construction of the Jiao Altar*

At Changzhou when construction work on the *jiao* Altar, the Deities' Altar, the Kitchen and the matshed Theatre was about to begin, there was no special ritual by the Taoists. At Longyuetou, however, as shown in Table 1, a total of three consecutive rituals were performed. This attempt to abide by the wishes of the heavens is again a characteristic of the *jiao* in the Cantonese tradition.

In the above comparison we have seen the simplicity of the Changzhou *jiao* preparations and the meticulousness of the Longyuetou *jiao* preparation.

(II) Rituals on the Day Before

In its orthodox form, on the day before the beginning of a *jiao* , the Altar is formally opened, the local deities are invited, and the Announcement is sent to Tiandi inviting his honoured presence. Changzhou and Longyuetou are similar in this basic structure but where the rituals are concerned, they are different.

(1)　*Drawing Water (qushui* 取 水 *) (Longyuetou)*

During the *jiao* , holy water taken from a nearby river to purify the *jiao* Altar is put below the Altar. This is a ritual that attaches importance to the water veins used in the irrigation of the fields. It is not performed at Changzhou, which is a fishing village. Perhaps it was never performed there in the first place.

(2)　*Inviting the Deities to the Altar (Yingshen* 迎 神 *)*

In the afternoon all sorts of local gods are invited to the Deities' Altar. At Changzhou they are the gods who occupy different parts of the island under and including the principal deity Beidi, namely Tianhou 天 后 , Hongsheng 洪 聖 , Guandi 關 帝 , etc.; their statues are put on sedan chairs and carried to the Deities' Altar from their respective temples. Small earth gods like Bogong 伯 公 etc. are not invited. At Longyuetou, besides the major gods like the principal deity, Tianhou, Wenchangdi 文 昌 帝 , Guandi etc. the Bogong of the many small shrines found all over the village are also invited. A group of Taoists and *yuanshou* take more than half a day to go round the village putting joss-sticks at the different earth shrines under big trees and stones. This entourage is called *xingxiang* 行 香 (*hang heung*). This invitation of all the gods is more meticulous than at Changzhou and it realizes the original purpose of the *jiao* .

(3)　*Setting Up the Altar (Putan* 舖 壇 or *qitan* 啓 壇 *) and Opening of the Eyes Ceremony (Kaiguang* 開 光 *)*

At Changzhou, the Setting Up of the Altar ritual is completed only with the Taoists reading the canon but the Opening of the Eyes Ceremony is carefully performed. One of the Taoist priests, as he chants the canon, takes a mirror to shine on the statues and paper images of the gods and makes several dots on the mirror with a brush. This rite can be said to be based on the Fukienese tradition, which has always emphasized the Opening of the Eyes Ceremony.[5] At Longyuetou, there is no Opening of the Eyes Ceremony but the Setting Up of the Altar ritual was Confucian and rather solemn and impressive. A feature of the *jiao* in the Cantonese tradition is that they have been influenced by Confucian Rites. First the Taoists and the *yuanshou* stand in a straight row facing each other. They re-arrange their robes and their skull-caps and then exchange wine cups. This is a rite modelled after the ceremony of *xiang yinjiu* 鄉 飲 酒 (an Official Banquet of Village Elders). Next the Taoist reads the canon announcing to Tiandi the names of the *yuanshou*.

(III) The First Day

(1) *Announcement to the Spirits (fazou* 發 奏 *)*

At Changzhou, after the Setting Up of the *Tan* (sacred area), Opening the Eyes and Inviting the Deities ritual, at a little past midnight an Announcement to the Spirits (*fazou*) ritual is performed where invitations are sent to the many gods in the heavens. One can say that performing the *fazou* in the early hours of the first day (before dawn) is a correct choice of time. However, the ritual is simplified.

A priest, accompanied by a *yuanshou* , faces a paper pavilion outside the *jiao* area, which symbolizes the three worlds: heaven, earth and water. He reads an official document, ties up the document to the horse and burns them (or sends them to heaven). At Longyuetou, in the evening before the first day of the *jiao* , soon after the Setting Up of the *Tan* ritual is over, the Announcement to the Spirits ritual which should be performed in the early hours of the next morning is performed. The reading of the official document and the rite of burning it with the horse are the same as at Changzhou but in the official document at Longyuetou, the names of all the villagers are written below those of the *yuanshou* . The reading of all these names takes the Taoists a long time. After that the Taoists perform a purification ritual by touching the villagers' names with the blood of a cock's comb. Then the document is tied to the horse and burnt. The whole ritual takes three hours. In the past two *jiao* , the names of all the villagers were sent to the heavens. And this time even with the increase in population, it is expected that no one's name is left out. Compared with Changzhou, the inclusion of every villager's name expresses the importance of the relationship between the villagers and the heavens.

(2) *Three Audiences (sanchao* 三 朝 *): Morning Audience and Raising of the Standards (zaochao yangfan* 早 朝 揚 旛 *), Noon Audience (wuchao* 午 朝 *), Evening Audience (wanchao* 晚 朝 *)*
The Taoists and *yuanshou* go round each and every standard three times a day: in the morning, afternoon and evening. Canons are read for the hungry ghosts who have gathered round the standards and food is offered. At Changzhou this rite is performed at all nine standards after which the whole entourage returns to the main altar immediately without visiting the Deities' Altar or the King of the Ghosts. This ritual is rather simplified. At Longyuetou, the entourage went round in the following order: 1) The *Tan* of the Heavenly Master, 2) The First Standard, 3) The Second Standard, 4) The Third Standard, 5) The Deities' *Tan* , 6) The King of the Ghosts, 7) The City God, 8) The Kitchen, 9) The *jiao Tan* .
This means that all the spirits (gods and ghosts) within the festival ground are visited; it is more meticulous than at Changzhou. In the morning of the first day there should be a separate Raising of the Standards ceremony before the morning audience but at neither Changzhou nor Longyuetou was this ceremony performed separately but in conjunction with the morning audience.

(3) *Reading of the Canon of Repentance (lichan* 禮 懺 *)*
Three times a day after each audience the Taoists and the *yuanshou* will go back to face the *jiao* Altar and announce the sins (offences) of the villagers to the heavens and read the Canon of Repentance to the Three Origins to pray for forgiveness. At Changzhou the Three Officials Canon and the North Pole Canon are also recited. This practice belongs to the *jiao* in the Fukienese tradition. At Longyuetou only the Canon of Repentance to the Three Origins is Read. This is the Cantonese style.

(4a) *Night Prayer to the Three Officials (wanchao* 晚 朝 *)*
At Changzhou, late in the night of the first day a ritual of Night Prayers to the Three Officials is performed, while at Longyuetou Lighting the New Fire, Military Training of Heavenly Soldiers and Purifying of the Altar are performed. One can say that the former is Fukienese style and the latter Cantonese style. At Changzhou the Night Prayers to the Three Officials is a ritual asking for forgive-

ness of the villagers' sins by sending a petition and other offerings to the Heavenly Emperor (Tianguan 天 官), Earth Emperor (Tiguan 地 官) and Water Emperor (Shuiguan 水 官), who respectively rule the realms of heaven, earth and water. It is believed that the peace of the place is disturbed by the increased number of hungry ghosts due to the sins of the villagers. Therefore they have to pray for forgiveness to the Three Officials who rule the three worlds. The red-head Taoists in Taiwan also read the Three Officials canon on the first day. It is therefore probably a ritual of *jiao* in the Fukienese tradition.

(4b) *Lighting of the New Fire (fendeng 分 燈), Military Training of the Heavenly Soldiers (dawu打武) and Purification of the Tan (jintan 禁 壇) at Longyuetou.*

On the first night at Longyuetou the rituals of the Lighting of the New Fire and the Purification of the *Tan* are performed. During the Lighting of the New Fire not only the Three Officials, but also those above them, the Nine Emperors, are invited and a fire (lantern) is offered. Fire symbolises Yang. It is believed that Yin and Yang are not balanced. As the element Yin becomes stronger there are more calamities. This ritual is performed in order to restore the balance, by bringing in an increase of the element Yang.

One can say that the Night Prayer to the Three Officials is to reduce the element Yin and the Lighting of the New Fire is to increase the element Yang. Thereupon the Purification of the *Tan* is performed. This is a ritual to shut out the evil spirits and demons from the sacred area and to purify the festival ground so that the invited heavenly spirits can be welcomed. In order to chase away the ghosts it is necessary to borrow the strength of the Heavenly Soldiers. The Taoists summon the Heavenly Soldiers and perform a kind of war training. This acrobatic performance is call *dawu* . When this is over, the Taoist shuts and guards all five directions East, West, South, North and Center. This is a ritual that strongly shows the power of Tiandi. One can say that this is an orthodox Taoist ritual.

(IV) Rituals on the Second Day

(1) *Three Audiences (sanchao)*
Generally the same as on day one.

(2) *Reading the Canon of Repentance (chan)*
At Changzhou it is the North Pole Canon that is mostly read. At Longyuetou the reading of the Canon of Repentance to the Three Origins is continued.

(3) *Noon Offering (wugong* 午 供 *) at Changzhou*
At Changzhou, on the second day a large scale *wugong* is performed together with the noon audience. Five tables are put out in each of the five directions. Food is offered on these tables to the gods of each direction. This Noon Offering ritual is a common feature of the *jiao* in Fukienese tradition seen in Taiwan. One can say that the Changzhou *jiao* also belongs to such a family. At the *jiao* in the Cantonese tradition in Longyuetou, there is no such ritual.

(4) *Public Presentation to Heaven of the Memorial with the List of the Villagers' Names (yingbang* 迎 榜 *)*
A public presentation to Heaven of the memorial with the list of the villagers' names is called *fabang* or *yingbang*. At Changzhou a representative of the villagers, on receiving the *bang* (memorial) from the Taoist priests, pastes it on a long notice board at the pier. Only the names of household heads are written on the memorial. No special ceremony accompanies this. One can say that no *fabang* is performed. At Longyuetou, however, there is an elaborate ceremony for the Taoists to hand over the memorial to the *yuanshou*. The Taoists and the *yuanshou* stand in a straight row facing each other as at the Setting Up of the *Tan* ritual. After exchanging greetings with each other they go to the mirror to "arrange their robes and skullcaps." Next there is a greeting through drinking. The chief among leaders-of-worship wears a red sash and cap. Next, the Taoists put the memorial on the table and one after another they sign their names at the end of the memorial. When that is completed, the Taoists all together hold the memorial high up and pass it into the hands of the *yuanshou*. At this time music is played. The *yuanshou* then take the memorial to the outside and paste it up on a notice board. In the memorial are not only the names of household heads but also the names of every family member including newborns. The memorial is about fifty metres long. After the whole memorial has been pasted up, the Taoists dot the names of the household with chicken blood for purification. This ritual, one of the most important rituals in the Cantonese style *jiao*, is as important as

the Invitation of the Jade Emperor ritual.

(5) *Feeding the Hungry Spirits in Water (shuiyou* 水 幽 *), Small Scale Feeding of the Hungry Spirits (xiaoyou* 小 幽 *)*

In the evening of the same day, at Changzhou, boats are sent off in water. The Taoists read the canons and food is thrown into the sea to comfort the souls who died in the waters. A corresponding ritual at Longyuetou is the Feeding of the Hungry Spirits in a small scale ritual. An altar is set up outside and a couple of judges and a husband and wife pair (selling miscellaneous goods) are carried out from the City God Matshed and put facing the altar. The canons are read and food is given to the hungry ghosts who have gathered around in the festival area. *Shuiyou* is probably a fishing village ritual while *xiaoyou* is probably an agricultural village ritual.

(6a) *Night Prayer to the Jade Emperor (wanchao* 晚 朝 *) at Changzhou*

On this night in Changzhou, a ritual to welcome the Jade Emperor, the highest ranking deity in the heavens is performed. This is the *wanchao*. Five Taoist priests invite the Ten Emperors from the heavens by offering incense, flowers, light, tea, cakes, food, water, treasures, jewels, and wine, a total of ten types of offering. Then a petition on blue paper (*qingci* 青 詞) with characters written in gold is presented to the emperors under the Jade Emperor. Following that a petition on yellow paper with bright red characters written on it, is presented to the Jade Emperor. This ritual can be seen in the *jiao* of the Fukienese Tradition in Taiwan. The reading of *qingci* is an old Taoist practice.

(6b) *Invitation of the Jade Emperor (yingsheng* 迎 聖 *) at Longyuetou*

On the second night at Longyuetou, an Invitation of the Jade Emperor ritual is performed. This invitation is a very important ritual. The basic idea is the same as the *wanchao* in Changzhou but the rituals are completely different. On a stage facing the *jiao Tan* (the *Tan* for the Three Pure Ones) the tablets of the Three Pure Ones are placed in the center with the image of Tianhou, the tablets with Deities' names, the image of the City God, and the paper box with the memorial to the North Pole Palace. The Taoist leads the *yuanshou* to bow and pray to the tablets of the Three Pure Ones and the petition asking for the Jade Emperor's presence is recited. Following that, a bridge-like structure is formed in front of the tablets of the Three Pure Ones. It is believed that the gods will come down over this bridge to enter the festival area. One Taoist then stands on the left of the bridge and reads the petition to the Nine Emperors. Having done that the petition is attached to a paper horse and burnt. Next, a Taoist reads the memorial to heaven briefly and, accompanied by a master of ceremonies, sits at the entrance of the bridge and sings a question-and-answer song alternating with a drummer. This is a song in praise of the Jade Emperor. When the Singing is over, a ceremony priest picks up flowers from a plate and scatters them on the bridge. Then one by one the tablets of the Three Pure Ones, and the Image of Tianhou Goddess, are made to cross the bridge and put back in their original places. This ritual symbolises the return of the gods. Here the presentation of the memorial to the North Pole Nine Emperors' Palace can be seen as a combination of *lidou* 禮斗 and *yingshen*. The effect of the Invitation of the Jade Emperor ritual is the same as the *wanchao* in Changzhou but the gathering of several gods, the presentation of offerings, the singing, the scattering of flowers, etc. is a beautiful devotional act which can be said to reflect clearly the idea of worshipping Heaven according to the *jiao* in the Cantonese tradition.

(V) Rituals on the Third Day

(1) *The Audiences, Taking down of the Standards (xiefan* 謝旛 *)*
Same as on the first day but Changzhou finishes after the Morning Audience during which the Standards are taken down, while the Noon Audience and the Evening Audience are skipped. At Longyuetou all three audiences are performed and during the Evening Audience the Standards are taken down.

(2) *Reading of the Canon of Repentance (chan)*
At Changzhou, the Canon of Repentance to the Stars is read before the break of daylight and the North Pole Canon is read in the Morning. At Longyuetou the reading of the Canon of Repentance to the Three Origins is continued until the Evening Audience when it ends.

(3) *Sending off the Paper Boat (qianchuan* 遺船 *) at Changzhou*

At Changzhou in the afternoon of the third day, a *qianchuan* ritual to chase away the Demon of Pestilence is performed. A Taoist issues orders to a Heavenly Envoy to carry off the boat (*yachuan shizhe* 押 船 使 者) and puts the Demon of Plague on a boat and then leaves it in the outer seas. The Heavenly Envoy, like the King of the Ghosts, has a fierce-looking face. It is an image of about one metre high and the boat is a small one of about one and half metres long. A Taoist lifts the Heavenly Envoy to a stage in the Matshed Theatre and then chants a question-and-answer song which instructs the Heavenly Envoy.[6]

Having finished that, the villagers then put the Heavenly Envoy into the boat loaded with offerings. This boat is then taken to the sea shore and left on the waters. This ritual can be seen as coming from the same family as the Fukienese King's Boat Rite (*wangchuan* 王 船) seen in Taiwan. There is no such ritual at Longyuetou but we can see traces of it in the Carrying off the Paper Boat ritual which will be described later.

(4) *Distributing the Pardon and Talismans to the Villagers (banshe* 頒 赦 *)*

In the afternoon, the *banshe* , a ritual of asking for amnesty from Tiandi for the sins of the villagers is performed. At Changzhou, talismans with certificates of pardon are put on the *jiao* Altar. Three Taoists perform a dance of ritual steps as they read the canon in front of the altar. When that is over, the talismans with certificates of pardon are distributed to the villagers. It is a simple ritual. At Longyuetou, however, it is more complicated. First, the heavenly messenger, the paper horse that was dispatched to Heaven, is put in front of the altar of the Heavenly Master. The Taoists read the canon, ordering the horse to start on its journey. Then a villager carries the paper horse on his shoulder and runs one round of the village before returning. This ritual symbolises the sending of pardon from Tiandi. Then a Taoist takes the Document of Pardon attached to the horse, climbs onto a high stage and reads it. In this Document of Pardon, like in the petition or memorial, the names of all the villagers are written down and therefore it takes one priest a long time to finish reading. The *yuanshou* are seated facing the Taoist, listening to the reading. When it is over, the names are dotted with chicken blood and the Document of Pardon is tied to the paper horse and burnt.

Here again, the reading of all the villagers' names to receive pardon is a feature of Cantonese rituals. One can even say that reporting all the villagers' names to Heaven is obligatory.

(5) *Procession of Local Deities for Patrol (huijing:* 會景 *) at Changzhou*

At Changzhou, all the deities under and including Beidi and the Deities' Altar are put on portable thrones and paraded round the whole island. Against the backdrop of the final moments of the *jiao* , it is a ritual to suppress the hungry ghosts by the gods going round the island. Longyuetou does not have such a ritual.

(6) *Freeing the Living Creatures (fangsheng* 放生 *)*

After the *banshe* (mentioned above), Freeing the Living Creatures is performed so as to extend amnesty to the animals. At Changzhou, during the *huijing* , Taoists free birds and turtles. At Longyuetou, in the evening after the Evening Audience, an altar is set up in an open space outside and birds are freed from cages.

(7) *Feeding the Hungry Ghosts on a Large Scale (dayou* 大 幽 *, chaoyou* 超幽 *)*

Late in the night on this day, a *chaoyou* ritual is performed during which food and clothing are given to the hungry ghosts who have gathered on the festival ground before sending them off to the underworld. It is on a much bigger scale than the Small Scale ritual Feeding of the Hungry Ghosts. This ritual is almost the same at both Changzhou and Longyuetou. First the villagers carry the image of the King of the Ghosts to an open space outside and an altar facing it is erected. Between the King of the Ghosts and the altar, many paper clothes and food are placed and two rows of candles are stuck in the earth, one on each side. The Taoists read the canons on the altar stage. When finished, the candles are lit and finally the large image of the King of the Ghosts is also lit and burnt. When all is over, it is already past midnight. At this point the three day *jiao* ritual is completed and what is left to do are only minor activities.

(VI) Rituals on the Day After

(1) *Sending off the Spirits of the Three Worlds (songshen 1* 送神 *)*

On this morning after the deities in the Deities' *Tan* and

the Matshed for the City God have been thanked, they are sent off in a ritual called *songshen*. At Changzhou this is performed immediately after the previous *dayou* which finished a little past midnight. One Taoist goes round every altar to read the canon. After each reading the paper images of the gods in the respective altars are burnt. At Longyuetou it is in the morning, after daybreak, that Taoists leading the *yuanshou* go round each altar to gather up the paper images in front of the ancestral hall and burn them.

(2) *Sending off the Local Deities to the Temple (songshen 2 送神)*

After the paper images have been burnt as mentioned above, the remaining tablets and images of local deities in the Deities' *Tan* are returned to their original temples or shrines. At Changzhou, as in the *huijing* of the previous day, the respective gods are put in portable shrines and a group surrounds each portable shrine. These groups parade one after another as the deities are sent back to their usual residences. This is called the Second Procession (*juijing*). At Longyuetou there is no such gaiety. The *yuanshou* carry the images of Tianhou and the deities' tablets from the Deities' *Tan* to the Tianhou temple and that is all.

(3) *Carrying off the Paper Boat (pachuan 扒船) at Longyuetou*

After *songshen* at Longyuetou the Taoists visit each home in the village distributing talismans. At the same time things which symbolise impurity, e.g. feathers, beans, etc., are collected in a boat made of paper. This is called Pulling off the Duck's Feathers and Carrying off the Paper Boat (*laya pachuan 拉鸭扒船*). The putting of impure things in a small boat and sending them off to the other world is probably a ritual related to the King's Boat Festival (*wangchuan*). After the Taoists and several young men have collected the impure things in a boat, they are all gathered in one place and burnt. At Changzhou the Sending off of the Paper Boat takes place but no other ritual is performed.

In the above I have compared the *jiao* as performed at Changzhou and at Longyuetou. To summarise briefly, one can say that the *jiao* at Changzhou is idolatrous while that at Longyuetou is a ritual with Confucian features in line with ancestral worship.

In the following paragraphs I shall further discuss the ritual with Confucian features of the *jiao* at Longyuetou, expand on several examples and attempt an analysis.

B. The Characteristics of the *Jiao* in Cantonese Tradition

What we have described above is a three-day *jiao* at Longyuetou. However, when a village is large or when several villages co-operate, there is a wider territorial coverage and substantial financial strength. In such a case, the same Cantonese Taoists are invited and often a five-day *jiao* is performed. In such a *jiao*, the Confucian characteristics proper to Cantonese ritual seen in the three-day *jiao* at Longyuetou appear even more clearly. The following table shows the change in the structure of rituals when a three-day *jiao* is expanded to a five-day one. See Table 2.

TABLE 2 Order for a Five-Day *jiao* Compared with that for a Three-Day *jiao*

A Three-Day *jiao*		A Five-Day *jiao*	
The Day Before	Drawing Water (*qushui*) Inviting the Local Deities (*yingshen*) Setting Up the *jiao* Altar (*qitan*) Announcing to the Spirits (*fabiao*)	The Day Before	Drawing Water (*qushui*) Inviting the Local Deities (*yingshen*) Setting Up the *jiao* Altar (*qitan*) Announcing to the Spirits (*fabiao*)
The 1st Day	Three Audiences (*sanchao*) Lighting the New Fire (*fendeng*) Military Training (*dawu*) Purifying the *jiao* Tan (*jintan*)	The 1st Day	Three Audiences (*sanchao*) Lighting the New Fire (*fendeng*) Military Training (*dawu*) Purifying the *jiao* Tan (*jintan*)
		The 2nd Day	Three Audiences (*sanchao*) Feeding the Hungry Spirits on a Small Scale (*xiaoyou*)

The 2nd Day	Three Audiences (*sanchao*) Feeding the Hungry Spirits on a Small Scale (*xiaoyou*) Inviting the Jade Emperor (*yingsheng*)	The 3rd Day	Three Audiences (*sanchao*) Inviting the Jade Emperor (*yingsheng*)
		The 4th Day	Three Audiences (*sanchao*) Prayer to North Pole and the Great Dipper (*lidou*)
The 3rd Day	Three Audiences (*sanchao*) Carrying a Pardon from Heaven (*zou sheshu*) Freeing the Living Creatures (*fang sheng*) Feeding the Hungry Spirits on a Large Scale (*dayou*)	The 5th Day	Three Audiences (*sanchao*) Carrying a Pardon from Heaven (*zou sheshu*) Freeing the Living Creature (*fang sheng*) Feeding the Hungry Spirits on a Large Scale (*dayou*)
The Day After	Sending Off the Spirits (*songshen*) Carrying the Paper Boat (*pachuan*)	The Day After	Sending Off the Spirits (*songshen*) Carrying the Paper Boat (*pachuan*)

As is clear from the table, in a five-day *jiao*, the two rituals of Feeding the Hungry Ghosts on a Small Scale and the Invitation of the Jade Emperor were cramped into the second day schedule of a three-day *jiao*. Now, the Invitation of the Jade Emperor ritual is pushed on to the third day and the Prayer to the North Pole and the Great Dipper is scheduled on the fourth day. There is not much change in the basic structure of the ritual, they are more carefully performed. Thus, the characteristics of the Cantonese ritual become very clear. Now let us have a look at the following points.

(1) *Inviting the Local Deities (yingshen* 迎神), *Offering Incense Procession (xingxiang* 行香)
At Longyuetou, during the Invitation of the Local Deities and

the Procession for Offering Incense rituals, the earth gods in the small shrines all over the village are visited and incense sticks burnt, it takes half a day to complete. However, in a large village, or when there is a co-operation of villages, owing to the larger geographical area, very often many days are needed. One can say that the reason for the expansion of a *jiao* to a five-day festival is because this Procession for Incense takes time. E.g. at each of the five-day *jiao* at Jintian (Kam Tin 錦田), Xiacun (Ha Tsuen 厦村) and Shatian (Sha Tin 沙田), three days were taken up for the *xingxiang* procession.[7] At these villages besides the Taoists and *yuanshou*, dragon dance troupes made up of young men also accompanied the Procession.

One can interpret this procession not so much as a show of respect to the gods through dedication of incense but as a demonstration of organization which strengthens inter-village (or intra-village, in the case of single large village) solidarity. This is an important function of the *jiao* in Cantonese villages.

(2) *The jiao Tan : Ten Magistrates of Hell*

In the case of a five-day *jiao*, the size of the *jiao* Tan is also enlarged, and often Pictures of the Ten Magistrates of Hell are put up. This was so at Jintian, Xiancun, Lincun (Lam Tsuen 林村), Shatian (Sha Tin) and Taiheng (Tai Hang 泰亨).[8] At Jintian and Xiacun, besides the Pictures of the Ten Magistrates, Paper Courts of Ten Magistrates are also built. The Pictures of the Ten Magistrates which show the bureaucratic hierarchy in the netherworld are there to warn the villagers not to act immorally but to do good. Especially in these pictures there is a scene depicting the magistrate in hell imposing punishments on the protagonists of a love story in Zhejiang 浙江, Liang Shanbo 梁山伯 and Zhu Yingtai 祝英台 for the sin of immoral union. From this conservatism, which is the exact opposite of local legends praising love, one can say that the conservative spirit which stresses social order, is clearly shown in the Cantonese style *jiao*.

(3) *Public Presentation of the Memorial to Heaven*

As mentioned above this is a ritual to announce to Tiandi the names of the villagers and to pray for protection. However, when a village has a large population, the memorial becomes enormously long. It can be seen as something to show off to other villages the size of the village's population and its financial strength, etc. In the memorial at Lincun there were 6,000 names, at Xiacun 10,000 names and at Shatian 12,000 names. In the Cantonese villages, where inter-village rivalry can be rather serious, this memorial functions as a demonstration of one's own village's power, influence and organized strength. Like the procession mentioned above, in the *jiao* of the Cantonese Tradition the organizational function of the ritual is strongly emphasized.

(4) *Inviting the Jade Emperor*

In a five-day *jiao*, the *yinshen* becomes a grand affair. E.g. at a three-day *jiao*, during the invitation of the Jade Emperor, the memorial to Heaven is read very briefly. At a five-day *jiao*, however, the Taoist takes the original text of the memorial to Heaven and takes a long time to finish reading all the names of the villagers. At Jintian it took four hours, finishing at about 2:00 a.m. There was also a special ritual to pray to the Three Pure Ones. The Taoists change to new robes and arrange their skull-caps in front of a mirror. Then at the Heavenly Stair in front of the Three Pure Ones, the Taoists following the orders given by a Master of ceremonies kneel three times. At each kneeling the Taoists will knock the ground three times with the head. Thus a total of nine knocks on the ground (3 kneelings and 9 kowtows) are performed. This method, like the method of worshipping ancestors in ancestral halls is Confucian in style. One can see how strongly the *jiao* in the Cantonese tradition is influenced by Confucian ideas of community ritual in this *yingshen* ceremony.

(5) *Prayer to the North Pole and the Great Dipper (lidou* 禮 斗 *)*

Scheduled on the fourth day of a five-day *jiao* is the *lidou* ritual. This is a ceremony of prayers to the Seven Stars of the Northern Dipper and the Queen Star of the Dipper, who are believed to determine life-spans, and will hopefully prolong the life of the villagers. Hanging from the roof are several tens of lanterns each with the name of a different Star written on it. On the altar different shapes of Seven Stars of the Northern Dipper, Six Stars of the Southern Dipper and the Chinese character *shou* 壽 are arranged with many candle dishes.

A Taoist facing the altar reads the North Pole Canon. Having finished, the lanterns and candle dishes are distributed to the villagers. However, only the *yuanshou* or influential families in the village can receive it. Here again the ruling order of the village is emphasized. One can say that *lidou* also has a strong feature of ritual which functions to organize the community.

(6) *Patrol of the King of the Ghosts (dashi chuyou* 大 士 出 遊 *)*

During Feeding the Hungry Ghosts on a large scale, the image of the King of the Ghosts is carried outside and burnt. However, when the village is territorially large, the King of the Ghosts is carried to several corners of the village and then rituals in front of the altar are performed, after which it is burnt and dispatched. It is believed that the ghosts who lurk in the nooks and corners of the village can be exposed by the King of the Ghosts. This patrol is rather frequent in the five-day *jiao* (Jintian, Taiheng, etc.). The role of carrying the King of the Ghosts is distributed between groups of young men of different districts who organize themselves to take turns to carry the King of the Ghosts. One can say that this is one type of ritual which stresses the social bonds of the community.

As described above, when a three-day *jiao* is extended to five days, there is no marked increase in the reading of canons or in rituals at the altar. The increased time and schedule is found in *xingxiang, yingbang, yingshen, lidou, dashi chuyou*, etc. which give the villagers a chance to participate. In other words, although three days are sufficient for the Taoist rituals, there is a strong tendency to extend the celebration to five days to include rituals which strengthen community bonds. One can thus safely say that in the *jiao* in the Cantonese tradition, the factor of ritual evincing communal solidarity, i.e. ritual in the village community and ritual in the ancestral hall, is important.

C. Conclusion

If the above account is accurate, what is the historical background from which the factor of ritual in the ancestral hall and ritual in the village community developed in the *jiao* in Cantonese tradition? I wish to consider the following points and make some suggestions.

Firstly, the factor of the ritual in the ancestral hall is probably strong because the village community of Guangdong, in which this region is located, developed from what was basically a lineage. A large part of the villages in the New Territories are single-lineage villages where all members belong to the same lineage, while there are some multi-lineage villages where several lineages live together. Therefore each lineage has its own ancestral hall. Twice every year, in spring and autumn, members gather at the ancestral hall to worship their ancestors. During that ceremony, all the participants will, under the command of a master of ceremonies kneel three times and knock their head on the ground at every kneeling, a ritual inspired by Confucian ideas and style. The village elders who manage the *jiao* frequently visit their own ancestral hall to pay respect to the ancestors. Their closeness to Confucian Rites enables a change in the Taoist ritual in the *jiao* to something similar to *Confucian Ritualistic Style in an ancestral hall.*

The second point. Due to the influence of rituals which stress the strengthening of community bonds during the *jiao*, one can also think of a historical background: the administrative and tax systems of the villages during the Ming period. At that time the *lijia* system 里甲制 prevailed, in which one hundred households constituted one *li* 里 (administrative unit). In each *li* a *xianglitan* 鄉厲壇 (altar for hungry ghosts) was built, where the unattached wandering souls were cared for. The *xianglitan* festival is managed by a headman of the *li* or one of the village elders. And the rituals are probably like rituals in ancestral halls: rituals with a Confucian idea and style. The *lijia* system was abolished in mid-Ming and the *xianglitan* also deteriorated.[9] During the Qing Period, the Altars for the hungry ghosts in the local villages disappeared. Except for Guangdong, the *tujia* system 圖甲制 which succeeded the *lijai* system continued to exist until the Qing period.[10] And a system based on this previous one, the *baojia* system 保甲制 (system for local security) still exists today. The reason why many of the *jiao* in the Cantonese tradition are celebrated once in ten years, can be seen as an influence from the *lijia* system where the household register is compiled once every ten years. One can see the similarity between the villagers' names register in the memorial, the original text of the memorial to Heaven and the official household register. Also, at the *jiao* at Lincun, the *baojia* organisation for village security or the Ten Tablets System (*shipai* 十牌) of the Qing period is still functional today.[11] A *qitan* or *yingbang*, the exchange of wine cups and polite greetings between the Taoists and the *yangshou* can be seen as an influence from *xiang yin jiuli* 鄉飲酒禮 (the ritual banquet held by village elders) practised in villages since Ming and Qing. That is to say that the *jiao* in the Cantonese tradition held today in the villages of the New Territories of Hong Kong, has a distinctly Confucian style, when compared with the *jiao* in the Fukienese tradition seen in Changzhou and in Taiwan. This is because the Taoist rituals of the Heavenly Master Sect which entered Hong Kong were influenced by Confucian ritual which supported the local lineage community and village community.

NOTES

[1] For more on the *jiao* festival in Changzhou, see Tanaka Issei 田中一成, *Chugoku saishi engeki kenkyu* 中國祭祀演劇研究 [*Ritual Theatres in China*], (University of Tokyo Press, 1981), pp. 94-99 and Tanaka Issei, *Chugoku no sozoku to engeki* 中國の宗教之演劇 [*Lineage and Theatre in China*], (University of Tokyo Press, 1985), pp. 227-302.

[2] For details on the *jiao* festival in Longyuetou, see Tanaka Issei, *Chugoku no sozoku to engeki* , pp. 609-720.

3 At the *jiao* in Taiwan, several bamboo poles are erected in front of the temple. See Liu Chih-wan 劉枝萬 *Taipei shih Sung-shan chi-an chien-chiao ji-dian* 台北市松山祈安建醮祭典 *["Great Propitiatory Rites of Petition for Beneficence at Sungshan, Taipei"]* (Taipei: Institute of Ethnology, Academia Sinica, Monograph No. 14, 1987), p. 81.

4 At the *jiao* in Taiwan, the paper pavilion for the Three Worlds is also used. See Liu Chih-wan, *"Great Propitiatory Rites of Petition for Beneficence at Sungshan"*, pp. 106-108.

5 Michael R. Saso, *Taoism and the Rite of Cosmic Renewal* (Washington University Press, 1972), p. 62.

6 Here in rituals of the past, the envoy and Taoist exchange questions and answers. See Ofuchi Ninji 大淵忍爾 , *Chugokujin no syuko girei* 中國人の宗教儀禮 *[The Religion Rituals of the Chinese]*, (Tokyo: Fukutake Shoten, 1983), p. 750.

7 For more information on the *xingxiang* (procession) at Xiacun, see Tanaka Issei, *Chugoku no sozuko to engeki* , p. 216.

8 See Tanaka Issei, *Chugoku no sozuko to engeki* , pp 356-528 for more on the Lincun *jiao* and Pictures of the Ten Magistrates in Hell.

9 Wada Hironori 和田博德 "Rikosei to rishidadan. Kyoreidan — Mindai no kyoson shiha to saishi" 里甲制と里社壇──郷属壇 "Lijia System and Lishe Tan (the Local Deities' Altar), Xiangli Tan (the Hungry Ghosts' Altar) — Control System and Rites in Village Society during the Ming period" in *Nishi to Higashi* 西 と 東, と *("West and East"*), Tokyo, Kyuko shoyin, 1985.

10 Katayama Tsuyoshi 片山剛 "Shinmatsu Kanton-sho shuko derata no zuko-hyo to soreo meguru shomondai — zeiryo, koseiki, dozoku" 清未廣東省デルタの圖甲表 ("Some Problems Concerning the T'u-chia Charts in the Pearl River Delta in Kwangtung province during the Late Ch'ing Period: Land Tax, Household Registers and Lineage"), *Shigaku Zasshi* 91-4, 1982.

11 At the Lincun *jiao* evidence of the Six-jia system can still be seen. See Tanaka Issei, *Chugoku no Sozoku to Engeki* , pp. 375-376.

296

Fig. 1: Layout of the site of the *jiao* in Changzhou

1. Bei Di Temple
2. The Altar for the Three Pure Ones
3. The Altar for Local Deities
4. Paper Image of the Mountain Deity
5. Paper Image of the Earth God
6. Paper Image of the King of Ghosts
7. Matshed Theatre
8. The Heroes' Hall

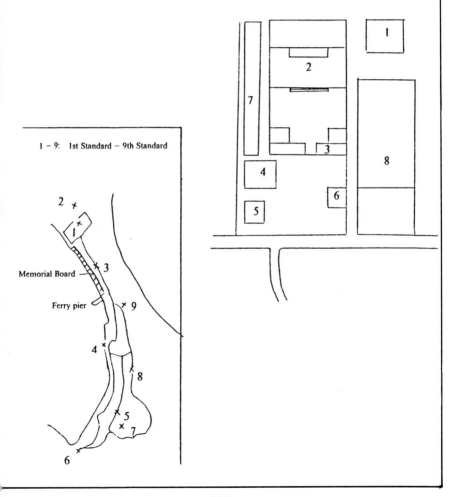

1 – 9: 1st Standard – 9th Standard

Memorial Board

Ferry pier

Fig. 2: Layout of the Site of the *jiao* in Longyuetou

1. Tianhou Temple
2. The Altar for the Three Pure Ones
3. The Altar for the Zhang Tianshi (Heavenly Master)
4. The Altar for Local Deities
5. The Paper Image of the King of Ghosts
6. Paper Image of the City God
7. The Kitchen for Vegetable Foods
8. Matshed Theatre

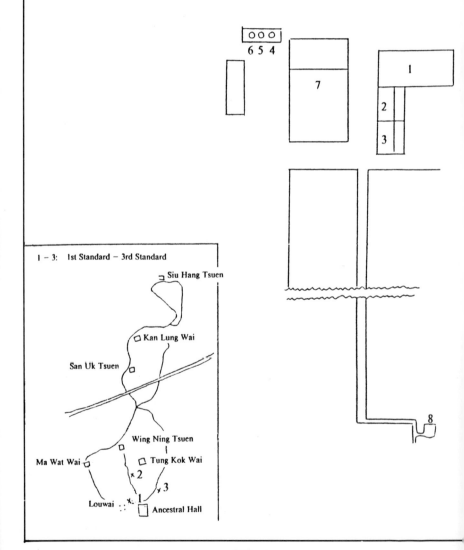

RECENT DEVELOPMENTS IN BUDDHISM
IN HONG KONG

Bartholomew P. M. Tsui

Introduction

The academic study of religion has not always been sufficiently sensitive to contemporary living religions. The reason for this is perhaps that the living specimen is usually so complex that it threatens to undermine the ideal picture which the academician may have formed of that religion from its scriptures. Buddhism in Hong Kong is such an example. There are altogether only two short articles written about some aspects of this religion.[1] This lack of attention is hardly justified when one considers the importance of certain developments in the past forty years or so. The Buddhist community has faced unprecedented challenges from several fronts, all brought about by the challenges from several directions, due to the unique circumstances of Hong Kong. One of these is the almost complete urbanization of the Territory. This had a great impact on the activities of the samgha, while the urban lifestyle indirectly created a chronic shortage of monks. Another factor is the presence of a substantial number of Christians within a basically Chinese population. A third factor is the proximity with China. For years, religion has practically been suppressed by the Communist government, and the Buddhist samgha in Hong Kong saw itself as one of the only two remnants of the Chinese samgha in the world. With the recent relaxation of control over religions in the mainland, monks in Hong Kong have eased their anxiety a little over the question of samgha recruitment, but they are kept in suspense by another problem: the future of Hong Kong after the year 1997 when Hong Kong will revert to China.

In the following I shall report on how Buddhism has faced up to these challenges, and in the process mention the priorities confronted by a living religion.

Factors Leading to Change

A convenient date to begin our story of accelerated pace of change is 1949, when the Communist regime took over the whole of China. Previous to that date, most Buddhist centres were scattered throughout isolated spots in the New Territories, where monks and nuns, mostly Cantonese, carried on their religious practices. After 1949 a large number of northern monks came with immigrants who tried to escape the Communist regime. For a time the number of monks was so great that they saturated the local temples. New monasteries were constructed and old ones enlarged. However, this situation did not last for long. Soon the monks discovered that they could easily emigrate to Taiwan, Southeast Asian countries, Australia and North America where Buddhism was gaining some acceptability. The recent uncertainty about Hong Kong's future has further accelerated emigration. According to an informed estimate, the present number of monks in Hong Kong cannot be more than 200.

This surge and decline in the number of monks coincided with a period of great population expansion, of urbanization and of unprecedented affluence. The resultant modern lifestyle is often blamed by the monks for the failure of recruitment into the samgha. Indeed, the lack of candidates for the monastery is only one effect of social change. The samgha itself is greatly affected by the wealth it is able to collect through donations and remunerations for the performance of rituals for the dead. With increased wealth many new temples are being built. There are altogether about 400 monasteries or places of worship. Such a disparity between the number of temples and the number of monks does not augur well for religious community life. The population increase and urbanization has the effect of reducing to zero the number of places of isolation within the Territory. Even the most isolated monastery, the Po Lin 寶蓮 on Lantao Island, is actively promoted as a tourist spot and is indeed invaded by numerous local and overseas visitors every day.

Paradoxically, industrialization and the adoption of a modern lifestyle do not lead to a decline in the need for spirituality, especially if the latter does not entail too great a sacrifice. For the majority of city people, life is as full of anxiety-creating factors as for the agricultural people. The general tone of the media such as the press and television is not critical of Chinese religious beliefs and customs. In fact, in their unexamined acceptance of Chinese beliefs, the media are actually promoting them. The number of lay worshippers has been on the increase. In order to impress the worshippers and visitors, temples have been greatly expanded and ornately decorated.

Temples are in demand for another reason. There is a policy of tight control over land use. A visitor to Hong Kong may see vast expanses of green mountains and hills, but as a general rule, no one may bury the dead anywhere except in designated cemeteries which are always near saturation point. This policy has driven the price of burial plots beyond the reach of ordinary mortals. One solution to this problem is to deposit the ashes or commemorative tablets of the dead within temples or pagodas. This arrangement has the added advantage of enabling the dead to share in the merits gained by prayers and rituals performed in the temple. The temples in turn charge a handsome fee for this practice.

Another factor which plays a part in the changes occurring in Buddhism is that it sees itself as a competitor with Christianity, which, although accounting for only about ten per cent of the population, exerts its presence far in excess of its numbers. The success of the Christians is thought to be most effective in education and social work. In order to prove their usefulness to society, Buddhists also build schools and develop various branches of social work. Of course, these activities are easily justified by the Mahayana ideal of rendering assistance to all sentient beings.

Finally, the changing policy of the Chinese government with regard to religion has a great impact on how the Buddhists perceive their future.

To sum up, the factors which led to changes in Buddhism were: 1. The fluctuating population of monks and nuns. 2. The urbanization of Hong Kong and the New Territories. 3. The increased popular demand for spirituality. 4. The substantial presence of Christianity. 5. The Chinese government policy towards religion.

The Monks

However much the lay Buddhists may have improved their lot within the Buddhist hierarchy, monastic life remains the essential element within Buddhism. Monks are the disciples who make full use of the means instituted by the Buddha. They are Buddhist par excellence. Hence, any change in the conception of monkhood is a serious change for Buddhism.

The full functioning of a monastic tradition can only be realized when there is a sufficient number of monks. Such a situation was found in the public monasteries in traditional China, and the daily life of the monks before 1949 is well described by Holmes Welch.[2] Although a few monasteries in Hong Kong are called public, it is doubtful whether full implementation of the monastic system has ever been carried out.[3] Nowadays, monks in any one monastery are so few in number that community life itself is nearly nonexistent, to say nothing of the implementation of traditional monastic life.

It is not entirely true to say that shortage of monks has caused the demise of monastic life. If the two hundred monks were willing to live in one monastery, they would make a rather sizeable community. It is rather a case of the monks' exposure to the unprecedented circumstances and opportunities offered by Hong Kong, which they have not learned to cope with. One of these unprecedented opportunities is wealth. I have already referred to the fact that temples are in demand to provide repositories for spirit tablets of the dead and that a handsome fee is collected. Just this reason alone is sufficient for the proliferation of temples. Secondly, it seems that there is nothing wrong in a monk aspiring to have his own *tao-ch'ang* (field of cultivation), which is an elegant way of referring to a temple. Hence when circumstances are favourable, each monk would aspire to become an accomplished master, each monk is an abbot in his own temple.

What are the effects of these changes? There are no longer common meditations in the meditation hall. Meditation has become a private affair left to the initiative of each individual monk. Morning and evening devotions appear to have become optional. These rituals are stipulated by the rules of most large centres but individual monks feel free to attend or not to attend.[4] I asked about the practice of *uposatha* , the bi-monthly confession of infringements of the rules. Most monks replied that this was suspended for lack of numbers. One monk at Tung Lam Nien Fah Tong 東林念佛堂 replied that it was carried out regularly. However, I could not find out what was done exactly. My impression was that it was little more than a formality.

In inverse proportion to traditional monastic practices are the rituals performed for the dead. With only 200 monks for a population of five and a half million, they are constantly called upon to perform lengthy funeral and commemorative rituals. One distinguished abbot complained that the involvement with this activity was detrimental to the spiritual health of the monastery.

The proliferation of temples mentioned earlier takes place in a society which is being progressively urbanized. The older monasteries find themselves gradually engulfed by buildings. Again, some new Buddhist centres are merely flats within a high-rise building. The result is that on the one hand, the quiet atmosphere of monastic life is shattered. On the other hand, monks come into greater contact with a mobile general population. It is only natural that sooner or later monks make a review of their purpose of life and the role they play in society.[5] However, even before the monks have thought about their role in society, Buddhism itself has been thinking about this question. In imitation of the Christians, Buddhists built schools, homes for the aged, hospitals and engaged in other forms of social work. So, when the monks came to reflect on their own role, they easily visualized themselves as wearing the shoes of teachers or of social workers. These sentiments were vocalized and brought to the surface in connection with the problem of the acute shortage of new recruits for the samgha. As the immigrant monks from China were getting older, they became increasingly worried that the lack of recruits could mean the extinction of Chinese monkhood. Buddhist monasteries in China, which had been the usual source of monks, were disbanded and could no longer provide monks. This situation triggered off a full scale self-examination. The monks asked themselves why monkhood was no longer appealing to young men. This led to an attempt to clarify their own role in which social involvement became the newly added element. The widespread recognition of the need of re-examination is shown by the almost universal response from the usually lethargic monks to a call for a samgha reformation voiced by an eminent monk in 1977.[6] Unfortunately, the fear of scandal smothered the flame for change, and the proposed conference to discuss reform never took place. This incident is a classic example of the inherent weakness of the Buddhist institution. Buddhist egalitarianism pays a high price for the lack of authority and a fear of change from a formalistic adherence to rules. The future shape of the Buddhist community in Hong Kong is far from clear. Since the debacle of the proposed reform, the monks have been unusually quiet. Are they following the adage "As long as porridge is served I may as well perform the Buddhist rite"? Recent relaxation of control over religions in China has provided a spark of hope for the local samgha. Perhaps young monks can again be imported into Hong Kong. More will be discussed below. However, the problem of the challenge of modern life, urbanization and the role of monkhood remains unsolved and cries out for resolution. If this is true, the samgha now faces a crucial transitional period in which any decision taken or omitted will have a profound effect on the future.

Temples

Over the centuries, various Chinese names have been coined in reference to Buddhist holy places of worship. These names may have originally reflected subtle differences in meaning, but they are lost when applied to Buddhist buildings in Hong Kong. In terms of function, Buddhist places of worship can be divided into two types. One type is the *tao-ch'ang*, especially set up so that a Buddhist master can claim the title of abbot. In this sense, this type of temple serves the need of a Buddhist monk. Otherwise, the elaborate images and other paraphernalia are not necessary to the needs of one monk. Temples of the other type serve the public more than anybody else. They are usually large and very ornate, and never at any stage of construction was their size considered in relation to the number of monks or nuns they hoped to serve. What is the point of building a temple which can house several hundreds of persons when there are at any time only a handful of monks and nuns? Temples serve the function to impress. They are intended to impress the visitor with the success of Buddhism. Again, building a temple is like building a pagoda. The larger and more grandiose the building is, the more merits the builders gain. As was noted by other scholars, Chinese temples, especially those of Taoism, also double as recreational spots. Many Buddhist temples in Hong Kong are only too eager to fulfill this function too. Summing up, public Buddhist temples in Hong Kong do not, and were never intended to, serve the needs of monks and nuns. They are rather the grand symbols of the presence of the Buddhist faith, and they are constructed in such a manner as to solicit this recognition.

Education and Social Work

While the monastic tradition has been subjected to the various challenges discussed above, Buddhism as a whole has been trying to project a new image in relation to the wider society. Although it is impossible to document the exact influence of Christianity in this regard, Buddhist involvement in education and social work must have followed the lead of the Christians. The Buddhists have not forgotten the success of Christian schools and hospitals in China.

The Buddhists are fortunate in being able to form a smoothly running Hong Kong Buddhist Association, a society consisting of dedicated laymen, monks and nuns. Formed in 1945, it has appropriated to itself all sorts of functions, from being a member of the World Fellowship of Buddhists to running a Buddhist cemetery. This organization had the vision of venturing into education and social work on a very large scale. Today it has about twenty-five schools of various levels under its care, with students numbering about thirty thousand. These make up about half the number of about fifty Buddhist schools in the Territory. It runs the only Buddhist hospital, seven homes for the aged and the sick, and three youth organizations. All these show that Buddhism has received generous financial support from the general population.

How successful is Buddhist education? Buddhism is taught as a subject in the middle schools, and students can sit for either one of two papers in the School Certificate Examinations. Thus, there is substantial amount of exposure to Buddhism for the students. One difficulty faced by the schools is to find suitable teachers. I think this situation is symptomatic of a deeper problem, the application of traditional Buddhist teaching to modern life. It is relatively easy to put up a school building. What is needed is merely sufficient capital. But to implement religious education, we need a model of an exemplary Buddhist in modern life. We also need to show how this exemplary modern Buddhist is guided by Buddhist principles. In Hong Kong, we do find considerable enthusiasm for the name of Buddhism. But probably very few people understand its essential elements, and still fewer put the teachings into practice. It is only after we have a living reality in modern life that we can translate it into relevant educational material.

The second difficulty faced by Buddhist education is the lack of suitable textbooks. Some efforts have been made in this direction but the result has been far from satisfactory. I think there has been a fundamental misunderstanding on the part of the writers of these textbooks as to the nature of educational material for middle schools. What is needed now is for the Buddhists to produce some educational specialists who may search for the correct formula in this matter.

Rituals

It has been widely observed by students of religion that there is nearly always a certain incongruity between the believers' explanation of their religion and the rituals they perform. Furthermore, the latter appear to be rather persistent and even autonomous, while in the former, there is always an attempt to adjust it to current trends of thought. What are the forces which account for the survival of the religion? The believer harbours no doubt that it is the superiority of the teaching. The historian of religion, on the other hand, cannot feel he has finished his job until he has also examined the unarticulated forces working under man's manifest actions. If this is true, many histories of religion are yet to be written. It may turn out that those rituals which are played down or even derided by certain enlightened believers are the reasons for the preservation of its historical existence. Whatever the outcome, the study of rituals cannot be ignored.

Buddhism in Hong Kong offers an example of the above paradox. When asked about rituals, believers invariably play down their importance, assigning them the function of mere external reminders. They say that the important thing in Buddhism is enlightenment or the understanding that "everything is a fabrication of the mind." Yet there is no sign of a decline in ritual devotion. Indeed, it seems that few opportunities for ritual celebration are ever missed. When there are not enough monks available to perform a calendrical festival such as the *Ullambana* (fifteenth day of the seventh month) in all the temples at once, the dates of celebration are varied so that the same monks may perform in several temples one after another.

In the major centres, the regular yearly celebrations are the six or seven birthdays of Buddhas and bodhisattvas, depending on the custom of individual temples, the commemoration of ancestors on the *Ch'ing Ming* festival, and the *Ullambana.* Each of these celebrations may last several days up to one week. The daily rituals consist of morning and evening prayers. Other celebrations are occasional ones. Those most frequently performed are rituals for the dead. Monks are also called upon to perform exorcism, to pray for peace and to pray for rain.

If these occasions for celebration are examined closely, one can see how Buddhism has bent itself to suit popular demands. The most frequently celebrated birthdays are those of Sakyamuni, Amitabha, Bhaisajyaguru (the Buddha of Healing), Mahasthamaprapta, Ksitigarbha, and Kuan-yin. Two other Kuan-yin festivals are also frequently celebrated. Nothing needs to be said about Sakyamuni, since he was the historical founder of Buddhism. Amitabha is the presiding Buddha of the Pure Land: a belief in his paradise is universally accepted by the Buddhist masses. Bhaisajyaguru is honoured for his willingness to heal the sick and to grant longevity; Mahasthamaprapta for his powerful help; and Ksitigarbha for deliverance from hell. Noted for being the most compassionate, Kuan-yin is the best loved divine figure among all the Buddhas and bodhisattvas, and her help extends to all circumstances.

The *Ch'ing Ming* festival is a purely Chinese feast in which deceased ancestors are commemorated. To show support for this Chinese custom, the Buddhists appropriated it to be their own. Likewise, the popularity of the *ullambana* festival is attributable to a legend concerning the filial piety of monk Mu-lien.

Thus, it is clear that the entire Buddhist calendar is geared towards the demands of the popular masses. Thus, it can be argued that Buddhism has been kept alive in Hong Kong not so much by its abstruse concepts of reality as by its support of popular religious demands.

One new addition to the body of Buddhist rituals is that of the marriage ceremony. It is difficult to conceive, given the Buddha's exhortation to go forth from home, how Buddhism could have anything positive to say about marriage or get involved in presiding over a marriage. For 2,500 years Buddhism had distanced itself from anything to do with marriage. Nowadays, Buddhist monks in Hong Kong have seen fit to invent a Buddhist marriage ritual. This is an implicit admission that Buddhism supports the established order of human life in *samsara*. It takes over the functions of the popular religion which seeks to sanctify momentous events in human life such as birth, maturity, marriage and death. In a sense then Buddhism has gone a full circle back to be identified with the Hindu dharma, upheld by Suddhodana, the Buddha's father and vehemently rejected by the Buddha.

Another innovation in Hong Kong is the short-term going-forth-from-home. This means that laymen and laywomen may taste the lifestyle of monks and nuns for a short period of time, at the end of which they go back to secular life. This practice is borrowed from the Buddhists in Thailand and has become very popular in Hong Kong. The monks, worried by the lack of recruitment, took this initiative in the hope that some might like the ascetic lifestyle and take on a permanent commitment.

When the contents of Buddhist rituals are examined, one is struck by the gulf between the practice of Buddhism and the philosophical understanding of Buddhist truth. Although there are individual differences according to the specific purpose of each ritual, nearly all Buddhist rituals have a similar structure. They normally contain the following parts: 1. Offering incense. 2. Addressing the names of Buddhas and bodhisattvas. 3. Recitation of sutras. 4. Recitation of mantras or dharanis. 5. Chanting the names of Buddhas. 6. Application of merit to specified beings.

What is striking in a Buddhist ritual is the overall dominance of the idea of merit, the power of recitation and the supreme importance of magical words, the mantras and dharanis. The mere calling of the names of Buddhas and bodhisattvas is meritorious. Sutras are chanted, not so much because of the meaning of the content but because the chanting of a powerful sutra brings powerful merits. Again sutras are usually accompanied by mantras and dharanis. The recitation of these magical words brings along all the fruits of the sutra. Then, there is usually a lengthy dharani or mantra specific to each ritual. These are supposedly transliterations of Sanskrit into Chinese. They are sounds whose meaning should not be investigated by any person. They are most powerful and extremely effective. They bring the ultimate fruit of Buddhism: instant enlightenment. There is also a universalistic aspect of Buddhist rituals. The merits gained are shared by all in imitation of the bodhisattva's compassion. At the evening ritual, one indispensable part is the feeding of the hungry ghosts. This part is performed even when the entire evening ritual is cut down to just ten minutes.

Lay Activities

One of the successful aspects of Buddhism in Hong Kong is the active participation of lay believers in Buddhist affairs. Without the assistance of lay Buddhists in the Buddhist Association, it is doubtful whether expansion into education and social work could have got off the ground. Thanks to a long association with Chinese culture, Buddhism is able to become a focus of spiritual yearning for a large number of people. Whether the full implication of Buddhism as a religion is understood or not is beside the point. What is important is that Buddhism successfully projects itself as containing profound wisdom. Again, Buddhism can be attractive to different people in different ways. To the less sophisticated, its simple doctrine, its beautiful rituals, and miraculous stories are satisfying. To those disillusioned with Western values, Buddhism is believed to be a spiritual haven. For the academically inclined, Buddhist abstruse treatises are reputed to be most profound. Consequently, lay participation is most active in rituals and in lectures. Advertisements of classes on Buddhism are frequently found in Buddhist magazines and newspapers.

Hong Kong Buddhists and China

There is no need to emphasize the fact that any policy China adopts with regard to Buddhism will have a great impact in Hong Kong. For years, organized Buddhism has practically been nonexistent in China, so that Hong Kong Buddhists have been left to develop on their own. Now that China has adopted a more lenient policy and allows institutionalized religions to function again, it is interesting to see how events in China have afffected Hong Kong Buddhists.

The first reaction of Hong Kong Buddhists to the re-emergence of Buddhist institutions in China was a feeling of delight. As explained to me by a prominent Buddhist, any revival of Buddhism is a positive step towards improvement, for there was absolutely nothing at all before. What are the examples of development which have delighted Hong Kong Buddhists? The most important thing is, of course, the recruiting of a new body of monks and nuns, and the opening of centres for the training of these aspirants. In Beijing, there is the Fa-yuan Szu 法 源 寺 where there are a few dozens of young novices. Sixty learners are found in the Shanghai Buddhist Institute (上 海 佛 學 院). Classes for nuns are found in Tz'u-hsiu an 慈 修 庵 . Twenty sramanera are found in Lung-hua Szu 龍 華 寺 . The Buddhist Institute in Fo Ch'ien initiated classes for nuns. Sichuan also has an institute for nuns with forty students. Some of these are trained for missionary purposes.

The rapid expansion of these institutions has put a great demand on a handful of survivors of past persecutions, and monks from Hong Kong have sometimes been asked to deliver lectures. Some Hong Kong monks did respond to this request and gave lectures for short periods. In fact, visits to China by Hong Kong monks have become easy and frequent. Visits in the opposite direction are more difficult to arrange. However, there have been courtesy visits. For example, there was the visit of the representatives of the Shanghai Buddhist Association. There was the visit of Master Hai-teng. Chinese Buddhists have also made a donation of a copy of the Lung Tsang 龍 藏 , a very rare edition of the Tripitaka, to Po Lin Temple. Thus, it can be seen that Buddhists on both sides have been friendly. However, the Buddhist Associations are separate, and there is no plan for any merger.

The developments in China have some psychological effects on Chinese monks in the free world. The pressing problem of samgha recruitment mentioned earlier is taking a breather. This sentiment was expressed by Master Ming-chih 敏智 when he visited Fa-yuan szu in Beijing.[7] On seeing the several dozens of novices, he was greatly pleased and said that the problem of finding successors to the samgha was now solved.

One Hong Kong monk tried to arrange for a young monk to come to Hong Kong. However, immigration control is strict and it is not known whether the application will be successful. Even if it were possible to arrange for immigration of monks, some Buddhists in Hong Kong have doubts about the wisdom of such arrangements. They feel that the Hong Kong environment is such that it would be extremely difficult for young monks from China to cope with it. Furthermore, no matter how fast Buddhism has been developing in China recently, Hong Kong Buddhists, many of whom have fled from a more repressive era, have no illusion that it is not entirely free from government control, and this is detrimental to the cause of Buddhism. This same attitude is revealed in another question: the future of Hong Kong after the year 1997.

Buddhist monks and nuns are worried not only whether, under the future government, they will be allowed to carry on their work in schools and other social works or not, but also whether the rather muddled question of property ownership will be changed or not. Right now, two Buddhist representatives are sitting on the Basic Law Drafting Committee and they are working towards ensuring the continuity of the freedom which the various religions are enjoying in Hong Kong.[8]

Conclusion

If it is felt necessary to link up the diverse topics discussed in this paper, the unifying element is that they are all aspects of the Buddhist response to modern city life and to particular circumstances. The key factors precipitating changes are the fluctuating numbers of monks, the population expansion and urbanization of the Territory, the increased popular demand for spirituality, the substantial presence of Christians in a Chinese society and the changing Chinese policy towards religion.

Some of the remarkable developments in the last forty years or so have been the building and running of Buddhist schools and colleges, the organization of charitable and social works, and a strong lay Buddhist interest and involvement.

This period is also a time of great transitions. The decline of the monastic population, the great increase in wealth and the proliferation of temples has led to the disappearance of traditional monastic life. This necessitates a search for a new lifestyle and an identity of the monk's role in society, a process which is as yet unfinished.

Our investigation also shows that the demands for popular Buddhism are undiminished. An analysis of the rituals indicates that the driving force of Buddhism may largely be due to a belief in their effectiveness. The ritual structure also shows the extent to which Buddhism has accommodated itself to native religious instincts. The function of the temple has also been delineated: they do not serve the needs of monks and nuns, but rather serve as symbols of the faith's presence within the Territory.

Our review of the development of Buddhism in Hong Kong may have a wider application. It concerns the assessment of Buddhism as a religion which has manifested itself in a history of 2,500 years. The true nature of this historical religion should not only be sought in the explicit accounts of scriptures and philosophical treatises, but also in its practical institutions which provide its driving force and which probably account for its long survival.

310

NOTES

1 Holmes Welch, "Buddhist Organizations in Hong Kong", *Journal of the Royal Asiatic Society, Hong Kong Branch* (1960-61), 98-114; and Bartholomew Tsui, "Self-Perception of Buddhist Monks in Hong Kong Today," *Journal of the Royal Asiatic Society, Hong Kong Branch* (1983), 23-40.

2 Holmes Welch, *The Practice of Chinese Buddhism 1900-1950.* Cambridge: Harvard University Press, 1967.

3 One logistical difficulty of full implementation is that monasteries in Hong Kong have the peculiar custom of housing both monks and nuns in the same monastery.

4 I witnessed two evening rituals in July 1986: one at Po Lin Temple in which seven monks and one nun attended. The monk with whom I was having an interview that day did not attend. The other one was at Tung Lan Nien Fah Tong in which two monks alone performed the ritual.

5 A more detailed account can be found in my article mentioned in note 1.

6 See especially the relevant articles found in the *Hong Kong Buddhist Journal* (香 港 佛 教 in Chinese), vols. 206, 208, 209. The last volume is almost entirely dedicated to the question of samgha reformation.

7 *Hong Kong Buddhist Journal*, vol. 310 (1986), p. 24.

8 *Ibid*., vol. 302 (1985), p. 37.

BIBLIOGRAPHY

As a service to our readers, I thought it was useful to attach two lists of bibliographic references: the first lists books and articles in Western languages, published after 1980; the second contains books and articles published by Chinese scholars from the People's Republic of China from 1977 to 1987.

The listed titles cover all areas of research within the field of Chinese religion, but only a few deal directly with the current situation of religion in China.

While the first catalogue merely lists the titles in alphabetical order of the author, the second one has categorized the titles into topical sections. I realize that this is not an ideal method, but if we wait for perfection, we would have to sacrifice two or three years.

I am grateful to the two compilers, Professors Alvin Cohen and Man Kam Leung, who sacrificed many days of summer to prepare their bibliographies. But I am sure many readers' appreciation will compensate their efforts.

The Editor

WESTERN LANGUAGE PUBLICATIONS ON CHINESE RELIGIONS, 1981-1987

Compiled by Alvin P. Cohen

Laurence G. Thompson's *Chinese Religion in Western Languages: A Comprehensive and Classified Bibliography of Publications in English, French and German through 1980* (Tucson: University of Arizona Press, 1985) was a great milestone in the study of Chinese religions. For the first time, a complete bibliography of English, French and German writings on Chinese religions was assembled into a single research source. The entire field of Chinese religious studies is indebted to Professor Thompson for this extraordinary feat of bibliographic research.

The following is an attempt to supplement *Chinese Religion in Western Languages* in order to bring it up to the present time. Due to limitations of time and library resources, I cannot possibly prepare a research tool anywhere near as comprehensive nor as cogently organized as Professor Thompson's bibliography. I therefore present merely a list of articles from a small number of common journals, monographs and collections, plus many references (which I could not verify) taken from published bibliographies and footnotes. A number of substantial articles in the recent *Encyclopedia of Religion* are also included. Although it is very incomplete, I hope it will be of some use until Professor Thompson, or another bibliographer of equal quality, provides us with the next instalment of *Chinese Religion in Western Languages* .

Abbreviations of Journal Titles and Collections

AA	*Arts of Asia* (Hong Kong).
AFS	*Asian Folklore Studies.*
BEFEO	*Bulletin de l'École Française d'Extrême-Orient.*
BIE	*Bulletin of the Institute of Ethnology, Academia Sinica* (Taipei).
BSOAS	*Bulletin of the School of Oriental and African Studies.*
BSSCR	*Bulletin of the Society for the Study of Chinese Religions* (starting from 1982 called *Journal of Chinese Religions*).
BTS	*Buddhist and Taoist Studies II* (see Chappell, 1987).

Chappell (1987)	Chappell, David W., ed. *Buddhist and Taoist Practice in Medieval Chinese Society: Buddhist and Taoist Studies II.*
Clammer (1983)	Clammer, John R., ed. *Studies in Chinese Folk Religion in Singapore and Malaysia.*
De Vos (1984)	De Vos, George A., and Takao Sofue, eds. *Religion and the Family in East Asia.*
Ebrey (1981)	Ebrey, Patricia Buckley, ed. *Chinese Civilization and Society: A Sourcebook.*
ER	*The Encyclopedia of Religion.* Ed. Mircea Eliade et al. New York: Macmillan Publishing Co., 1987.
Gimello (1983)	Gimello, Robert M., and Peter N. Gregory, eds. *Studies in Ch'an and Hua-yen.*
Gregory (1986)	Gregory, Peter N., ed. *Traditions of Meditation in Chinese Buddhism.*
Gregory (1987)	Gregory, Peter N., ed. *Sudden and Gradual Approaches to Enlightenment in Chinese Thought.*
HJAS	*Harvard Journal of Asiatic Studies.*
HR	*History of Religions.*
Hsieh (1985)	Hsieh Jih-chang and Chuang Ying-chang, eds. *The Chinese Family and Its Ritual Behavior.*
JA	*Journal Asiatique.*
JAAR	*Journal of the American Academy of Religion.*
JAOS	*Journal of the American Oriental Society.*
JAS	*Journal of Asian Studies.*
JCP	*Journal of Chinese Philosophy.*
JCR	*Journal of Chinese Religions.*
JHKBRAS	*Journal of the Hong Kong Branch of the Royal Asiatic Society.*
JIABS	*Journal of the International Association of Buddhist Studies.*
Johnson (1985)	Johnson, David, et al., eds. *Popular Culture in Late Imperial China.*
JRAS	*Journal of the Royal Asiatic Society* (London).
Naundorf (1985)	Naundorf, Gert, et al., eds. *Religion und Philosophie in Ostasien*
PEW	*Philosophy East and West.*
Rosemont (1984)	Rosemont, Henry, Jr., ed. *Explorations in Early Chinese Cosmology*

Strickmann (1981-83) Strickmann, Michel, ed. *Tantric and Taoist Studies in Honour of R.A. Stein.* 2 vols. (Melanges chinois et bouddhiques, 20-21).
TP *T'oung Pao.*

Listings

Ackerman, Susan E., and Raymond L. M. Lee. "Pray to the Heavenly Father: A Chinese New Religion in Malaysia." *Numen* 29 (1982), 62-77.

Adelmann, Frederick J., ed. *Contemporary Chinese Philosophy.* The Hague: Martinus Nijhoff, 1982.

Ahern, Emily M. *Chinese Ritual and Politics.* Cambridge: Cambridge University Press, 1981.

Ahern, Emily M. "The Thai Ti Kong Festival." In *The Anthropology of Taiwanese Society.* Eds. Emily M. Ahern and Hill Gates (Stanford: Stanford University Press, 1981), pp. 397-425, 469-470.

Allan, Sarah. *The Heir and the Sage: Dynastic Legend in Early China.* San Francisco: Chinese Materials Center, 1981.

Allan, Sarah. "Sons of Suns: Myth and Totemism in Early China." *BSOAS* 44 (1981), 290-326.

Allen, Joseph Roe, III. "The Myth Studies of Wen I-to: A Question of Methodology." *Tamkang Review* 13 (1982), 137-160.

Anagnost, Ann S. "Polities and Religion in Contemporary China." *Modern China* 13 (1987), 40-62.

Asian Comparative Collection Newsletter (University of Idaho, Moscow, Idaho). Edited by Priscilla Wegars. Vol. 1 no. 1 (March 1984).

Aune, David E. "Oracles." *ER* XI, 81-87.

Baker, Hugh D. R. "Ancestral Images: A Bibliographical Note" *JHKBRAS* 23 (1983), 221-232.

Baldrian-Hussein, Farzeen. "Lu Tung-pin in Northern Sung Literature" *Cahiers d'Extrême-Asie* 2(1986), 133-169.

Baldrian-Hussein, Farzeen. *Procédés Secrets du Joyau Magique: Traité d'Alchimie Taoiste du XIe siècle.* Paris: Les Deux Oceans, 1984.

Baldrian-Hussein, Farzeen. "Taoism: An Overview." *ER* XIV, 288-306.

Baldrian-Hussein, Farzeen. "Yueh-yang and Lu Tung-pin's *Ch'in-yuan ch'un*: A Sung Alchemical Poem." In Naundorf (1985), pp. 19-31.

Banck, Werner. *Das Chinesische Tempelorakel, Teil II: Ubersetzung und Analysen.* (Asiatische Forschungen, Band 90). Wiesbaden: Harrassowitz, 1985.

Barnes, Nancy J. "Women in Buddhism." In *Women in World Religions*. Ed. Arvind Sharma (Albany: State University of New York Press, 1987).

Barrett, T.H. "Taoist and Buddhist Mysteries in the Interpretation of the *Tao-te-ching.*" *JRAS* 1 (1982), 35-43.

Barrett, T.H. "Taoism: History of Study." *ER XIV,* 329-332.

Bays, Daniel H. "Popular Religious Movements in China and the United States in the Nineteenth Century." *Fides et Historia* 15 (1982), 24-38.

Bechert, Heinz. "Samgha: An Overview." *ER* XIII, 36-40.

Bechert, Heinz. "Etienne Lamotte (1903-1983)" *JIABS* 88 (1985), 151-156.

Beh, Leo Juat, and John Clammer. "Confucianism as Folk Religion in Singapore: A Note." In Clammer (1983), pp. 175-178.

Benn, Charles. "Religious Aspects of Emperor Hsuan-tsung's Taoist Ideology" in Chappell (1987), 127-145.

Bergeron, Marie-Ina. *Wang-Pi, Philosophe du "non-avoir ".* (Varietes Sinologiques, new series, no.69) Taipei and Paris: Ricci Institute, 1986.

Berling, Judith A. "Religion and Popular Culture: The Management of Moral Capital in *The Romance of the Three Teachings .*" In Johnson (1985), pp. 188-218.

Berthrong, John H. "The Problem of Mind: Mou Tsung-san's Critique of Chu Hsi." *JCR* 10 (1982), 39-52.

A Bibliography of Bibliographies of Asian Religious Studies. Stony Brook, NY: the Institute for Advanced Studies of World Religions, 1986.

Bielefeldt, Carl. "Ch'ang-lu Tsung-tse's *Tso-Ch'an I* and the 'Secret' of Zen Meditation." In Gregory (1986), 129-161.

Birnbaum, Raoul. "Avalokitesvara." *ER* II, 11-14.

Birnbaum, Raoul. "Buddhist Meditation Teachings and the Birth of 'Pure' Landscape Painting in China." *BSSCR* 9 (1981), 42-58.

Birnbaum, Raoul. "Seeking Longevity in Chinese Buddhism: Long Life Deities and Their Symbolism." *JCR* 13-14 (1985-86), 143-176.

Birnbaum, Raoul. "Thoughts on T'ang Buddhist Mountain Traditions and Their Context." *T'ang Studies* 2 (1984), 5-23.

Bishop, Donald H. "Confucianism in Contemporary Taiwan." *Chinese Culture* 24 (1983), 69-83.

Bock, Felicia G., trans. *Classical Learning and Taoist Practices in Early Japan. With Translation of Books XVI and XX of the Engi-Shiki.* Tempe: Center for Asian Studies, Arizona State University, 1985.

Bohr, P. Richard. "The Heavenly Kingdom in China: Religion and the Taiping Revolution, 1837-1853." *Fides et Historia* 17 (1985), 38-52.

Bokenkamp, Stephen R. "Sources of the Ling-pao Scriptures." In Strickmann (1983), 434-486

Bokenkamp, Stephen R. "The 'Peach Flower Font' and the Grotto Passage." *JAOS* 106 (1986), 65-77.

Boltz, Judith M. "Opening the Gates of Purgatory: A Twelfth-century Taoist Meditation Technique for the Salvation of Lost Souls." In Strickmann (1983), 487-511.

Boltz, Judith M. "In Homage to T'ien-fei." *JAOS* 106 (1986), 211-232.

Boltz, Judith M. "Lao-tzu." ER VIII, 454-459.

Boltz, Judith M. *A Survey of Taoist Literature, Tenth to Seventeenth Centuries.* (China Research Monographs, no. 32). Berkeley: Institute of East Asian Studies, University of California, 1987.

Boltz, Judith M. "Taoism: Taoist Literature." *ER* XIV, 317-329.

Bond, George D. "The Development and Elaboration of the Arahant Ideal in the Theravada Buddhist Tradition." *JAAR* 52 (1984), 227-242.

Brandon, James R. "Drama: East Asian Dance and Theater." *ER* IV, 459-462.

Brereton, Joel P. "Lotus." *ER* IX, 28-31.

Broman, Sven, *Chinese Shadow Theatre*. (Ethnographical Museum of Sweden, Monograph Series, No. 15). Stockholm: Ethnographical Museum of Sweden, 1981.

Broughton, Jeffrey. "Early Ch'an Schools in Tibet." In Gimello (1983), 1-68.

Brown, Chappel. "Inner Truth and the Origin of the Yarrow Stalk Oracle". *JCP* 9 (1982), 190-211.

Buswell, Robert E., Jr. "Chinul's Systemization of Chinese Meditative Techniques in Korean Son Buddhism." In Gregory (1986), 199-242.

Buswell, Robert E., Jr. "The 'Short-cut' Approach of *K'an-hua* Meditation: The Evolution of a Practical Subitism in Chinese Ch'an Buddhism." In Gregory (1987), 321-377.

Cabezon, Jose I. "The Concepts of Truth and Meaning in the Buddhist Scriptures." *JIABS* 4 (1981), 7-23.

Cahill, James, "Tung Ch'i-ch'ang's 'Southern and Northern Schools' in the History and Theory of Painting: A Reconsideration." In Gregory (1987), 429-456.

Cahill, Suzanne. "Beside the Turquoise Pond: The Shrine of the Queen Mother of the West in Medieval Chinese Poetry and

Religious Practice." *JCR* 12 (1984), 19-32.

Cahill, Suzanne. "Sex and the Supernatural in Medieval China: Cantos on the Transcendent who Presides over the River." *JAOS* 105 (1985), 197-220.

Cahill, Suzanne. "Performers and Female Taoist Adepts: Hsi Wang Mu as the Patron Deity of Women in Medieval China." *JAOS* 106 (1986), 155-168.

Cahill, Suzanne. "Reflections of a Metal Mother, Tu Kuang-t'ing's Biography of Hsi Wang Mu." *JCR* 13-14 (1985-86), 127-142.

Caldarola, Carlo, ed. *Religions and Societies : Asia and the Middle East* . Berlin: Walter de Gruyter, 1982.

Cammann, Schuyler. "Some Early Chinese Symbols of Duality." *HR* 24 (1985), 215-254.

Cass, Victoria B. "Female Healers in the Ming and the Lodge of Ritual and Ceremony." *JAOS* 106 (1986), 233-240.

Cedzich, Ursula-Angelika. "Wu-t'ung. Zur bewegten Geschichte eines Kultes." In Naundorf (1985), 33-60.

Chan, Hok-Lam, and William Theodore de Bary, eds. *Yuan Thought: Chinese Thought and Religion Under the Mongols* . New York: Columbia University Press, 1982.

Chan, Sin-wai. "Buddhism and the late Ch'ing Intellectuals". *Journal of the Institute of Chinese Studies of the Chinese University of Hong Kong* 16 (1985), 97-109.

Chan, Wing-tsit, Trans. and Ed. *New-Confucian Terms Explained* (The *Pei-hsi tzu-i* by Ch'en Ch'un (1159-1223)). New York: Columbia University Press, 1986.

Chan, Wing-tsit. "Chinese Religion: Religious and Philosophical Texts." *ER* III, 305-312.

Chan, Wing-tsit. "Confucian Thought: Foundations of the Tradition." *ER* IV, 15-24.

Chan, Wing-tsit. "Confucian Thought: Neo-Confucianism." *ER* IV, 24-36.

Chang, Chi-yun. "Confucius' Religious Philosophy." *Chinese Culture* 23 (1982), 39-63.

Chang, Garma C.C. General Editor. *A Treasury of Mahayana Sutras. Selections from the Maharatnakuta Sutra* . Translated from the Chinese by the Buddhist Association of the United States. University Park & London: Pennsylvania State Univ. Press, 1983. (Published in cooperation with IASWR, New York).

Chang, K.C., *Art, Myth, and Ritual: The Path to Political Authority in Ancient China*. Cambridge, MA: Harvard University Press, 1983.

Chao, Paul. "The Chinese Natural Religion: Confucianism and Taoism." *Chinese Culture* 24 (1983), 1-14.

Chappell, David W., ed. *T'ien-t'ai Buddhism. An Outline of the Fourfold Teachings.* Tokyo: Daiichi-Shobo (distributed by University of Hawaii Press), 1983.

Chappell, David W. "From Dispute to Dual Cultivation: Pure Land Responses to Ch'an Critics." In Gregory (1986), 163-197.

Chappell, David W., ed. *Buddhist and Taoist Practice in Medieval Chinese Society: Buddhist and Taoist Studies II.* (Asian Studies at Hawaii, no. 34). Honolulu: University of Hawaii Press, 1987.

Chappell, David W. "Ching-t'u." *ER* III, 329-333.

Che, Choong Ket. "Chinese Divination: An Ethnographic Case Study." In Clammer (1983), 49-97.

Ch'en, Kenneth K.S. "Religious Changes in Communist China." In *Chinese Political Science.* Eds. Sah Mong-wu, et al. (Taipei: China Academy, 1980), 347-353.

Cheng, Hsueh-Li. "The Roots of Zen Buddhism." *JCP* 8 (1981), 451-478.

Ch'ien, Edward T. "The Concept of Language and the Use of Paradox in Buddhism and Taoism." *JCP* 11 (1984), 375-399.

Ch'ien, Edward T. "The Neo-Confucian Confrontation with Buddhism: A Structural and Historical Analysis." *JCP* 9 (1982), 307-328.

Ching, Julia. "Confucius." *ER* IV, 38-42.

Chiu, Milton M. "Taiwanese Religions." *ER* XIV, 252-255.

Chiu, Milton M. *The Tao of Chinese Religion.* Lanham, MD: University Press of America, 1984.

Chu, Hai-yuan. "The Impact of Different Religions on the Chinese Family in Taiwan." In Hsieh (1985), 221-231, 302-303.

Clammer, John R., ed. *Studies in Chinese Folk Religion in Singapore and Malaysia.* (Contributions to Southeast Asian Ethnography, no.2). Singapore: Department of Sociology, National University of Singapore, 1983.

Cleary, J.C. *Zen Dawn: Early Zen Texts from Tun Huang.* Boston: Shambala, 1986.

Cleary, Thomas, *Entry into the Inconceivable: An Introduction to Hua-yen Buddhism.* Honolulu: University of Hawaii Press, 1983.

Cohen, Alvin P. "Chinese Religion: Popular Religion," *ER* III, 289-296.

Cohen, Alvin P. "Completing the Business of Life: The Vengeful Dead in Chinese Folk Religion." In *Folk Culture, 2, Folkways in Religion, Gods, Spirits, and Men.* (Cuttack, India: Institute of Oriental and Orissan Studies, 1983), 59-66.

Cohen, Alvin P. "A Taiwanese Puppeteer and His Theatre." *AFS* 40 (1981), 33-49.

Cohen, Alvin P., trans. *Tales of Vengeful Souls: A Sixth Century Collection of Chinese Avenging Ghost Stories*. (Varietes Sinologiques, n.s. 68) Paris & Taipei: Institut Ricci, Centre d'Etudes Chinoises, 1982.

Collcutt, Martin. "Monasticism: Buddhist Monasticism." *ER* X, 41-44.

Collins, Steven. "Soul: Buddhist Concepts." *ER* XIII, 443-447.

Company, Robert. "Cosmogony and Self-cultivation: Demonic and the Ethical in Two Chinese Novels." *Journal of Religious Ethics* 14 (1986), 81-112.

Corless, Roger J. "T'an-luan: Taoist Sage and Buddhist Bodhisattva" in Chappell (1987), 36-45.

Covell, Ralph R., ed. "Religion in China." *Missiology* 11 (1983), 258-360.

Coyle, Mark, trans. "Book of Rewards and Punishments." In Ebrey (1981), 71-74.

Cummings, Mary. *The Lives of the Buddha in the Art and Literature of Asia*. (Michigan Papers on South and Southeast Asia, no. 20). Ann Arbor: Center for South and Southeast Asian Studies, University of Michigan, 1982.

Dalia, Albert A. "The 'Political Career' of the Buddhist Historian Tsan-ning" in Chappell (1987), 146-180.

Davis-Friedmann, Deborah, *Long Lives: Chinese Elderly and the Communist Revolution*. (Harvard East Asian Series, 100). Cambridge, MA: Harvard University Press, 1983. Pp. 60-70, "Funerals and Filial Piety".

Dean, Kenneth. "Field Notes on Two Taoist *Jiao* Observed in Zhangzhou, December 1985." *Cahiers d'Extreme-Asie* 2 (1986).

Debon, Gunther. "Randbemerkungen zum sechsten Kapitel des *Chuang-tzu* ." In Naundorf (1985), 401-413.

Demiéville, Paul, *Buddhism and Healing: Demieville's Article 'Byo' from Hóbógirin* . Trans. Mark Tatz. Landham, MD: University Press of America, 1985.

Demiéville, Paul. "The Mirror of the Mind." In Gregory (1987), 13-40.

Despeux, Catherine, "Les Lectures Alchimiques du *Hsi-yu chi.*" In Naundorf (1985), 61-75.

De Vos, George A., and Takao Sofue, eds. *Religion and the Family in East Asia.* Berkeley : University of California Press, 1984.

De Woskin, Kenneth J. *Doctors, Diviners and Magicians of Ancient China : Biographies of Fang-shih.* New York : Columbia

University Press, 1983.

De Woskin, Kenneth J. "A Source Guide to the Lives and Techniques of Han and Six Dynasties *Fang-shih.*" *BSSCR* 9 (1981), 79-105.

Donner, Neal. "Chih-i's Meditation on Evil" in Chappell (1987), 49-64.

Donner, Neal. "Sudden and Gradual Intimately Conjoined: Chih-i's T'ien-t'ai View." In Gregory (1987), 201-226.

Dove, Victor. "Temples, Tombs and Gardens in Szechwan" *AA*, 15 (1985), #3, 72-79.

Dumoulin, Heinrich. "Ch'an." *ER* III, 184-192.

Eberhard, Wolfram. *A Dictionary of Chinese Symbols : Hidden Symbols in Chinese Life and Thought.* Trans. from German by G.L Campbell. London : Routledge & Kegan Paul, 1986 (English trans. of *Lexikon* ... Köln , 1983).

Eberhard, Wolfram. *Dictionnaire des Symboles Chinois.* Paris : Editions Seghers, 1984 (French trans. of *Lexikon* ... Köln, 1983).

Eberhard, Wolfram. *Lexikon chinesischer Symbole : Geheime Sinnbilder in Kunst und Literatur, Leben und Denken der Chinesen.* Koln: Eugen Diederichs Verlag, 1983.

Ebert, Jorinde, Barbara M. Kaulbach, and Martin Kraatz. *Religiose Malerei aus Taiwan.* (Veroffentlichungen der Religionskundlichen Sammlung der Philipps-Universitat Marburg, Nr. 1). Koln: Museum fur Ostasiatische Kunst, 1981. (Catalog of exhibition, Oct. 12 to Nov. 23, 1980, at Universitatsmuseum fur Kunst und Kulturgeschichte, Marburg).

Ebrey, Patricia Buckley, ed. *Chinese Civilization and Society: a Sourcebook.* New York, The Free Press, 1981.

Ebrey, Patricia Buckley, trans. "Exhortations on Ceremony and Deference." In Ebrey (1981), 204-207.

Ebrey, Patricia Buckley, trans. "Local Cults." In Ebrey (1981), 38-40.

Eichhorn, Werner. "Das *Tung-ming chi* des Kuo Hsien." In Naundorf (1985), 291-300.

Eilert, Hakan. "A Brief Outline of Pure Land Buddhism in India and in Early China." *Japanese Religions* 14 (1985) 1-12.

Fang, Keli. "The Categories *Ti* and *Yong* in Chinese Philosophy." *Social Sciences in China 6 (1985),* 115-141.

Faure, Bernard. "Le Maître de Dhyana Chih-ta et le 'Subitisme' de l'Ecole du Nord." *Cahiers d'Extrême-Asie,* 2 (1986), 123-132.

Faure, Bernard. "The Concept of One-Practice Samadhi in Early Ch'an." In Gregory (1986), 99-128.

Faure, David, and Lee Lai-mui. "The Po Tak Temple in Sheung

Shui Market." *JHKBRAS* 22 (1982), 271-279.

Franke, Herbert. "Sha-lo-pa (1259-1314), a Tangut Buddhist Monk in Yuan China." In Naundorf (1985), 201-222.

Franke, Wolfgang. "Notes on Chinese Temples and Deities in Northwestern Borneo." In Naundorf (1985), 267-289.

Frodsham, J.D., trans. *Goddesses, Ghosts, and Demons: The Collected Poems of Li He (790-816).* San Francisco: North Point Press, 1983 (revised ed. of *The Poems of Li Ho* , Oxford, 1970).

Fry, C. George. "Confucianism." In *Great Asian Religions.* Eds. C. George Fry, et al. (Grand Rapids: Baker Book House, 1984), pp. 87-110.

Fry, C. George, ed. *Great Asian Religions.* Grand Rapids: Baker Book House, 1984.

Fu, Charles Wei-Hsun. "Chinese Buddhism as an Existential Phenomenology." In *Phenomenology of Life in a Dialogue Between Chinese and Occidental Philosophy.* Ed. Anna-Teresa Tymieniecka (Dordrecht: D. Reidel, 1984), 229-251.

Fu, Pei-jung. "On Religious Ideas of the Pre-Chou China." *Chinese Culture* 26 (1985), 23-39.

Gates, Hill. "Money for the Gods." *Modern China* 13 (1987), 259-277.

Gaulier, Simone, and Robert Jera-Bezard. "Iconography: Buddhist Iconography." *ER* VII, 45-50.

Gernet, Jacques. "Techniques de Recueillement, Religion, Philosophie: A propos du jingzuo Néo-Confucéen." *BEFEO* 69 (1981), 289-305.

Gibbs, Nancy, trans. "Funeral Processions". In Ebrey (1981), 289-293.

Gimello, Robert M. "Hua-yen." *ER* VI, 485-489.

Gimello, Robert M., and Peter N. Gregory, eds. *Studies in Ch'an and Hua-yen.* (Kuroda Institute Studies in East Asian Buddhism, no. 1). Honolulu: University of Hawaii Press, 1983.

Gimello, Robert M. "Li T'ung-hsuan and the Practical Dimensions of Hua-yen." In Gimello (1983), 321-389.

Girardot, Norman J. "Chinese Religion: History of Study." *ER* III, 312-323.

Girardot, Norman J . "Chinese Religion: Mythic Themes." *ER* III, 296-305.

Girardot, Norman J. *Myth and Meaning in Early Taoism: The Theme of Chaos (Hun-tun).* Berkeley: University of California Press, 1983.

Girardot, Norman J., and John S. Major. "Introduction [Myth and Symbol in Chinese Tradition]." *JCR* 13-14 (1985-86), 1-14.

Gjertson, Donald E. "The Early Chinese Buddhist Miracle Tale: A Preliminary Survey." *JAOS* 101 (1981), 287-301.

Gomez, Luis O. "The Direct and the Gradual Approaches of Zen Master Mahayana: Fragments of the Teachings of Mo-ho-yen." In Gimello (1983), 69-167.

Gomez, Luis O. "Language: Buddhist Views of Language." *ER* VIII, 446-451.

Gomez, Luis O. "Purifying Gold: The Metaphor of Effort and Intuition in Buddhist Thought and Practice." In Gregory (1987), 67-165.

Goodrich, Anne Swann. *Chinese Hells: The Peking Temple of Eighteen Hells and Chinese Conceptions of Hell.* St. Augustin, Germany: Monumenta Serica, 1981.

Gordon, Arvan. "Religion and the New Chinese Constitution." *Religion in Communist Lands* 11 (Summer 1983), 130-134.

Grafflin, Dennis. "Geomantic Cliché and Geomagnetic Puzzle." *JAOS* 105 (1985), 315-316.

Graham, A.C. *Chuang-tzu: The Seven Inner Chapters and Other Writings from the Book of Chuang-tzu.* London: Allen and Unwin, 1981.

Grangereau, Philippe. "Résurrection du Bouddha en Chine: La Réouverture de Temples." *Sudestasie* 13 (1981), 64-67.

Gregory, Peter N. "Chinese Buddhist Hermeneutics : The Case of Hua-yen." *JAAR* 51 (1983), 231-249.

Gregory, Peter N., ed. *Sudden and Gradual Approaches to Enlightenment in Chinese Thought.* (Kuroda Institute Studies in East Asian Buddhism, no. 5). Honolulu: University of Hawaii Press, 1987.

Gregory, Peter N. "Sudden Enlightenment Followed by Gradual Cultivation: Tsung-mi's Analysis of Mind." In Gregory (1987). 279-320.

Gregory, Peter N. "The Teaching of Men and Gods: The Doctrinal and Social Basis of Lay Buddhist Practice in the Hua-yen Tradition." In Gimello (1983), 253-319.

Gregory, Peter N., ed. *Traditions of Meditation in Chinese Buddhism.* (Kuroda Institute Studies in East Asian Buddhism, no. 4). Honolulu: University of Hawaii Press, 1986.

Griffiths, Paul. "Concentration or Insight: The Problematic of Theravada Buddhist Meditation-Theory." *JAAR* 49 (1981), 605-624.

Gutheinz, Luis, and Edmond Tang. "Economic Development and Quality of Life: A Socio-religious Survey of Taiwan." *Pro Mundi Vita Asia-Australasia* Dossier no. 33 (1985), 1-29.

Haboush, JaHyun Kim. "Confucianism in Korea." *ER* IV, 10-15.

Hahn, Thomas. "Field Work in Daoist Studies in the People's Republic." *Cahiers d'Extreme-Asie* 2 (1986).

Hanson-Barber, A.W. " 'No-thought' in Pao-T'ang Ch'an and Early Ati-Yoga." *JIABS* 8 (1985), 61-73.

Hardacre, Helen. "Ancestors: Ancestor Worship." *ER* I, 263-268.

Harper, Donald. "A Chinese Demonography of the Third Century B.C.." *HJAS* 45 (1985), 459-498.

Harper, Donald. "Magic: Magic in East Asia." *ER* IX, 112-115.

Harrell, Stevan. "The Concept of Fate in Chinese Folk Ideology." *Modern China* 13 (1987), 90-109.

Harrell, Stevan. "Domestic Observances: Chinese Practices." *ER* IV, 410-414.

Hart, James A. "The Speech of Prince Chin: A Study of Early Chinese Cosmology." In Rosemont. *JAAR Thematic Studies* 50 (1984), 35-65.

Hase, Patrick H. "Old Hau Wong Temple, Tai Wai, Sha Tin." *JHKBRAS* 23 (983), 233-240.

Hase, Patrick H. "Traditional Funerals." *JHKBRAS* 21 (1981), 192-196.

Hayes, J.W. "The Kwun Yam: Tung Shan Temple of East Kowloon 1840-1940." *JHKBRAS* 23 (1983), 212-218.

Hayes, James. "Specialists and Written Materials in the Village World" in Johnson (1985), 75-111.

Heinze, Ruth-Inge. "Automatic Writing in Singapore." In Clammer (1983), 146-160.

Heinze, Ruth-Inge. "The Nine Imperial Gods in Singapore." *AFS* 40 (1981), 151-157.

Henderson, John B. *The Development and Decline of Chinese Cosmology.* New York: Columbia University Press, 1984.

Hendrischke, Barbara. "Chinese Research into Daoism after the Cultural Revolution." *Asiatische Studien/Etudes Asiatiques* 38 (1984), 25-42.

Hendrischke, Barbara. "How the Celestial Master Proves Heaven Reliable." In Naundorf (1985), 77-86.

Henricks, Robert G. "The Philosophy of Lao-tzu Based on the Ma-wang-tui Texts: Some Preliminary Observations." *BSSCR* 9 (1981), 59-78.

Hirakawa, Akira. "Buddhist Literature: Survey of Texts." *ER* II, 509-529.

Holzman, Donald. "The Cold Food Festival in Early Medieval China. " *HJAS* 46 (1986), 51-79.

Hoshino, Eiki. "Pilgrimage: Buddhist Pilgrimage in East Asia." *ER* XI, 349-351.

Howard, Angela Falco. "Heavenly Mounts — Horses and Elephants — in Chinese Buddhist Art." *Oriental Art* 28 (1982-83), 368-381.

Howard, Angela Falco. "Planet Worship: Some Evidence, Mainly Textual, in Chinese Esoteric Buddhism." *Asiatische Studien* 37 (1983), 103-119.

Howard, Angela Falco. *The Imagery of the Cosmological Buddha.* (Studies in South Asian Culture, 13), Leiden: E.J. Brill, 1986.

Howard, Jeffrey A. "Concepts of Comprehensiveness and Historical Change in the *Huai-nan-tzu.*" In Rosemont (1984), 35-65.

Hsieh, Jiann, and Ying-hsiung Chou. "Public Aspirations in the New Year Couplets: A Comparative Study Between the People's Republic and Taiwan." *AFS* 40 (1981), 125-149.

Hsieh, Jih-chang and Chuang Yin-chang, eds. *The Chinese Family and Its Ritual Behavior.* (Institute of Ethnology Monograph Series B, no. 15). Taipei: Institute of Ethnology, Academia Sinica, 1985.

Hsing, Fu-Chuan. "The General Development of Taiwanese Buddhism." *Chinese Culture* 22 (1981), 79-84.

Hua, Cecilia Ng Siew. "The Sam Poh Neo Neo Keramat: A Study of a Baba Chinese Temple." In Clammer (1983), 98-131.

Huey, Ju Shi. "Chinese Spirit Mediums in Singapore: An Ethnographic Study." In Clammer (1983), 3-48.

Hurvitz, Leon, trans. *A History of Early Chinese Buddhism: From its Introduction to the Death of Hui-yüan.* See Tsukamoto, Zenryu.

Iida, Shotaro. "The Three Stupas of Ch'ang An." In *Papers of the First International Conference of Korean Studies, 1979,* (Seongnam-Si, Kyeonggi-Do: Academy of Korean Studies, 1980), 484-497.

Jan, Yün-hua. "The Bodhisattva Idea in Chinese Literature: Typology and Significance." In *The Bodhisattva Doctrine in Buddhism.* Ed. Leslie S. Kawamura (Waterloo, Ontario: Wilfred Laurier University Press, 1981), 125-152.

Jan, Yün-hua. "Buddhism in Ta-tu: The New Situation and New Problems." In *Yuan Thought: Chinese Thought and Religion Under the Mongols.* Eds. Hok-Lam Chan and William Theodore de Bary (New York: Columbia University Press, 1982), 375-417.

Jan, Yün-hua. "The Chinese Understanding and Assimilation of Karma Doctrine." In *Karma and Rebirth: Post Classical Developments.* Ed. Ronald W. Neufeldt (Albany: State University of New York Press, 1986), pp. 145-168.

Jan, Yün-hua. "The Mind as the Buddha-nature: The Concept of the Absolute in Ch'an Buddhism." PEW 31 (1981), 467-477.

Jan, Yün-hua. "Political Philosophy of the *Shih Liu Ching* Attributed to the Yellow Emperor Taoism." *JCP* 10 (1983), 205-228.

Jan, Yün-hua. "The Religious Situation and the Studies of Buddhism and Taoism in China: An Incomplete and Imbalanced Picture." *JCR* 12 (1984), 37-64.

Jenner, W.J.F. *Memories of Loyang: Yang Hsüan-chih and the Lost Capital (493-534).* Oxford: Clarendon Press, 1981.

Jochim, Christian. *Chinese Religions: A Cultural Perspective.* (Prentice-Hall Series in World Religions). Englewood Cliffs, NJ: Prentice-Hall, 1986.

Johnson, David. et al., eds. *Popular Culture in Late Imperial China.* Berkeley: University of California Press, 1985.

Johnson, David. "The City-God Cults of T'ang and Sung China." *HJAS* 45 (1985), 363-457.

Jordan, David K. "Sworn Brothers: A Study in Chinese Ritual Kinship (with an Appendix on Sworn Brotherhood and Folk Law)." In Hsieh (1985), 232-262; 303-304.

Jordan, David K. "Taiwanese *poe* divination: Statistical Awareness and Religious Belief." *Journal for the Scientific Study of Religion* 21 (1982), 114-118.

Jordan, David K., and Daniel L. Overmyer. *The Flying Phoenix: Aspects of Chinese Sectarianism in Taiwan.* Princeton: Princeton University Press, 1986.

Kao, Karl S.Y., ed. *Classical Chinese Tales of the Supernatural and the Fantastic: Selections from the Third to the Tenth Century.* Bloomington: Indiana University Press, 1985.

Kasoff, Ira E. *The Thought of Chang Tsai (1020-1077).* Cambridge University Press, 1984.

Kawamura, Leslie, S., ed. *The Bodhisattva Doctrine in Buddhism.* Waterloo, Ontario: Wilfred Laurier University Press, 1981.

Keightley, David N. "Late Shang Divination: The Magico-Religious Legacy." *JAAR Thematic Studies* 50 (1984), 11-34.

Kelleher, Terry. "Women in Confucianism." In *Women in World Religions.* Ed. Arvind Sharma (Albany: State University of New York Press, 1987).

Kendall, Laurel. "Korean Shamanism: Women's Rites and a Chinese Comparison." In De Vos (1984), 57-73.

Kendall, Laurel. *Shamans, Housewives, and Other Restless Spirits: Women in Korean Ritual Life.* Honolulu: University of Hawaii Press, 1985.

King, James R. "The Taoist Tradition." In *Great Asian Religions.* Eds. C. George Fry, et al. (Grand Rapids: Baker Book House, 1984), 111-138.

King, Winston L. "Meditation: Buddhist Meditation." *ER* IX,

331-336.

Kirkland, J. Russell. "Chang Kao — Noteworthy Tang Taoist?" *T'ang Studies* 2 (1984), 31-35.

Kirkland, J. Russell. "The Last Taoist Grand Master at the T'ang Imperial Court: Li Han-kuang and T'ang Hsüan-tsung." *T'ang Studies* 4 (1986), 43-67.

Kirkland, J. Russell. "The Roots of Altruism in the Taoist Tradition." *JAAR* 54 (1986), 59-77.

Kloetzli, W. Randolph. "Cosmology: Buddhist Cosmology."*ER* IV, 113-119.

Knaul, Livia. "Chuang Tzu and the Chinese Ancestry of Ch'an." *JCP* 13 (1986), 411-428.

Knaul, Livia. *Leben und Legende des Ch'en T'uan.* (Würzburger Sino-Japonica, Bd. 9). Frankfurt am Main: Verlag Peter Lang, 1981.

Knaul, Livia. "Lost *Chuang Tzu* Passages." *JCR* 10 (1982), 53-79.

Knaul, Livia. "The Habit of Perfection: A Summary of Fukunaga Mitsuji's Studies on the *Chuang tzu* Tradition." *Cahiers d'Extrême-Asie* 1 (1985): 71-85

Kohn, Livia (née Knaul). "A Textbook of Physiognomy: The Tradition of the *Shenxiang quanbian." AFS* 45 (1986), 227-258.

Kojima, Hajime. "Die Jodo-Lehre Shinrans und der Taoismus in China." In Naundorf (1985), 233-238.

Koseki, Aaron K. "Chi-tsang's *Sheng-man pao-k'u*: The True Dharma Doctrine and the Bodhisattva Ideal." *PEW* 34 (1984), 67-84.

Koseki, Aaron K. "Later Madhyamika in China: Some Current Perspectives on the History of Chinese Prajnaparamita Thought." *JIABS* 5 (1982), 53-62.

Kramers, Robert P. "On Religion and Religious Values in China Today." *Ching Feng* 27 (1984), 196-203.

Kroll, Paul W. "In the Halls of the Azure Lad." *JAOS* 105 (1985), 75-94.

Kroll, Paul W. "Li Po's Rhapsody on the Great P'eng-bird."*JCR* 12 (1984), 1-17.

Kroll, Paul W. "Li Po's Transcendent Diction." *JAOS* 106 (1986), 99-117.

Kroll, Paul W. "Notes on Three Taoist figures of the T'ang Dynasty." *BSSCR* 9 (1981), 19-41.

Kroll, Paul W. "Verses from on High: The Ascent of T'ai Shan." *TP* 69 (1983). 223-260.

Lagerwey, John. "The Oral and the Written in Chinese and Western Religion." In Naundorf (1985), 301-322.

Lagerwey, John. "Priesthood: Taoist Priesthood." *ER* XI, 547-

550.

Lagerwey, John. "Taoism: The Taoist Religious Community." *ER* XIV, 306-317.

Lagerwey, John. "Worship and Cultic Life: Taoist Cultic Life. *ER* XV, 482-486.

Lagerwey, John, trans. *Wu-shang pi-yao: Somme Taöiste du VIe siècle*. (Publications de l'Ecole Française d'Extrême-Orient, 74). Paris: Adrien-Maisonneuve, 1981.

Lagerwey, John. "La Conférence Taöiste à Hong Kong, decembre 1985." *Cahiers d'Extrême-Asie* 2 (1986): 185-190.

Lagerwey, John. *Taoist Ritual in Chinese Society and History*. New York: Macmillan Publishing Co., 1987.

Lai, Whalen. "A Different Religious Language: The Tian-tai Idea of the Triple Truth." *Ching Feng* 25 (1982), 67-78.

Lai, Whalen, and Lewis R. Lancaster, eds. *Early Ch'an in China and Tibet*. Berkeley: Asian Humanities Press, 1983.

Lai, Whalen. "The Early Chinese Buddhist Understanding of the Psyche: Chen Hui's Commentary on the *Yin chih ju ching*." *JIABS* 9 (1986), 85-103.

Lai, Whalen. "Emperor Wu of Liang on the Immortal Soul, *Shen Pu Mieh*." *JAOS* 101 (1981), 167-175.

Lai, Whalen. "Faith and Wisdom in the Tien-t'ai Buddhist Tradition: A Letter by Ssu-ming Chih-li," *Journal of Dharma* 6 (1981), 283-298.

Lai, Whalen. "From Sakyamuni to Amitabha: The Logic Behind Pure Land Devotion." *Ching Feng* 24 (1981), 156-174.

Lai, Whalen. "Kang Yuwei and Buddhism: From Enlightenment to Sagehood." *Ching Feng* 26 (1983), 14-34.

Lai, Whalen. "The Mahaparinirvana-Sutra and Its Earliest Interpreters in China: Two Prefaces by Tao-lang and Tao-sheng." *JAOS* 102 (1982), 99-105.

Lai, Whalen. "Ma-tsu Tao-i and the Unfolding of Southern Zen." *Japanese Journal of Religious Studies* 12 (1985), 173-192.

Lai, Whalen. "Popular Moral Tracts and the Chinese Personality." *Ching Feng* 25 (1982), 22-31.

Lai, Whalen. "Sinitic Speculations on Buddha-nature: The Nirvana School." *PEW* 32 (1982), 135-149.

Lai, Whalen. "Yao Hsing's Discourse on Mahayana: Buddhist Controversy in Early Fifth-century China." *Religious Traditions* 4 (1981), 35-58.

Lai, Whalen, and Lily Hwa, trans. "Precepts of the Perfect Truth Taoist Sect." In Ebrey (1981), 75-78.

Lai, Whalen. "The Earliest Folk Buddhist Religion in China: *T'i-wei Po-li Ching* and its Historical Significance." In Chappell

(1987), 11-35.

Lai, Whalen. "Tao-sheng's Theory of Sudden Enlightenment Re-examined." In Gregory (1987), 169-200.

Lancashire, Douglas, trans. *Chinese Essays on Religion and Faith*. (Chinese Materials Center, Asian Library Series, 26). San Francisco: Chinese Materials Center, 1981.

Lancaster, Lewis R. "The Bodhisattva Concept: A Study of the Chinese Buddhist Canon." In *The Bodhisattva Doctrine in Buddhism*. Ed. Leslie S. Kawamura (Waterloo, Ontario: Wilfred Laurier University Press, 1981), 153-163.

Lancaster, Lewis R. "Buddhism and Family in East Asia." In De Vos (1984), 139-151.

Lancaster, Lewis R. "Buddhist Literature: Canonization." *ER* II, 504-509.

Lancaster, Lewis R. "Buddhist Studies." *ER* II, 554-560.

Lancaster, Lewis R. "Elite and Folk: comments on the Two-Tiered Theory." In De Vos (1984), 87-95.

Lancaster, Lewis R. "Maitreya." *ER* IX, 136-141.

Lanciotti, Lionello, ed. *Incontro di religioni in Asia tra il III e il X secolo d C.* Florence, Italy: Leo S. Olschki Editore, 1984.

Langlois, John D., Jr., and Sun K'o-k'uan. "Three Teachings Syncretism and the Thought of Ming T'ai-tsu." *HJAS* 43 (1983), 97-139.

Larre, Claude. *Le Traité VII du Houai Nan Tseu: Les Esprits Légers et Subtils Animateurs de l'Essence.* (Variétés Sinologiques, n.s. 67). Paris and Taipei: Institut Ricci, Centre d'Etudes Chinoises, 1982.

Lau, Theodora. *The Handbook of Chinese Horoscopes.* London: Arrow, 1981.

Law, Joan, and Barbara E. Ward. *Chinese Festivals.* Hong Kong: South China Morning Post, 1982.

Le Blanc, Charles."A Re-Examination of the Myth of Huang-ti."*JCR* 13-14 (1985-86), 45-63.

Le Blanc, Charles, *Huai Nan Tzu, Philosophical Synthesis in Early Han Thought.* Hong Kong University Press, 1985.

Ledderose, Lothar. "Some Taoist Elements in the Calligraphy of the Six Dynasties." *TP* 70 (1984), 246-278.

Lee, Cyrus. "Life, Death and Reincarnation: A Comparative Study on Hanshan Tzu and Thomas Merton." *Chinese Culture* 22 (1981), 111-120.

Lee, Peter K.H. "A Bird's Eye View of Ideological Contours in Hong Kong Today." *Ching Feng* 25 (1982), 16-21.

Lee, Peter K.H. "A Brief Presentation of Chinese Religion: An Exercise in Communication." *Ching Feng* 25 (1982), 79-91.

Lee Peter K.H. "A Chinese New Year Interreligious Celebration." *Ching Feng* 25 (1982). 32-36

Lee, Peter K.H. "Theology and Myth: A Reflection on the Lady Flying to the Moon and the Archer Shooting Down Nine Suns." *East Asia Journal of Theology* 3 (1985), 228-242.

Lee, Raymond L. M. "Dancing with the Gods: A Spirit Medium Festival in Urban Malaysia." *Anthropos* 78 (1983), 355-368.

Lee, Raymond L.M., and S.E. Ackerman. "Ideology, Authority and Conflict in a Chinese Religious Movement in West Malaysia." In Clammer (1983), 132-145.

Leslie, Donald Daniel. "Persian Temples in T'ang China." *Monumenta Serica* 35 (1981-83), 275-303.

Leung, Koon-Loon. "An Algebraic Truth in Divination." *JCP* 9 (1982), 243-258.

Levering, Miriam. "Ta-hui and Lay Buddhists: Ch'an Sermons on Death" in Chappell (1987), 181-206.

Levi, Jean. "L'Abstinence des Céréales chez les Taoistes." *Etudes Chinoises* 1 (1983).

Levi, Jean. "Dong Yong le Fils Pieux et le Mythe Formosan de l'Origine des Singes." *JA* 272 (1984), 82-132.

Levi, Jean. "Les Fonctionaires et le Divin: Luttes de Pouvoirs entre Divinites et Administrateurs dans les Contes des Six Dynasties et des Tang." *Cahiers d'Extrême-Asie* 2 (1986), 81-110.

Li, Jung-hsi, trans. *Biographies of Buddhist Nuns: Pao-chang's Pi-chiu-ni chuan.* Osaka: Tohokai, 1981.

Li, Yih-yuan. "On Conflicting Interpretations of Chinese Family Rituals." In Hsieh (1985), 263-281; 304-306.

Lievens, Bavo. *The Recorded Sayings of Matsu.* Introduced, translated into Dutch and annotated by Bavo Lievens. Translated from the Dutch by J.F. Pas. New York: E. Mellen Press, 1987 (*Studies in Asian Thought & Religion,* vol.6).

Lim, Guek Eng. "Christianity Encounters Ancestor Worship in Taiwan." *Evangelical Review of Theology* 8 (1984), 225-235.

Lip, Evelyn. *Chinese Temples and Deities.* Singapore: Times Books International, 1981.

Lip, Evelyn. *Chinese Temple Architecture in Singapore.* Singapore: Singapore University Press, 1983.

Liu, James J.Y. "Poetry: Chinese Religious Poetry." *ER* XI, 378-380.

Liu, Ming-Wood. "The P'an-chiao system of the Hua-yen School in Chinese Buddhism." *TP* 67 (1981), 10-47.

Liu, Ming-Wood. "The Three-Nature Doctrine and its Interpretation in Hua-yen Buddhism." *TP* 68 (1982), 181-220.

Liu, Ts'un-yan, and Judith Berling. " 'The Three Teachings' in the

Mongol-Yüan Period." In *Yüan Thought: Chinese Thought and Religion Under the Mongols*. Eds. Hok-Lam Chan and William Theodore de Bary (New York: Columbia University Press, 1982), 479-512.

Loewe, Michael. *Chinese Ideas of Life and Death: Faith, Myth and Reason in the Han Period* (202 BC — AD 220) London: George Allen & Unwin, 1982.

Loewe, Michael, and Carmen Blacker, eds. *Oracles and Divination*. Boulder : Shambhala, 1981.

Louton, John. "Concepts of Comprehensiveness and Historical Change in the *Lü-shih ch'un-ch'iu*." In Rosemont. *JAAR Thematic Studies* 50 (1984), 105-117.

Lowe, H.Y. *The Adventures of Wu. the Life Cycle of a Peking Man* (2 vols in 1). Princeton Univ. Press, 1983 (original edition by the Peking Chronicle Press, 1940-41).

Luo, Tzong-tao, "Au Sujet du Terme *bian* : Les Procédés d'Adaptation des Textes Bouddhiques aux *Bianwen*." *JA* 269 (1981), 151-157.

Lynn, Richard John. "The Sudden and Gradual in Chinese Poetry Criticism: An Examination of the Ch'an Poetry Analogy." In Gregory (1987), 381-427.

Ma, Qicheng. "A Brief Account of the Early Spread of Islam in China". *Social Sciences in China* 4 (1983), 97-113.

MacInnis, Donald E. "Secularism and Religion in China: The Problem of Transcendence." In *Contemporary Chinese Philosophy*. Ed. Frederick J. Adelmann (The Hague: Martinus Nijhoff, 1982), 117-133.

Magnin, Paul. "Deux prières Liturgiques pour célébrer l'anniversaire de l'empereur." *Cahiers d'Extrême-Asie* 3 (1987).

Mair, Victor H., Ed., *Experimental Essays on Chuang-tzu*. Honolulu: University of Hawaii Press, 1983.

Mair, Victor H. "Language and Ideology in the Written Popularizations of the *Sacred Edict*." In Johnson (1985), 325-359.

Mair, Victor H. "The Origins of an Iconographical Form of the Pilgrim Hsüan-tsang." *T'ang Studies* 4 (1986), 29-41 and 7 plates.

Mair, Victor H. "Records of Transformation Tableaux (*pien-hsiang*)." *TP* 72 (1986), 3-43.

Mair, Victor H. "Wandering in and Through the *Chuang-tzu*." *JCR* 11 (1983), 106-117.

Major, John S. "The Five Phases, Magic Squares and Schematic Cosmography." In Rosemont. *JAAR Thematic Studies* 50 (1984), 133-166.

Major, John S. "New Light on the Dark Warrior." *JCR* 13-14 (1985-86), 65-86.

Mak, Lau-Fong. "Chinese Secret Societies: Criminologically Defined." *BIE* 59 (1985), 143-161.

Mak, Lau-Fong. *The Sociology of Secret Societies: A Study of Chinese Secret Societies in Singapore and Peninsular Malaysia.* Kuala Lumpur: Oxford University Press, 1981.

Malek, Roman. *Das Chai-chieh lu: Materialien zur Liturgie im Taoismus* . (Wurzburger Sino-Japonica, Band 14). Frankfurt/M: Peter Lang, 1985.

Maliszewski, Michael. "Material Arts." *ER* IX, 224-228.

Maraldo, John C. "Is There Historical Consciousness Within Ch'an?" *Japanese Journal of Religious Studies* 12 (1985), 141-172.

Martin, Helmut. *Cult and Canon. The Origins and Development of State Maoism.* Armonk, N.Y. : M.E. Sharpe, Inc., 1982.

Martinson, Paul V. "Some Theological Reflection on Current Marxist Studies of the Confucian Tradition." *Theology and Life* 4 (1982), 60-74.

Maspero, Henri. *Taoism and Chinese Religion* . Trans. Frank A. Kierman, Jr. Amherst: University of Massachusetts Press, 1981.

Mather, Richard B. "The Bonze's Begging Bowl: Eating Practices in Buddhist Monasteries of Medieval India and China." *JAOS* 101 (1981), 417-424.

Mather, Richard B. "The Impact of the *Nirvana Sutra* in China." In *Literature of Belief: Sacred Scripture and Religious Experience.* Ed. Neal E. Lambert (Salt Lake City: Religious Studies Center, Brigham Young University, 1981), 155-173.

Mather, Richard B. "The Life of the Buddha and the Buddhist Life: Wang Jung's (468-93) 'Songs of Religious Joy' (*Fa-le tz'u*)." *JAOS* 107 (1987), 31-38.

Mather, Richard B. "Wang Jung's 'Hymns on the Devotee's Entrance into the Pure Life." *JAOS* 106 (1986), 79-98.

Mathieu, Rémi. *Etude sur la Mythologie et l'Ethnologie de la Chine Ancienne.* Paris, 1983.

McRae, John R. "The Ox-head School of Chinese Ch'an Buddhism: From Early Ch'an to the Golden Age." In Gimello (1983), 169-252.

McRae, John R. *The Northern School and the Formation of Early Ch'an Buddhism* . (Studies in East Asian Buddhism, no. 3). Honolulu: University of Hawaii Press, 1986.

McRae, John R. "Shen Hui and the Teaching of Sudden Enlightenment in Early Ch'an Buddhism." In Gregory (1987), 227-

278.
Meister, Michael W., and Nancy Shatzman Steinhardt. "Temple: Buddhist Temple Compounds." *ER* XIV, 373-380.

Meyer, Jeffrey F. "The Sacred Truth Society." *Studies in Religion* 11 (1982), 285-298.

Morgan, Carole. "Dog Divination from a Dunhuang Manuscript." *JHKBRAS* 23 (1983), 184-193.

Morgan, Carole. "Nouvelle Etude sur la Divination d'après le Cri du Corbeau dans les mss de Dunhuang." *Cahiers d'Extrême-Asie* 3 (1987).

Moses, Larry. "Triplicated Triplets: The Number Nine in the *Secret History of the Mongols*." *AFS* 45 (1986). 287-294.

Murphy, Laurence T. "Religion and Politics in China." *Journal of Dharma* 7 (1982), 46-55.

Myers, John T. "Traditional Chinese Religious Practices in an Urban-industrial Setting: The Example of Kwun Tong." In *Social Life and Development in Hong Kong*. Eds. Ambrose Y.C. King and Rance P.L. Lee (Hong Kong: Chinese University Press, 1981), 275-288.

Nakamura, Hajime. "Bodhisattva Path." *ER* II, 265-269.

Naquin, Susan. *Shantung Rebellion: The Wang Lun Uprising of 1774*. New Haven: Yale University Press, 1981.

Naquin, Susan. "The Transmission of White Lotus Sectarianism in Late Imperial China." In Johnson (1985), 255-291.

Naundorf, Gert, et al., eds. *Religion und Philosophie in Ostasien: Festschrift für Hans Steininger zum 65 Geburtstag*. Würzburg: Konigshausen & Neumann, 1985.

Naundorf, Gert. "Raumvorstellungen in den *chiao*-Opfern bis zur Han-Zeit." In Naundorf (1985), 323-330.

Needham, Joseph, and Lu Gwei-djen. *Science and Civilisation in China*. Cambridge: Cambridge University Press, 1983, vol. 5, part 5, "Alchemy and Chemistry".

Nivison, David S. "Chinese Philosophy." *ER* III, 245-257.

Nivison, David S. "Meng-tzu." *ER* IX, 373-376.

Nivison, David S. "Tao and Te." *ER* XIV, 283-286.

Nosco, Peter. "Confucianism in Japan." *ER* IV, 7-10.

Okada, Takehiko, and Rodney L. Taylor. "*Zazen to Seiza*: A Prefatory Discussion." *JCR* 10 (1982), 19-38.

Orzech, Charles D. "Chen-yen." *ER* III, 233-238.

Overmyer, Daniel L. "Attitudes Toward the Ruler and State in Chinese Popular Religious Literature: Sixteenth and Seventeenth Century *Pao-chüan*." *HJAS* 44 (1984), 347-379.

Overmyer, Daniel L. "Chinese Religion: An Overview." *ER* III, 257-289.

Overmyer, Daniel L. *Religions of China : The World as a Living System.* San Francisco: Harper and Row, 1986.

Overmyer, Daniel L. "Values in Chinese Sectarian Literature: Ming and Ch'ing *Pao-chüan.*" In Johnson (1985), 219-254.

Overmyer, Daniel L., and David K. Jordan. *The Flying Phoenix: Aspects of Chinese Sectarianism in Taiwan.* Princeton: Princeton University Press, 1986.

Ozaki, Masaharu. "The Taoist Priesthood: From Tsai-chia to Ch'u-chia." In De Vos (1984), 97-109.

Pachow, W. "Chinese Ch'an: A Transformation of Indian Dhyana." *Chinese Culture* 26 (1985), 45-58.

Pachow, W. "A Hermeneutical Approach to the Supernatural Phenomena in Buddhist History." *Chinese Culture* 27 (1986), 15-36; 73-100.

Pachow, W. "The Tooth Relic of the Buddha: A Bridge to International Buddhist Unity." *Chinese Culture* 25 (1984), 81-90.

Palmer, Martin, ed. and trans., with Mak Hin Chung, Kwok Man Ho, and Angela Smith. *T'ung shu: The Ancient Chinese Almanac.* Boston: Shambhala, 1986.

Paludan, Ann. *The Imperial Ming Tombs.* New Haven: Yale University Press, 1981.

Paper, Jordan. "From Shamanism to Mysticism in the *Chuang-tzu.*" *Scottish Journal of Religious Studies* 3 (1982), 27-45.

Paper, Jordan. "Riding on a White Cloud: Aesthetics as Religion in China." *Religion* 15 (1985), 3-27.

Paper, Jordan. "The *Feng* in Protohistoric Chinese Religion." *HR* 25 (1986), 213-235.

Park, Ynhui. "The Concept of Tao: A Hermeneutical Perspective." In *Phenomenology of Life in a Dialogue Between Chinese and Occidental Philosophy.* ed. Anna-Teresa Tymieniecka (Dordrecht: D. Reidel, 1984), 203-213.

Parkin, Harry. "Postscript: Chinese Religious Studies Today." In Clammer (1983), 161-174.

Pas, Julian F. "Chuang Tzu's Essays on 'Free Flight into Transcendence' and 'Responsive Rulership' (Chapters One and Seven of the *Chuang Tzu*)." *JCP* 8 (1981), 479-496.

Pas, Julian F. "Religion in China Today." *Canada-China Journal* 5 (1985), 18, 21.

Pas, Julian F., comp. *Select Bibliography on Taoism.* Stony Brook, N.Y. : Institute for the Advanced Study of World Religions, 1988.

Pas, Julian F. "Yin-yang Polarity: A Binocular Vision of the World." *Asian Thought & Society* 8 (1983), 188-201.

Pas, Julian F. "Dimensions in the Life and Thoughts of Shan-tao

(613-681)." in Chappell (1987), 65-84.

Pas, Julian F. trans. *The Recorded Sayings of Matsu*. Introduced, translated into Dutch and annotated by Bavo Lievens. Translated from the Dutch by J.F. Pas. New York: E. Mellen Press, 1987 (Studies in Asian Thought & Religion, Vol. 6).

Paul, Diana Y. *Philosophy of Mind in Sixth-Century China: Paramartha's 'Evolution of Consciousness'*. Stanford: Stanford University Press, 1984.

Peterson, Willard. "Making Connections: Commentary on the Verbalizations of the Book of Changes," *HJAS*, 42 (1982), 67-116.

Piediscalzi, Nicholas, "China's New Policy on Religion." *The Christian Century* 102 (June 19-26, 1985), 611-614.

Pokora, Timoteus. " 'Living Corpses' in Early Medieval China — Sources and Opinions." In Naundorf (1985), 343-357.

Pong, Raymond. "China: Religion in Revolutionary Society." In *Religions and Societies: Asia and the Middle East*. Ed. Carlo Caldarola (Berlin: Walter de Gruyter, 1982), 551-577.

Powell, William. "More Laughter at Tiger Creek: Impressions of Buddhism in Modern China." *JCR* 12 (1984), 77-87.

Powell, William, Trans. *The Record of Tung-shan*. Honolulu: University of Hawaii Press, 1986.

Pruden, Leo M. "T'ien-t'ai." *ER* XIV, 510-519.

Quack, Anton. *Priesterinnen, Heilerinnen, Schamaninnen? Die Poringao der Puyuma von Katipol (Taiwan)*. (Collectanea Instituti Anthropos, Bd. 32). Berlin: Dietrich Reimer, 1985.

Raguin, Yves. "L'Eschatologie dans le Monde Chinois." *Studia Missionalia* 32 (1983), 181-194.

Rand, Christopher C. "*Chuang-tzu* : Text and Substance." *JCR* 11 (1983), 5-58.

Reat, N. Ross. "Some Fundamental Concepts of Buddhist Psychology." *Religion* 17 (1987), 15-28.

Reiter, Florian C. "Der Name *Tung-hua ti-chun* und sein Umfeld in der Taoistischen Tradition." In Naundorf (1985), 87-101.

Reiter, Florian C. "Some Observations Concerning Taoist Foundations in Traditional China." *Zeitschrift der Deutschen Morgenlandischen Gesellschaft* 133 (1983), 363-376.

Renaud, Rosario. *Le Diocèse de Süchow (Chine): Champ Apostolique des Jésuites Canadiens de 1918 à 1954*. Montréal: Ed. Bellarmin, 1981.

Reynolds, Frank E., John Holt, and John Strong. *Guide to Buddhist Religion*. Boston: G.K. Hall, 1981.

Rhie, Marylin N. "Late Sui Buddhist Sculpture: A Chronology and Regional Analysis." *Archives of Asian Art* 35 (1982), 27-54.

Riegel, Jeffrey K. "Early Chinese Target Magic." *JCR* 10 (1982), 1-18.

Robinet, Isabelle. "Le *Ta-tung chen-ching:* Son authenticite et sa place dans les textes du *Shang-ch'ing ching.* " In Strickmann (1983), 394-433.

Robinet, Isabelle. "*Chuang-tzu* et le Taoisme Religieux." *JCR* 11 (1983), 59-105.

Robinet, Isabelle. "Notes Préliminaires sur quelques Antinomies Fondamentales entre le Bouddhisme et le Taoïsme." In *Incontro di Religioni in Asia tra il III e il X secolo d C.* Ed. Lionello Lanciotti (Florence, Italy: Leo S. Olschki Editore, 1984), 217-242.

Robinet, Isabelle. " La notion de *hsing* dans le Taoisme et son rapport avec celle du Confucianisme." *JAOS* 106 (1986), 183-196.

Robinet, Isabelle. *La Revelation du Shangqing dans l'Histoire du Taoisme* . 2 vols. (Publications de l'Ecole Francaise d'Extrême-Orient, 137). Paris: L'Ecole Française d'Extrême-Orient, 1984.

Robinet, Isabelle, "The Taoist Immortal: Jesters of Light and Shadow, Heaven and Earth." *JCR* 13-14 (1985-86), 87-105.

Robinet, Isabelle. "L'Unité Transcendante des Trois Enseignements selon les Taoïstes des Sung et des Yüan." In Naundorf (1985), 103-126.

Rosemont, Henry, Jr., Ed. *Explorations in Early Chinese Cosmology. JAAR Thematic Studies.* 50.2 (1984), 1-171.

Rosner, Erhard. "Frauen als Anfuhrerinnen Chinesischer Sekten." In Naundorf (1985), 239-246.

Rossabi, Morris. "Islam: Islam in China." *ER* VII, 377-390.

Roth, Harold D. "Fang-shih." *ER* V, 282-284.

Roth, Harold D. "Liu An." *ER* IX, 1-3.

Rubin, Vitaly A. "Ancient Chinese Cosmology and *Fa-chia* Theory." *JAAR Thematic Studies* 50 (1984), 95-104.

Rubin, Vitaly A . "The Concepts of Wu-Hsing and Yin-Yang." *JCP* 9 (1982), 131-157.

Sangren, P. Steven. "Female Gender In Chinese Religious Symbols: Kuan Yin, Ma Tsu, and the 'Eternal Mother'." *Signs* 9 (1983), 4-25.

Sangren, P. Steven. "Orthodoxy, Heterodoxy, and the Structure of Value in Chinese Rituals." *Modern China* 13 (1987), 63-89.

Sankar, Andrea. "Sisters and Brothers, Lovers and Enemies: Marriage Resistance in Southern Kwangtung." *Journal of Homosexuality* 11 (1985), 69-81.

Saso, Michael. "Taiwan: Old Gods and Modern Society." In *Religions and Societies: Asia and the Middle East.* Ed. Carlo

Caldarola (Berlin: Walter de Gruyter, 1982), 579-605.

Sawatzky, Sheldon. "State-Church Conflict in Taiwan: Its Historical Roots and Contemporary Manifestations." *Missiology* 9 (1981), 449-463.

Schafer, Edward H, "Empyreal Powers and Chthonian Edens: Two Notes on T'ang Taoist Literature." *JAOS* 106 (1986), 667-677.

Schafer, Edward H. *Mirages on the Sea of Time: The Taoist Poetry of Ts'ao T'ang.* Berkeley: University of California Press, 1985.

Schafer, Edward H. "The Princess Realized in Jade." *T'ang Studies* 3 (1985), 1-23

Schafer, Edward H, " The Snow of Mao Shan: A Cluster of Taoist Images." *JCR* 13-14 (1985-86), 107-126.

Schafer, Edward H. "Transcendent Elder Mao." *Cahiers d'Extreme-Asie* 2 (1986), 111-122.

Schafer, Edward H. "Two Taoist Bagatelles." *BSSCR9* (1981), 1-18.

Schafer, Edward H. "Wu Yun's Stanzas on 'Saunters in Sylphdom'." *Monumenta Serica* 35 (1981-83), 1-37.

Schafer, Edward H. "The Cranes of Mao Shan." In Strickmann (1983), 372-394.

Schipper, Kristofer. *Le Corps Taoïste, Corps Physique — Corps Social.* (Collection "L'espace Intérieur" 25). Paris: Artheme Fayard, 1982.

Schipper, Kristofer. *Index du Yunji qiqian.* 2 vols. (Publications de l'Ecole Française d'Extrême-Orient). Paris: L'Ecole Francaise d'Extrême-Orient, 1981.

Schipper, Kristofer. "Le Monachisme taoiste." In *Incontro di religioni in Asia tra il III e il X secolo d C.* Ed. Lionello Lanciotti (Florence, Italy: Leo S. Olschki Editore, 1984), 199-216.

Schipper, Kristofer. "Seigneurs royaux, Dieux des epidemies." *Archives de Sciences Sociales des Religions* 30 (59) (1985), 31-40.

Schipper, Kristofer. "Taoist Ordination Ranks in The Tunhuang Manuscripts." In Naundorf (1985), 127-148.

Schipper, Kristofer. "Vernacular and Classical Ritual in Taoism." *JAS* 45 (1985), 21-57.

Schirokauer, Conrad. "Chu Hsi." *ER* III, 469-472.

Schluchter, Wolfgang, Ed. *Max Webers Studie über Konfuzianismus und Taoismus: Interpretation und Kritik.* Frankfurt: Suhrkamp Verlag. 1983.

Schmidt, Hans-Hermann. "Die Hundertachtzig Vorschriften von Lao-chun." In Naundorf (1985), 149-159.

Schmidt-Glintzer, Helwig. "Eine Ehrenrettung fur den Suden. Pao-chih (418/25-514) und Fu Hsi (497-569) — Zwei Heilige aus dem unteren Yangtse-Tal." In Naundorf (1985), 247-265.

Schmidt-Glintzer, Helwig. *Die Identität der Buddhistischen Schulen und die Kompilation Buddhistischer Universalgeschichten in China.* (Munchener Ostasiatische Studien, Bd. 26). Wiesbaden: Franz Steiner, 1981.

Schneider, R. "Les Pèlerinages du Wutaishan d'après les Manuscrits de Dunhuang." *Cahiers d'Extrême-Asie* 3 (1987).

Schopen, Gregory. "Filial Piety and the Monk in the Practice of Indian Buddhism: A Question of 'Sinicization' Viewed from the Other Side." *TP* 70 (1984), 110-126.

Schuster, Nancy. "Striking a Balance: Women and Images of Women in Early Chinese Buddhism." In *Women, Religion, and Social Change*. Ed Yvonne Yazbeck Haddad and Ellison Banks Findly (Albany: State University of New York Press, 1985), 87-112.

Schuyler, Cammann. "Some Early Chinese Symbols of Duality" *HR* 24 (1985), 215-254.

Schwartz, Benjamin I. *The World of Thought in Ancient China.* Cambridge, Mass. & London: the Belknap Press of Harvard University Press, 1985.

Scott, David A. "Christian Responses to Buddhism in Pre-medieval Times." *Numen* 32 (1985), 88-100.

Seah, Ingram S. "Nestorian Christianity and Pure Land Buddhism in T'ang China." *Taiwan Journal of Theology* 6 (1984), 75-92.

Seaman, Gary. " The Divine Authorship of *Pei-yu chi* (Journey to the North)." *JAS* 45 (1986), 483-497.

Seaman, Gary. "In the Presence of Authority: Hierarchical Roles in Chinese Spirit Medium Cults." In *Normal and Abnormal Behavior in Chinese Culture.* Eds. Arhur Kleinman and T. Y. Lin (Dordrecht: Reidel, 1981), 61-74.

Seaman, Gary. "Only Half-Way to Godhead: The Chinese Geomancer as Alchemist and Cosmic Pivot." *AFS* 45 (1986), 1-18.

Seaman, Gary. *Pei Yu Chi: Journey to the North.* Berkeley: University of California Press, 1987.

Seaman, Gary, "The Sexual Politics of Karmic Retribution." In *The Anthropology of Taiwanese Society*. Eds. Emily M. Ahern and Hill Gates (Stanford: Stanford University Press, 1981), 381-396; 468-469.

Seaman, Gary. "Spirit Money: An Interpretation." *JCR* 10 (1982), 80-91.

Seidel, Anna. "Afterlife: Chinese Concepts." *ER* I, 124-127.

Seidel, Anna. "Geleitbrief an die Unterwelt: Jenseitsvorstellungen in den Graburkunden der Späteren Han Zeit." In Naundorf (1985), 161-183.

Seidel, Anna. "Taoist Messianism." *Numen* 31 (1984), 161-174.

Seidel, Anna. "Imperial Treasures and Taoist Sacraments: Taoist Roots in the Apocrypha." In Strickmann (1983), 291-371.

Seiwert, Hubert. "Religious Response to Modernization in Taiwan: The Case of I-kuan Tao." *JHKBRAS* 21 (1981), 43-70.

Seiwert, Hubert. *Volksreligion und Nationale Tradition in Taiwan. Studien zur Regionalen Religionsgeschichte einer Chinesischen Provinz.* (Münchener Ostasiatische Studien, no. 38). Wiesbaden & Stuttgart: Franz Steiner Verlag, 1985.

Sellmann, James D. "The 'Cosmic Talisman' of Liturgical Taoism: An Analysis of the Structure and Content ot the *Ling Pao Chen Wen.* " *Chinese Culture* 24 (1983), 57-69.

Sellmann, James D. "From Myth to Festival: A Structural Analysis of the Chinese New Year Celebration." *Chinese Culture* 23 (1982), 41-58.

Shek, Richard. "Millenarianism: Chinese Millenarian Movements." *ER* IX, 532-536.

Shim, Jae-ryong. "Faith and Practice in Hua-yen Buddhism: a Critique of Fa-tsang (643-712) by Li T'ung-hsuan (646-750)" in Chappell (1987), 109-124.

Shiratori, Yoshiro, ed. *The Dragon Boat Festival in Hong Kong .* Tokyo: Sophia University, 1985.

Sivin, Nathan. "Alchemy: Chinese Alchemy." *ER* I, 186-190.

Skaja, Henry G. "Li (Ceremonial) as a Primal Concept in Confucian Spiritual-Humanism." *Chinese Culture* 25 (1984), 1-26.

Snellgrove, David L. "Celestial Buddhas and Bodhisattvas." *ER* III, 133-144.

Song, Arhtur. "Chinese Religion: The Chinese Community in Southern Africa." *Religion in South Africa* 3 (198), 19-30.

Southhard, Samuel, and Donna Southhard. "Demonizing and Mental Illness: The Problem of Identification, Hong Kong." *Pastoral Psychology* 33 (1985), 173-188.

Soymie, Michel. "Notes d'iconographie bouddhique — des Vidyárája et Vajradhara de Touen-houang." *Cahiers d'Extrême-Asie* 3 (1987), 9-26.

Spae, J. "China's New Constitution and Its Impact on Religion." *China Update* 3 (spring 1983), 31-42.

Sponberg, Alan. "Meditation in Fa-hsiang Buddhism." In Gregory (1986), 15-43.

Sponberg, Alan. "A Report on Buddhism in the People's Republic of China." *JIABS* 5 (1982), 109-177.

Sponberg, Alan. "The Study of Buddhism in China: Some Observations on the Chinese Buddhist Association and its Seminary." *JCR* 12 (1984), 65-76.

Stein, Rolf A. "Avalokitesvara Kouan-yin, un Exemple de Transformation d'un Dieu en Déesse." *Cahiers d'Extrême-Asie* 2 (1986), 17-77.

Stein, Rolf A. "Sudden Illumination or Simultaneous Comprehension: Remarks on Chinese and Tibetan Terminology." In Gregory (1987), 41-65.

Stevenson, Daniel B. "The Four Kinds of Samadhi in Early T'ien-t'ai Buddhism." In Gregory (1986), 45-97.

Strickmann, Michel, ed. *Tantric and Taoist Studies in Honour of R. A. Stein* , 2 vol. (Melanges Chinois et Bouddhiques, 20-21), Brussels: Institut Belge des Hautes Etudes Chinoises, 1981-83 (vol. 3, forthcoming).

Strickmann, Michel. *Le Taoisme du Mao Chan: Chronique d'une Révélation* . (Memoires de l'Institut des Hautes Etudes Chinoises, 17). Paris: Presses Universitaires de France, 1981.

Strickmann, Michel. "Therapeutische Rituale und das Problem des Bösen im frühen Taoismus." In Naundorf (1985), 185-200.

Strong, J. " Filial Piety and Buddhism: The Indian Antecedents to a 'Chinese' Problem." In *Traditions in Contact and Change* . Eds. Peter Slater and Donald Wiebe (Waterloo, Ontario: Wilfried Laurier University Press, 1983), 171-186.

Suenari, Michio. "The 'Religious Family' Among the Chinese of Central Taiwan." In De Vos (1984), 169-184.

Sun, K'o-k'uan, "Yu Chi and Southern Taoism during the Yüan Period." In *China Under Mongol Rule* . Ed. John D. Langlois (Princeton: Princeton University Press, 1981), 212-253.

The Sutra of Contemplation of the Buddha of Immeasurable Life. Kyoto: Ryukoku University Translation Centre, 1984.

Sutton, Donald. "Pilot Surveys of Chinese Shamans, 1875-1945: A Spatial Approach to Social History." *Journal of Social History* 15 (1981), 39-50.

Swanson, Gerald. "The Concept of Change in the *Great Treatise.* " In Rosemont (1984), 67-93.

Swanson, Paul L. "T'ien-t'ai Studies in Japan." *Cahiers d'Extrême-Asie* 2 (1986), 219-232.

Sweater, Donald K. "Buddhist Religious Year." *ER* II, 547-554.

Swearer, Donald K. "Folk Religion: Folk Buddhism." *ER* II, 540-547.

Tan, Chee-Beng. "Chinese Religion in Malaysia: A General View." *AFS* 42 (1983), 217-252.

Tan, Chee-Beng. *The Development and Distribution of Dejiao As-*

sociations in Malaysia and Singapore: A Study on a Chinese Religious Organization . (Occasional Paper No. 79). Singapore: Institute of Southeast Asian Studies, 1985.

Tao Fong Shan Work Group on Chinese Society. "Religion and Chinese Society: A Quarterly Review." *Ching Feng* (1981-) [a regular feature of the journal].

Tatz, Mark. *Buddhism and Healing* . See Demiéville, Paul.

Taylor, Rodney L. *The Way of Heaven: An Introduction to the Confucian Religious Life* . (Iconography of Religions, XII/3). Leiden: E.J. Brill, 1986.

Teiser, Stephen F. "Engulfing the Bounds of Order: The Myth of the Great Flood in *Mencius.*" *JCR* 13-14 (1985-86), 15-43.

Teiser, Stephen F. "Ghosts and Ancestors in Medieval Chinese Religion: The Yü-lan-p'en Festival as Mortuary Ritual." *HR* 26 (1986), 47-67.

Teiser, Stephen F. "T'ang Buddhist Encyclopedias: An Introduction to *Fa-yuan chu-lin* and *Chu-ching yao-chi.*" *T'ang Studies* 3 (1985), 109-128.

Teiser, Stephen F. *The Ghost Festival in Medieval China.* Princeton: Princeton University Press, 1988.

Teng, Ssu-yu. *Protest and Crime in China: A Bibliography of Secret Societies, Popular Uprisings, Peasant Rebellions* . New York: Garland, 1981.

Thompson, Laurence G. *Chinese Religion in Western Languages: A Comprehensive and Classified Bibliography of Publications in English, French, and German through 1980* . (Association for Asian Studies Monograph, XLI). Tucson: University of Arizona Press, 1985.

Thompson, Laurence G. "Chinese Religious Year." *ER* III, 323-328.

Thompson, Laurence G. "Confucian Thought: The State Cult." *ER* IV, 36-38.

Thompson, Laurence G. "Iconography: Taoist Iconography." *ER* VII, 50-54.

Thompson, Laurence G. "The Moving Finger Writes: A Note on Revelation and Renewal in Chinese Religion." *JCR* 10 (1982), 92-147.

Thompson, Laurence G. "Obiter Dicta on Chinese Religion as Play." In *Chung-yang yen-chiu-yuan kuo-chi Han-hsueh-hui i-lun wen-chi* . Taipei: Chung-yang yen-chiu-yuan, 1981, 59-72.

Thompson, Laurence G. "Observations on Religion in Communist China: Introductory Comments." *JCR* 12 (1984), 33-36.

Thompson, Laurence G. "Popular and Classical Modes of Ritual in a Taiwanese Temple." *BSSCR* 9(Fall 1981), 106-122.

Thompson, Laurence G. "Taoism: Classic and Canon." In *The Holy Book in Comparative Perspective*. Eds Frederick M. Denny and Rodney L. Taylor (Columbia, SC: University of South Carolina, 1985), 204-223.

Thompson, Laurence G. "T'ien." *ER* XIV, 508-510.

Till, Barry and Paula Swart. "Funerary Sculpture of the Northern Sung Dynasty." *AA* 14 (1984), 81-89.

Tirone, Gail. "A Gathering of Deities Occasions Gala Celebrations." *Free China Review* 36 (July 1986): 36-41 (The Matsu festival at Peikang, 1986).

Tober, Linda M., and F. Stanley Lusby. "Heaven and Hell." *ER* VI, 237-243.

Tong, Fung-Wan. "Understanding of the Social Ethical Dimensions of Buddhism and Christianity." Trans. G. G. Caldwell. *Taiwan Journal of Theology* 7 (1985), 151-165.

Tsai, Kathryn A. "The Chinese Buddhist Monastic Order for Women: The First Two Centuries." In *Women in China: Current Directions in Historical Scholarship*. Eds Richard W. Guisso and Stanley Johannesen (Youngstown, NY: Philo Press, 1981), 1-20.

Tsao, Pen-yeh, ed. *Taoist Ritual and Music*. Hong Kong: University of Hong Kong Press (forthcoming).

Tsao, Pen-yeh. "Variation Technique in the Formal Structure of the Music of Taoist Jiao-shi in Hong Kong." *JHKBRAS* 23 (1983), 172-183.

Tsui, Bartholomew P.M. "The Self-Perception of Buddhist Monks in Hong Kong Today." *JHKBRAS* 23 (1983), 23-40.

Tsukamoto, Zenryu. *A History of Early Chinese Buddhism: From its Introduction to the Death of Hui-yüan*. 2 vols. Trans. Leon Hurvitz. Tokyo: Kodansha International; New York: Harper & Row, 1985.

Tu, Wei-ming. "Soul: Chinese Concepts." *ER* XIII, 447-450.

Twitchett, Denis. "Law and Religion: Law and Religion in East Asia." *ER* VIII, 469-472.

Tymieniecka, Anna-Teresa, ed. *Phenomenology of Life in a Dialogue Between Chinese and Occidental Philosophy*. Dordrecht: D. Reidel, 1984.

Ulin, Robert C. "Peasant Politics and Secret Societies: The Discourse of Secrecy." *Anthropological Quarterly* 59 (1986), 28-39.

Unno, Taitetsu. "Worship and Cultic Life: Buddhist Cultic Life in East Asia." *ER* XV, 467-472.

Unschuld, Paul U. *Medicine in China: A History of Ideas*. Berkeley: University of California Press, 1985.

Van der Loon, Piet. *Taoist Books in the Libraries of the Sung Period: A Critical Study and Index.* (Oxford Oriental Institute Monographs, no.7). London: Ithaca Press, 1984.

Veith, Ilza. "Medicine: Medicine and Religion in Eastern Traditions." *ER* IX, 312-319.

Walls, Jan, and Yvonne Walls, ed. and trans. *Classical Chinese Myths*. Hong Kong: Joint Publishing Co., 1984.

Wang, Yi-t'ung, trans. *A Record of Buddhist Monasteries in Loyang*. By Yang Hsuan-chih. Princeton: Princeton University Press, 1984.

Watson, Burton. "Chuang-tzu." *ER* III, 467-469.

Watson, James L. "Standardizing the Gods: The Promotion of T'ien Hou ("Empress of Heaven") Along the South China Coast, 960-1960." In Johnson (1985), 292-324.

Wayman, Alex. "Male, Female, and Androgyne: Per Buddhist Tantra, Jacob Boehme, and the Greek and Taoist Mysteries." In Strickmann (1983), 592-631.

Wechsler, Howard J. *Offerings of Jade and Silk: Ritual and Symbol in the Legitimation of the T'ang Dynasty*. New Haven: Yale University Press, 1985.

Weinstein, Stanley. "Buddhism, Schools of: Chinese Buddhism." *ER* II, 482-487.

Weinstein, Stanley. *Buddhism Under the T'ang*. Cambridge: Cambridge University Press, 1987.

Weller, Robert P. "Bandits, Beggars, and Ghosts: The Failure of State Control over Religious Interpretation in Taiwan." *American Ethnologist* 12 (1985), 46-61.

Weller, Robert P. "The Politics of Ritual Disguise: Repression and Response in Taiwanese Religion." *Modern China* 13 (1987), 17-39.

Weller, Robert P. "Sectarian Religion and Political Action in China." *Modern China* 8 (1982), 463-483.

Weller, Robert P. *Unities and Diversities in Chinese Religion*. Seattle: University of Washington Press, 1987.

Weller, Robert P., and Hill Gates. "Hegemony and Chinese Folk Ideologies." *Modern China* 13 (1987), 3-16.

Wilson, B.D. "Notes on Some Chinese Customs in the New Territories." *JHKBRAS* 23 (1983), 41-61.

Wilson, Richard W., Sidney L. Greenblatt, and Amy Auerbacher Wilson, Eds. *Moral Behavior in Chinese Society*. New York: Praeger, 1981.

Wong, David. "Taoism and the Problem of Equal Respect." *JCP* 11 (1984), 165-183.

Wong, Isabel. "Music: Music and Religion in China, Korea, and

Tibet." *ER* X 195-203.

Wright, Dale S. "The Significance of Paradoxical Language in Hua-yen Buddhism." *PEW* 32 (1982), 325-338.

Wu Kuang-Ming. "Trying Without Trying: Toward a Taoist Phenomenology of Truth." *JCP* 8 (1981), 143-167.

Wu, Yi. "On Chinese ch'an in Relation to Taoism." *JCP* 12 (1985), 131-154.

Yang, Winston L. Y. "From History to Fiction: The Popular Image of Kuan Yu." *Renditions* 15 (1981), 67-79.

Yao, Tao-chung. "Ch'iu Ch'u-chi and Chinggis Khan." *HJAS* 46 (1986), 201-219.

Yu, Anthony C. "Two Literary Examples of Religious Pilgrimage: The *Commedia* and *The Journey to the West.*" *HR* 22 (1983), 202-230.

Yu, Chun-fang. "Chinese Buddhist Responses to Contemporary Problems." *Journal of Dharma* 10 (1985), 60-74.

Yu, Chun-fang. "Chung-feng Ming-pen and Ch'an Buddhism in the Yuan." In *Yuan Thought: Chinese Thought and Religion Under the Mongols* . Eds. Hok-Lam Chan and William Theodore de Bary (New York: Columbia University Press, 1982), 419-477.

Yu, Chun-fang. *The Renewal of Buddhism in China: Chu-hung and the Late Ming Synthesis.* (IASWR series), New York: Columbia University Press, 1981.

Yu, Clara, trans. "Ancestral Rites." In Ebrey (1981), 79-83.

Yu, David C. "The Creation Myth and Its Symobolism in Classical Taoism." *PEW* 31 (1981), 479-500.

Yu, David C. *Guide to Chinese Religion* . Boston: G. K. Hall, 1985.

Yu, David C. "The Mythos of Chaos in Ancient Taoism and Contemporary Chinese Thought." *JCP* 8 (1981), 325-348.

Yu, Ying-shih. "New Evidence on the Early Chinese Conception of Afterlife: A Review Article." *HJAS* 41 (1981), 81-85.

Zeuschner, Robert. "The Understanding of Karma in Early Ch'an Buddhism." *JCP* 8 (1981), 399-425.

Zeuschner, Robert B. "Awakening in Northern Ch'an". In Chappell (1987), 85-108.

Zhao, Puchu. "A Recapitulation of Buddhism in China in the Last Thirty Years." Trans. D. MacInnis. *Ching Feng* 24 (1981), 108-118.

Zhou, Guidian, and Guo Shaoming. "Liu Zongyuan's Concept of Heaven." *Social Sciences in China* 6 (1985), 142-165.

Zürcher, Erik. "Amitabha." *ER* I, 235-237.

Zürcher, Erik. "Buddhism: Buddhism in China." *ER* II, 414-421.

Zürcher, Erik. "Eschatology and Messianism in Early Chinese Buddhism." In *Leyden Studies in Sinology: Papers Presented at the Conference Held in Celebration of the Fiftieth Anniversary of the Sinological Institute of Leyden University, December 8-12, 1980* . Ed. W. L. Idema (Leiden: E. J. Brill, 1981), 34-56.

Zurcher, Erik. " Perspectives in the Study of Chinese Buddhism." *JRAS* (1982), 161-176.

Zürcher, Erik. " 'Prince Moonlight', Messianism and Eschatology in Early Medieval Chinese Buddhism." *TP* 68 (1982), 1-75.

Zürcher, Erik. "The *Sangha* in China." In *The World of Buddhism* . Eds. H. Bechert and R. Gombrich (London: Thames and Hudson, 1984), 193-212.

Zürcher, Erik. "The Lord of Heaven and the Demons — Strange Stories from a Late Ming Christian Manuscript." In Naundorf (1985), 359-376.

1977－1987年中國出版有關中國宗教之書籍及論文索引

梁文金[*]

編者引言：

1. 本目錄收錄1977至1987年十年間中國學者在國內以中文發表有關中國宗教問題之專書及論文，其翻譯自外文者，概不收入。
2. 本目錄未收有關宗教藝術方面之資料。
3. 本目錄所收之專書及論文，均經編者過目，其編者未寓目者，概不列入。
4. 本目錄排列次序，先專書，次論文，其編排次序，以出版時間之先後爲序。
5. 重印書或論文，其有重要學術價值者，亦列入。

* Man Kam Leung

目錄

宗教一般

宗教一般：研究宗教之目錄

1. 《國內報刊有關宗教問題部份文章目錄索引（1979-1980）》，《世界宗教資料》，1981年第 2 期（1981年 6 月），58-65頁。
2. 《1981年國內報刊有關宗教部份文章目錄索引》，《世界宗教資料》，1982年第 2 期（1982年 5 月），59-65頁。
3. 《1982年國內報刊有關宗教問題部份文章目錄索引》，《世界宗教資料》，1983年第 2 期（1983年 5 月），59-64頁。
4. 《1983年國內報刊有關宗教問題部份文章目錄索引》，《世界宗教資料》，1984年第 2 期（1984年 5 月），58-65頁。
5. 《1984年國內報刊有關宗教問題部份文章目錄索引》，《世界宗教資料》，1985年第 2 期（1985年 5 月），59-65頁。
6. 《1985年國內報刊有關宗教問題部份文章目錄索引》，《世界宗教資料》，1986年第 2 期（1986年 6 月），58-65頁。

宗教一般：論文索引

1. 世界宗教研究所圖書資料室編，《《全唐文》宗教類篇目分類索引》，《世界宗教研究》，1981年第 4 期（總第 6 期）（1981年12月）128-158頁
2. 世界宗教研究所圖書資料室編，《《全上古三代秦漢三國六朝文》宗教類篇目分類索引》，《世界宗教研究》，1982年第 2 期（總第 8 期）（1982年 5 月），115-153頁
3. 世界宗教研究所圖書資料室編，《《東方雜誌》（1904-1948）宗教問題資料索引》，《世界宗教研究》，1982年第 3 期（總第 9 期）(1982年 8 月)，152-154頁；1982年第 4 期（總第10期）（1982年11月），147-152頁

宗教一般：宗教參考書

1. 任繼愈主編《宗教辭典》，上海，上海辭書出版社，1981年，1450頁。
2. 中國社會科學院哲學研究所編，《中國哲學年鑑》，上海，中國大百科全書出版社

1982年	436頁
1983年	615頁
1984年	630頁
1985年	579頁
1986年	568頁

3. 袁珂 《中國神話傳說辭典》，上海，上海辭書出版社，1985年，580頁。
4. 馮契等編 《哲學大辭典-中國哲學史卷》，上海，上海辭書出版社，1985年，862頁。
5. 方克立等編 《中國哲學史論文索引》，第1冊（1900-1949），北京，中華書局，1986年，573頁。

宗教一般：通論之部（專書）
1. 朱天順《原始宗教》 上海，上海人民，1964年初版，1978年等二次印刷，116頁。
2. 黃心川，戴康生等《世界三大宗教》，北京，三聯，1979年，138頁。
3. 牙含章《無神論和宗教問題》，上海，上海人民，1979年，136頁。
4. 三友三《中國無神論史綱》，上海，上海人民，1982年初版，376頁；修訂本，1986年，522頁。
5. 朱天順《中國古代宗教初探》，上海，上海人民，1982年，319頁。
6. 三友三編，顧曼君，馬俊南註《中國無神論史資料選編》（先秦編），北京，中華，1983年170頁。
7. 牙含章《民族問題與宗教問題》，成都，中國社會科學出版社，四川民族出版社，1984年，371頁。
8. 三友三編，顧曼君，馬俊南註《中國無神論史資料選編》（兩漢編），北京，中華書局，1985年，245頁。
9. 曹琦，彭耀《世界三大宗教在中國》，北京，中國社會科學出版社，1986年，149頁。
10. 陳麟書《宗教學原理》（高等大學文科教材），成都，四川大學，1986年361頁。
11. 雲南社會科學院宗教研究所編，《宗教論稿》，昆明，雲南人民出版社，1986年，281頁。

宗教一般： 通論之部（論文）
1. 丁寶蘭《中國原始宗教與無神論的萌芽》，《中山大學學報》《哲學社會科學》，1978年第4期（1978年7月），2-13頁。
2. 陳垣《談談宗教史研究的問題》，《中國哲學》第6輯（1981年5月），25-28頁。
3. 張繼安《對"宗教是人民的鴉片"這個論斷的初步理解》，《世界宗教研究》，1981年第2期（總4期）（1981年6月），1-12頁。
4. 三友三《中國無神論同有神論在鬥爭中起伏消長變化的規律》，《世界宗教研究》，1981年第3期（總5期）（1981年9月），78-87頁。

5. 蕭漢明《論中國古史上兩次"絕對天通"》，《世界宗教研究》，1981年3期（1981年9月），88-98頁。

6. 余敦康《中國原始宗教的演變》，《世界宗教研究》，1981年4期（總6期）（1981年12月），94-100頁。

7. 蔡家麒《自然、圖騰、祖先（原始宗教初探）》，《哲學研究》，1982年4期（1982年4月），54-61頁。

8. 夏之乾《略談我國不同經濟類型民族中原始宗教的差異》，《世界宗教研究》，1982年2期（總8期），（1982年5月），95-103頁。

9. 蔡家麒《關於原始宗教的研究》，《思想戰綫》，1982年4期（總46），（1982年8月），77-79頁。

10. 三方三《宗教，無神論與自然科學》，《世界宗教研究》，1982年3期，（總9期），（1982年8月），121-132頁。

11. 方德紹《試論宗教的社會作用》，《思想戰綫》，1983年4期，（總52期），（1983年8月），65-71及64頁。

12. 周星《中國古代岩畫中所見的原始宗教》，《世界宗教研究》，1984年1期，（總15期），（1984年2月），113-122頁。

13. 王友三《中國無神論在反神學的鬥爭中解決的幾個問題及其局限》，《世界宗教研究》，1984年1期（總15），（1984年2月），144-151頁。

14. 張繼安《試論我國社會主義時期的宗教》，《世界宗教研究》，1985年4期，（總22期），（1985年12月），58-69頁。

15. 周慶基《古代宗教觀念中靈魂與肉體的關係》，《世界宗教研究》，1985年4期，（總22），（1985年12月），70-75頁。

16. 龔學增《關於我國社會主義時期宗教方面的幾個問題》，《世界宗教研究》，1986年1期，（總23），（1986年3月），134-142頁。

17. 牟鍾鑒《中國宗教的歷史特點》，《世界宗教研究》，1986年2期，（總24期），（1986年6月），36-40頁。

18. 于本源《宗教信仰自由與無神論宣傳》，《世界宗教研究》，1986年3期，（總25期），（1986年9月），131-139頁。

宗教一般（論文集）

1. 湯敬昭編《中國無神論思想論文集》，江蘇人民出版社，1980年，273頁。

2. 中國無神論學會編《中國無神論文集》，湖北人民出版社，1982年，336頁。

3. 中國無神論學會編《宗教與無神論》，福州，福建人民出版社，1985年，318頁。

佛教

佛教目錄
  呂徵《新編漢文大藏經目錄》，濟南，齊魯書社，1980年，176頁。

佛教參考書
  中國佛教學會編《中國佛教》，北京，知識出版社；
    第1輯，1980年，399頁。
    第2輯，1982年，395頁。

佛教論文集
  1. 向達《唐代長安與西域文明》，北京，三聯，1957年第一版，
     1979年第2次印，670頁。
  2. 陳寅恪《寒柳堂集》，上海，上海古籍出版社，1980年，186頁。
  3. 陳寅恪《金明館叢稿初編》，上海，上海古籍出版社，1980年，
     370頁。
  4. 陳寅恪《金明館叢稿二編》，上海，上海古籍出版社，1980年，
     316頁。
  5. 任繼愈《漢唐佛教思想論集》，北京，人民出版社，1973年初版，
     1981年再版，349頁。
  6. 方立天《魏晉南北朝佛教論叢》，北京、中華書局，1982年，
     311頁。
  7. 季羨林《中印文化關係史論文集》，北京、三聯書店，1982年，
     488頁。
  8. 湯用彤《湯用彤學術論文集》，北京，中華書局，1983年，436
     頁。
  9. 任繼愈、季羨林、蔡尚思等《中國佛學論文集》，西安，陝西人
     民出版社，1984年，432頁。

佛教哲理（專書之部）
  嚴北溟《中國佛教哲學簡史》，上海，上海人民出版社，1985年，
    235頁。

佛教哲理（論文之部）
  1. 李富華《略論禪宗的形成》，《世界宗教研究》，第1集（1979
     年），284-300頁。
  2. 郭朋《三階教略論》，《世界宗教研究》，第2集，（1980年），
     24-42頁。
  3. 嚴北溟《論佛教哲學對某些唯物論者或進步思想家的影響》，《中
     國哲學》，第3輯（1980年8月），202-219頁。

4. 石峻，方立天《論魏晉時代佛學與玄學的異同》，《哲學研究》，1980年第10期，（1980年10月），31-41頁。

5. 虞愚《玄奘對因明的貢獻》，《中國社會科學》，1981年1期（總7期），（1981年1月），197-208頁。

6. 郭朋《禪宗五家-禪宗思想研究之二》，《世界宗教研究》，第3期（1981年2月），40-51頁。

7. 許抗生《略論兩晉時期的佛教哲學思想》，《中國哲學》，第6輯（1981年5月），29-60頁

8. 孫實明《唯識宗唯心主義哲學簡述》，《中國哲學》，第6輯（1981年5月），105-118頁。

9. 楊廷福《玄奘在國外的傳承》，《中國哲學》，第6輯（1981年5月），119-137頁。

10. 嚴北溟《論佛教的美學思想》，《復旦學報》（社會科學），1981年第3期（1981年5月），41-51頁。

11. 翁志鵬《試論智顗》，《世界宗教研究》，1981年第2集，（總4期）（1981年6月），44-53頁。

12. 杜繼文《略論康僧會佛學思想的特色》，《世界宗教研究》，1981年第2集，（總4期）（1981年6月），37-43頁。

13. 李富華《惠能和他的佛教思想》，《世界宗教研究》，1981年第3集，（總5期）（1981年9月），107-120頁。

14. 嚴北溟《論佛教哲學在思想史上的挑戰》，《哲學研究》，1982年第2期，（1982年2月），44-51頁。

15. 陳士強《佛教"格義"法的起因》，《復旦學報》（社會科學），1981年第3期（1982年5月），78-81頁。

16. 戒圓《試論大乘佛教對於"眞如"的不同看法》，《法音》，1982年第4期（總8期）（1982年7月），32-36頁。

17. 張春波《略論我國初傳的佛教》，《中國哲學史研究集刊》，第2輯，（1982年7月），295-306頁。

18. 文丁《早期佛教般若學和貴無派玄學的關係》，《中國哲學史研究集刊》，第2輯，（1982年7月），307-323頁。

19. 石峻，方立天《論隋唐佛教宗派的思想特點》，《中國哲學史研究集刊》，1984年第4期（總9期）（1982年10月），36-45及61頁。

20. 崔大華《說"陽儒陰釋"》-理學與佛學的聯繫和差別》，《中國哲學史研究集刊》，1982年第4期（總9期）（1982年10月），62-697頁。

21. 張建木《"風幡不動"解》，《法音》，1982年第6期（總10期）（1982年11月），6-7頁。

22. 郭朋《慧能的思想與《壇經》的演變》，《中國哲學史研究集刊》，1983年第1期（總10期）（1983年1月），33-40頁。

23. 潘桂明《從智圓的《閑居編》看北宋佛教的三教合一思想》,《世界宗教研究》,1983年第1期,(總11期)(1983年2月),78-94頁。

24. 李富華《宗密和他的禪宗》,《世界宗教研究》,1983年第1期(總11期)(1983年2月)95-106頁。

25. 溫玉成《禪宗北宗初探》,《世界宗教研究》,1983年第2期(總12期)(1983年5月)23-36頁。

26. 潘桂明《臨濟宗思想初探》,《世界宗教研究》,1983年第3期(總13)(1983年8月)1-15頁。

27. 呂希晨《評熊十力的《新唯識論》》,《世界宗教研究》,1983年第3期(總13期)(1983年8月)26-36頁。

28. 游有維《天台宗講要》,《法音》(上),1983年第5期(總15期)(1983年9月)8-13頁。(下),1983年16期(總16期)(1983年12月),69-74頁及68頁。

29. 楊曾文《佛教《般若經》思想與玄學的比較》,《世界宗教研究》,1983年第4期(總14期)(1983年11月),63-76頁。

30. 呂澂,熊十力《辨佛學根本問題》,《中國哲學》,第11輯,(1984年1月),169-199頁。

31. 任繼愈《唐宋以後的三教合一思潮》,《世界宗教研究》,1984年第1期(總15期)(1984年2月)1-6頁。

32. 陳士強《論吉藏的佛學思想》,《世界宗教研究》,1984年第1期(總15期)(1984年2月),60-77頁。

33. 余敦康《六家七宗-兩晉時期的佛教般若學思潮》,《世界宗教研究》,1984年第2期(總16期)(1984年5月),49-62頁。

34. 陳士強《中國早期佛教形神論與其他形神論之比較研究》,《中國哲學史研究》,1984年第4期(總17期)(1984年10月),70-77頁。

35. 廖明活《天台智顗對儒家道家的態度》,《世界宗教研究》,1984年4期(總18期)(1984年11月),25-35頁。

36. 樂壽明《論佛教對於"神"的理解》,《中國哲學》,第13輯,(1985年4月),213-226頁。

37. 溫玉成《禪宗北宗續探》,《世界宗教研究》,1985年第2期,(總25期)(1985年6月),72-79頁。

38. 馬序《法藏一多思想辨析》,《世界宗教研究》,1985年第4期(總22期)(1985年12月),98-104頁。

39. 樓宇烈《佛學與中國近代哲學》,《世界宗教研究》,1986年第1期,(總23期)(1986年3月),1-17頁。

40. 郭朋《南朝"佛性"論思想略述》,《世界宗教研究》,1986年第1期(總23期),(1986年3月),18-26頁。

41. 方立天《華嚴宗哲學范疇體系簡論》,《世界宗教研究》,1986

353

年第 2 期（總24期）（1986年 6 月），24-27頁。

42. 羅炤《禪宗評述》，《世界宗教研究》，1986年第 3 期（總25期）
（1986年 9 月），15-30頁。

43. 方立天《佛教，佛法，佛學與佛教哲學》，《世界宗教研究》，
1986年第 4 期（總26期）（1986年12月）60-65頁。

佛教歷史（專書之部）

1. 范文瀾《唐代佛教-附隋唐五代佛教大事年表》，北京人民出版
社，1979年，311頁。

2. 呂澂《中國佛學源流略講》，北京中華書局，1979年，340頁。

3. 郭朋《隋唐佛教》，濟南齊魯書社，1980年，653頁。

4. 郭朋《宋元佛教》，福州，福建人民，1981年，209頁。

5. 任繼愈主編《中國佛教史》，北京，中國社會科學出版社
第 1 卷，1981年，601頁。
第 2 卷，1985年，786頁。

6. 郭朋《明清佛教》，福州，福建人民，1982年，343頁。

7. 湯用彤《隋唐佛學史稿》，北京，中華書局，1982年，317頁。

8. 湯用彤《漢魏兩晉南北朝佛教史》，北京，中華書局，上下冊，
1983年，672頁。

9. （唐）道宣著，范祥雍校點《釋迦方誌》（中外交通史藉叢刊），
北京，中華書局，1983年，130頁。

10. （唐）慧立，彥悰著，孫毓棠，謝方校點《大慈恩寺三藏法師
傳》，（中外交通史籍叢刊），北京，中華書局，1983年，245
頁。

11. （東晉）法顯著，章巽校注《法顯傳校注》，上海，上海古籍出
版社，1985年，244頁。

12. （唐）玄奘，辯機著，季羨林等校點，《大唐西域記校注》（中
外交通史籍叢刊），北京，中華書局，1985年，1284頁。

13. 季羨林等《大唐西域記今譯》，西安，陝西人民出版社，1985年，
449頁。

14. 邱明洲《中國佛教史略》，成都，四川省社會科學院出版社，
1986年，177頁。

15. （日）圓仁著，顧承甫，何泉達校《入唐求法巡禮行記》，上海，
上海古籍出版社，1986年，225頁。

佛教歷史（論文之部）

1. 鞏紹英《從唐朝中葉以前反佛教鬥爭談到佛教在中國的發展道
路》，《中國哲學》，第 1 集（1979年 8 月），251-287頁。

2. 丁明夷《玄中寺》，《世界宗教研究》，第 1 集（1979年 8 月），
301-307頁。

3. 汪向榮《鑒眞在日本佛教史中的作用》，《世界宗教研究》，第 2集（1980年8月），11-19頁。

4. 李富華《中國封建統治者的興佛與廢佛》，《世界宗教研究》，第2集（1980年8月），82-92頁。

5. 湯用彤《五代宋元明佛教事略》，《中國哲學》，第5輯（1981年1月），456-472頁。

6. 李斌城《唐代佛道之爭研究》，《世界宗教研究》，1981年第2期（總4期）（1981年6月），99-108頁。

7. 溫玉成《少林寺與"孔門禪"》，《世界宗教研究》，1981年第2期（總4期）（1981年6月），136-147頁。

8. 徐梵登《韋陀教神壇與大乘菩薩道概觀》，《世界宗教研究》，1981年第3期（總5期）（1981年9月），63-77頁。

9. 業露華《北魏的僧祇戶和佛圓戶》，《世界宗教研究》，1981年第3期（總5期）（1981年9月）1138-143頁。

10. 方立天《梁武帝蕭衍與佛教》，《世界宗教研究》，1981年第4期（總6期）（1981年12月），16-33頁。

11. 何茲全《佛教經律關於寺院財產的規定》，《中國史研究》，1982年第1期，（總13期）（1982年3月，68-78頁。

12. 郭朋《明太祖與佛教》，《世界宗教研究》，1982年第1期（總7期）（1982年2月），43-69頁。

13. 周子美《近百年來我國佛學研究拾零》，《華東師範大學學報（哲學社會科學）》1982年1期（總39期），1982年2月），92-94頁。

14. 魏承恩《唐代長安和佛教》，《法音》，1982年第2期（總6期）（1982年3月），38-42頁。

15. 夏應元《中國禪僧東渡日本及其影響》，《歷史研究》，1982年3期，（1982年6月），131-192頁。

16. 談狀飛《名僧智凱死之疑》，《中國哲學史研究集刊》，第2輯，（1982年7月），376-390頁。

17. 李文生《中日佛教史上洛陽的佳話》，《世界宗教研究》，1982年第3期（1982年8月），63-64頁。

18. 蘇淵雷《略論我國近代學者研究佛學主要傾向和成就》，《法音》，1982年第5期（總9期）（1982年9月），5-10頁及142頁。

19. 游有維《中國淨土宗弘傳的歷史》，《法音》，1982年5期（總9期）（1982年9月），21-22頁。

20. 張英莉，戴禾《義邑制度述略-兼論南北朝佛道混合之原因》，《世界宗教研究》，1982年第4期（總10期）（1982年11月），48-55頁。

21. 張建木《玄奘法師的翻譯事業》，《法音》，1983年第2期（總12期）（1983年3月），8-13頁，（續一）1983年第3期（總113

期）（1983年5月），8-13頁。(續完)1983年第4期（總14期）
（1983年7月），12-16頁。

22. 蘇淵雷《論佛教在中國的演變及其對社會文化各方面的深刻影
響》，《華東師範大學學報》（哲學社會科學）》（上）1983年
第4期（總48期）（1983年8月），34-42頁（中）1983年第5
期（總49期）（1983年10月），60-66頁，（下）1983年第6期
（總50期）（1983年12月），62-67頁。

23. 張春波《佛教與中國的佛學研究》，《中國哲學史研究》，1983
年第4期，（總13期）（1983年10月），78-87頁。

24. 白文固《南北朝隋唐僧官制度探究》，《世界宗教研究》，1984
年第1期（總15期）（1984年2月），53-59頁。

25. 業露華《北魏的僧官制度》，《世界宗教研究》，1984年第2期
（總16期）（1984年5月），67-71頁。

26. 李玉昆《從龍門造像銘記看北朝的佛教》，《世界宗教研究》，
1984年第2期（總16期）（1984年5月），72-77頁。

27. 馬書田《明成祖的政治與宗教》，《世界宗教研究》，1984年第
3期（總17期）（1984年9月），35-51頁。

28. 孔繁《從《世說新語》看名僧和名士相交游》，《世界宗教研究》，
1984年第4期（總18期）（1984年11月），15-24頁。

29. 蘇晉仁《佛教傳記綜述》，《世界宗教研究》，1985年第1期
（總19期）（1985年3月），1-28頁。

30. 張新鷹《《雪竇寺志》淺考》，《世界宗教研究》，1985年第2
期（總20期）（1985年6月），91-96頁。

31. 蕭平漢《衡山寺院經濟試探》，《世界宗教研究》，1985年第2
期（總20期）（1985年6月），97-106頁。

32. 李玉昆《從龍門造像銘記看唐代佛教》，《世界宗教研究》，
1985年第3期（總21期）（1985年9月），34-39頁。

33. 白文固《明代的僧官制度》，《世界宗教研究》，1985年第4期
（總22期）（1985年12月），89-97頁。

34. 陳慶英《元帝師八思巴年譜》，《世界宗教研究》，1985年第4
期（總22期）（1985年12月），105-1223頁。

35. 任繼愈《中國佛教的特點》，《世界宗教研究》，1986年第2期，
（總24期）（1986年6月），6-9頁。

36. 杜繼文《中國佛教和中國文化》，《世界宗教研究》，1986年2
期，（總254期）（1986年6月），20-23頁。

37. 樓宇烈《中國近代佛學的振興者-楊文會》，《世界宗教研究》，
1986年第2期（總24期）（1986年6月），28-31頁。

38. 楊曾文《隋唐時期的中日佛教文化交流》《世界宗教研究》，1986
年第2期（總24期）（1986年6月），32-35頁。

39. 張羽新《清朝統一新疆與喇嘛教》，《世界宗教研究》，1986年

第 2 期（總24期）（1986年 6 月），105-118頁。

40. 謝重光《晋-唐僧官制度考略》，《世界宗教研究》，1986年第 3 期，（總25期）（1986年 9 月），31-46頁。

41. 程民生《略論宋代的僧侶與佛教政策》，《世界宗教研究》，1986年 4 期，（總26期）（1986年12月），49-59頁。

佛教文學與音樂（專書部）

常任俠　《佛經文學故事選》，上海，上海古籍出版社，1982年170頁。

佛教文學與音樂（論文之部）

1. 龍晦　《敦煌變文《雙恩記》本事考索》，《世界宗教研究》，1984年第 3 期，（總17期），（1984年 9 月），52-63頁。

2. 田青　《佛教音樂的華化》，《世界宗教研究》，1985年 3 期，（總21期），（1985年 9 月），1 -20頁。

3. 孔繁　《魏晉玄學、佛學和詩》，《世界宗教研究》，1986年 3 期，（總26期），（1986年 9 月），1 -14頁。

4. 龍晦　《論敦煌詞典所見之禪宗與淨土宗》，《世界宗教研究》，1986年第 3 期，（總25期），（1986年 9 期），59-67頁。

佛教典籍（專書之部）

1. 石峻，樓宇烈，方立天，許抗生，樂壽明編《中國佛教思想資料選編》
   第 1 卷，北京，中華書局，1981年，460頁
   第 2 卷 4 冊，北京，中華書局，1983年，1708頁

2. （唐）慧能著，郭朋校釋《壇經校釋》，（中國佛教典籍選刊），北京，中華書局，1983年，167頁。

3. （唐）法藏著，方立天校釋《華嚴金師子章校釋》，（中國佛教典籍選刊），北京，中華書局，1983年，211頁。

4. （宋）普濟著，蘇淵雷點校《五燈會元》（中國佛教典籍選刊），北京，中華書局，1984年，全 3 冊，1528頁。

5. 《中華大藏經》編輯局整理《中華大藏經（漢文部份）》，北京，中華書局
   第 1 冊-第 5 冊，1984年
   第 6 冊-第15冊，1985年
   第16冊-第21冊，1986年
   （附）《中華大藏經》編輯局《中華大藏經》（漢文部份）概論》
   《世界宗教研究》，1984年第 4 期，（總18期）（1984年11月），1 -14頁。

佛教典籍（論文之部）

1. 龍晦《大足佛教石刻《父母恩重經變像》跋》，《世界宗教研究》，
   1983年第 3 期（總13期）（1983年8月），16-25頁
2. 溫玉成《記新出土的荷澤大師會塔銘》，《世界宗教研究》，
   1982年第 2 期，（總16期）（1984年 5 月），78-79頁
3. 章瑋、方廣錩、金志良《元代官刻大藏經考證》，《世界宗教研
   究》，1986年 3 期，（總25期）（1986年9月），47-58頁

藏族宗教和文化（專書之部）

1. 王輔仁《西藏佛教史略》，西寧，青海人民出版社，1982年，
   318頁
2. 彭英全主編《西藏宗教概說》，拉薩，西藏人民出版社，1983年
   150頁
3. 牙含章編著《達頼喇嘛傳》，北京，人民出版社，1984年，368
   頁
4. 黃奮生編著，吳均校訂《藏族史略》，北京，民族出版社，1985
   年，442頁

藏族宗教和文化（論文之部）

1. 黃心川《沙俄侵略蒙藏與喇嘛教》，《世界宗教研究》，第1集
   （1979年 8 月），244-263頁
2. 王堯《西藏喇嘛教的形成》，《中國哲學》第 2 輯（1980年3 月），
   200-224頁
3. 王輔仁《關於西藏黃教寺院集團的幾個問題》，《世界宗教研究》，
   第 2 集（1980年 8 月），46-55頁
4. 黃文煥《河西吐蕃經卷目錄跋》，《世界宗教研究》，第 2 集，
   （1980年 8 月份，56-62頁
5. 湯池安《德格印書院》，《世界宗教研究》，第 2 集（1980年 8
   月），63-72頁
6. 楊化羣《藏傳因明學發展概況》，《世界宗教研究》，1981年第
   2 期，（總 4 期）（1981年 6 月）13-19頁
7. 陳那養，法尊譯《集量論頌》，《世界宗教研究》，1981年第 2
   期，（總 4 期）（1981年 6 月），20-28頁
8. 丁漢儒，渴華，唐景福《喇嘛教形成的特點問題》，《世界宗教
   問題》，1981年第 2 期（總 4 期）（1981年 6 月）54-63頁
9. 王堯《吐蕃佛教述略》，《世界宗教研究》，1981年第 2 期，
   （總 4 期）（1981年 6 月）64-74
10. 孫爾康《本教初探》，《世界宗教研究》，1981年第 3 期（總 5

期），（1981年9月），21-131頁

11. 王輔仁《關於西藏佛教（喇嘛教）及其教派的形成時間問題-兼與王堯同志商榷》，《世界宗教研究》，1981年第3期（總5期）（1981年9月），132-137及120頁

12. 黃顥、吳碧雲《六世達賴倉央嘉措生平考略》，《西藏研究》，1981年第1期，（1981年12月），105-109頁

13. 拉薩市政協文史資料組，恰白、次且平撥著《大昭寺史事述略》，《西藏研究》，1981年1期（1981年12月），36-50頁

14. 李安宅《從拉卜楞寺的護法諸神看佛教的象徵主義-兼談印藏佛教簡史》，《西藏研究》，1981年第1期，（1981年12月），23-35頁

15. 丁漢儒《宗喀巴宗教思想探討》，《世界宗教研究》，1982年第1期，（總7期），（1982年2月）76-83頁

16. 黃文煥《河西吐蕃卷式寫經目錄併後記》，《世界宗教研究》，1982年1期（總7期）（1982年2月），84-102頁

17. 黃顥《略述北京地區的西藏文物》，《西藏研究》，1982年1期，（總2期）（1982年3月），71-82頁

18. 李克誠《普寧寺的建立及其歷史作用》，《世界宗教研究》，1982年3期（總9期），（1982年8月），114-120頁

19. 強巴洛卓《甘丹寺及其創建者宗喀巴》，《西藏研究》，1982年2期，（總3期）（1982年8月），45-50頁

20. 張駿《西藏文化豐富了敦煌寶窟》，《西藏研究》，1982年2期，（總3期），（1982年8月），121-123頁

21. 賈湘雲、何宗英《關於向居寺創建者及始建年代問題》，《西藏研究》，1982年2期（總3期），（1982年8月），124-128頁

22. 喬鴻達《論清代的尊孔和崇奉喇嘛教》《社會科學輯刊》，1982年5期，（總22期）（1982年9月），107-113頁

23. 蒲文成《關於西藏佛教前後弘期歷史分岐》，《西藏研究》，1982年3期，（總4期）（1982年11月），127-132頁

24. 楊鶴書《廣東南華寺發現八思巴字藏文重要文物》，《中山大學學報》（哲學社會科學版），1982年2期（總83期），（1982年），40-48頁

25. 段克興《西藏原始宗教一本教簡述》，《西藏研究》，1983年1期，（總5期），（1983年2月），76及72頁

26. 王册《塔爾寺概述》，《西藏研究》，1983年1期，（總5期）（1983年2月），81-90頁

27. 蕭帶岩節錄注釋《元明漢族史家筆下的八思巴》，《西藏研究》，1983年1期，（總5期）（1983年2月），91-97頁

28. 唐景福、溫華《試論西藏佛教薩迦派的歷史及其作用》，《世界宗教研究》，1983年3期（總13期），（1983年3月）37-44頁

29. 王森《宗喀巴年譜》,《世界宗教研究》,1983年2期(總12期),(1983年5月),1-22頁

30. 楊李政《普米族的汗歸教》,《世界宗教研究》,1983年2期(總12期)(1983年5月)73-83頁

31. 楊許浩《簡析西藏的"政教合一"制度》《西藏研究》,1983年2期,(總6期)(1983年5月),85-90頁

32. 江道元《世界屋脊的明珠-布達拉宮》,《西藏研究》,1983年2期(總6期)(1983年5月),103-111頁

33. 胡繼歐《阿底峽的佛教思想及其對西藏佛教的影响》,《法音》,1983年4期(總14期)(1983年7月),26-28頁

34. 羅炤《漢藏合璧《聖腥慧到彼岸功德寶集偈》考慮》,《世界宗教研究》,1983年4期(總14期),(1983年11月),4-36頁

35. 班班多杰譯《聽宗喀巴大師講八難題備忘記錄》,《世界宗教研究》,1983年4期(總14期),(1983年11月),37-45頁

36. 阿沛、晋美《書松德贊時期漢地佛教在吐蕃的傳播和影响》,《世界宗教研究》,1983年4期(總14期)(1983年11月)46-62頁

37. 李家瑞《西北藏族地區本教的歷史及其特點初探》,《世界宗教研究》,1984年3期,(總17期)(1984年9月),144-151頁

38. 閻清,張羽新《康熙對西藏佛教的政策》,《世界宗教研究》,1985年1期(總19期)(1985年3月),43-54頁

39. 王堯,陳踐《吐蕃時期的占卜研究-敦煌藏文寫卷P.T.1047,1055號譯釋》,《世界宗教研究》,1985年3期(總21期),(1985年9月),91-102頁

40. 李冀誠《對西藏佛教的形成及其稱謂問題上的一切淺見》,《世界宗教研究》,1986年第1期(總23期)(1986年3月),43-55頁

41. 格勒,祝啓源《藏族本教的起源與發展問題探討》,《世界宗教研究》,1986年第2期(總24期)(1986年6月),124-133頁

42. 王堯,陳踐《從一張借契看宗教的社會作用-P.T.1297號敦煌吐蕃文書譯釋》,《世界宗教研究》,1986年第4期(總26期)(1986年12月),66-71頁

43. 楊明《州西牧區藏族游牧部落的本教》,《世界宗教研究》,1986年第4期(總26期)(1986年12月),128-134頁

道教

研究道教之目錄及索引

1. 楊光文輯《全國部份報刊道教論文目錄索引（1905-1983）》，《宗教學研究論集》（四川大學），第24期（1984），132-144頁
2. 和光　《解放後關於道家，道教，玄學部份論文索引（1949-1984）》，《中國哲學》第11輯（1984年1月），505-525頁

道教思想（專書之部）

1. 卿希泰　《中國道教思想史綱》
   第一卷　漢魏兩晉南北朝時期
   　成都　四州人民出版社，1980年，348頁
   第二卷　隋唐五代北宋時期
   　成都　四州人民出版社，1985年，519頁
2. 熊鐵基　《秦漢新道家略論稿》
   上海　，上海人民出版社，1984年，210頁
3. 王明　《道家和道教思想研究》
   北京　中國社會科學出版社，1984年，386頁
4. 吳光　《黃老之學通論》
   杭州　浙江人民出版社，1985年266頁
5. 趙明　《道家思想與中國文化》
   長春　吉林大學出版社，1986年260頁

道教思想（論文之部）

1. 湯一介　《略論早期道教關於生死神形問題的理論》，《哲學研究》1981年1期（1981年1月），50-59頁及75頁
2. 卿希泰　《從葛洪論儒道關係看神仙道教理識特點》，《世界宗教研究》，1981年第1期（總3期）（1981年2月），110-115頁
3. 朱越利　《炁氣二字異同辨》，《世界宗教研究》，1982年1期（總7期）（1982年2月），50-58頁
4. 陳俊民　《略論全眞道的思想源流》，《世界宗教研究》，1983年3期（總13期）（1983年8月），83-98頁
5. 徐西華　《浮明教與理學》，《思想戰綫》，1983年3期，（總51期）（1983年6月），35-40頁及34頁
6. 陳兵　《略論全眞道三教合一說》，《世界宗教研究》，1984年第1期（總15期），（1984年2月），7-21頁
7. 韓秉方　馬西沙　《林兆恩三教合一思想與三一教》，《世界宗教研究》，1984年第3期（總17期）（1984年9月），64-83頁
8. 馬序　盛國倉　《劉一明道教哲學思想初探》，《世界宗教研究》，

1984年第 3 期（總17期）（1984年 9 月）102-112頁

9. 劉國梁　《略論《周易》"三才"思想對早期道教的影响，《世界宗教研究》，1985年第 1 期（總19期），（1985年 3 月）98-106頁

10. 鐘肇鵬　《嚴遵的《老子撥歸》及其哲學和政治思想》，《世界宗教研究》，1985年第 2 期（總20期），（1985年 6 月）33-47頁

11. 李遠國　《試從陳搏的宇宙生成論》，《世界宗教研究》，1985年 2 期（總20期），（1985年 6 月）48-61頁

道教歷史（專書之部）

李遠國　《四川道教史話》，成都，四川人民出版社，1985年，107頁

道教歷史（論文之部）

1. 卿希泰　《道教產生的歷史條件和思想淵源》，《世界宗教研究》　第 2 集（1980年），106-114頁

2. 卞孝萱．《佛道之爭與鑒眞東渡》，《中國史研究》，1980年第 1 期，（1980年 3 月），29-36頁

3. 趙克堯，許道勛　《論黃巾起義與宗教的關係》，《中國史研究》，1980年第 1 期，（1980年 3 月），45-56頁

4. 蒙文通　《道教史瑣談》，《中國哲學》　第 4 輯（1980年10月），308-324頁

5. 卿希泰　《有關五斗米道的幾個問題》，《中國哲學》，　第 4 輯（1980年10月，325-336頁

6. 王明　《論陶弘景》，《世界宗教研究》，1981年第 1 集（總 3 集）（1981年 2 月），10-21頁

7. 劉琳　《論東晉南北朝道教的變革和發展》，《歷史研究》，1981年 5 期，（1981年10月），110-130頁

8. 王國軒　《關於道教研究的幾個問題》，《中國哲學史研究》，1982年 2 期（總 7 期）（1982年 4 月），97-103頁

9. 徐西華　《道家，神仙和道教》，《中國哲學史研究集刊》，第 2 輯（1982年 7 月），274-294頁

10. 周休照　《龍虎山上清宮沿革建置初探》，《中華文史論叢》，1982年2期，（總23期）（1982年8月），87-112頁

11. 湯一介　《論早期道教的發展》，《世界宗教研究》，1982年4期，（總10期）（1982年11月）1 -15頁

12. 龍晦　《全眞教三論》，《世界宗教研究》，1982年 1 期，（總 7 期）（1982年 2 月）27-36頁

13. 中國道教協會研究室　《中國道教史提綱》，《中國哲學史研究》，1983年1期（總10期），（1983年 1 月）41-50頁

14. 鍾國發　《前期天師道史論略》，《中國史研究》，1983年 2 期，
　　（總18期）（1983年 5 月），100-113頁
15. 李斌城　《茅山宗初探》，《中國史研究》，1983年 2 期，（總
　　18期）（1983年 5 月），114-127頁
16. 郭旃　《全真道的興起及其與金王朝的關係》，《世界宗教研究》，
　　1983年 3 期（總13期）（1983年 8 月），99-107頁
17. 龍顯昭　《論曹魏道教與兩晉政局》，《世界宗教研究》，1985
　　年 1 期，（總19期）（1985年 3 月），79-97頁
18. 聶長振，齊未了　《道教傳入日本及其對神道的影响》，《世界
　　宗教研究》，1985年2期，（總20期）（1985年 6 月）62-71頁
19. 柳存仁　《許遜與蘭公》，《世界宗教研究》，1985年 3 期，
　　（總21期）（1985年 9 月），40-59頁
20. 羊華榮　《宋徽宗與道教》，《世界宗教研究》，1985年 3 期，
　　（總21期），（1985年 9 月），70-79頁
21. 陳兵　《金丹派南宗淺探》，《世界宗教研究》，1985年 4 期
　　（總22期），（1985年12月），35-49頁
22. 陳兵　《元代江南道教》，《世界宗教研究》，1986年 2 期，
　　（總24期）（1986年 6 月）65-80頁
23. 陳智超　《真大道教新史料-兼評袁國藩《元代真大道教考》》，
　　《世界宗教研究》，1986年 4 期（總26期）（1986年12月）
　　16-24頁
24. 卿希泰　《關於道教齋醮及其形成問題初探》，《世界宗教研究》，
　　1986年 4 期，（總26期）（1986年12月）25-33頁

道教音樂與文學（專書之部）.
　　周沐照　倪少才　《道教與龍虎山傳說》，南昌，江西人民出版社，
　　1986年175頁

道教音樂與文學（論文之部）
1. 陳國符　《明清道音樂考稿(1)》，《中華文史論叢》，1981年
　　2 期（總18期）（1981年 5 月），1 -28頁
2. 劉守華　《道教與中國民間故事傳說》，《思想戰綫》，1983年
　　2 期，（總50期）（1983年 4 月），37-47頁
3. 龍晦　《論敦煌道教文學》，《世界宗教研究》，1985年 3 期，
　　（總21期），（1985年 9 月），60-69頁
4. 馬曉宏　《呂洞賓神仙信仰溯淙》，《世界宗教研究》，1986年
　　3 期（總25期）（1986年 9 月），79-95頁

道教科技與醫學（專書之部）
1. 張覺人　《中國煉丹術與丹藥》，成都，四川人民出版社，1981

年143頁

2. 周士一　潘啓明　《周易參同契新探》　長沙，湖南人民出版社，1981年，112頁

道教科技與醫學（論文之部）

1. 魏啓鵬　《《太平經》與東漢醫學》，《世界宗教研究》　1981年第 1 集（總 3 期）（1981年 2 月），101-109頁

2. 郭起華　《從葛洪和陶弘景看道教對古代醫學的影响》，《世界宗教研究》，1982年 1 期（總 7 期）（1982年 2 月），37-42頁

3. 孟乃昌　《中國煉丹術的基本理論是汞論》，《世界宗教研究》，1986年第 2 期（總24期）（1986年 6 月），81-91頁

道教典籍（專書之部）

1. 王明　《太平經合校》，北京，中華書局，1960年初版，1979年重印，782頁

2. 陳國符　《道藏源流考》，北京，中華書局，1963年初版，1985年第 2 次印刷，上下册，523頁。
（附）陳國符　《道藏源流續考》，香港，里仁書局，1983年
（又）台北，明文書局，1983年，496頁

3. 王明　《抱朴子內篇校釋》，北京，中華書局，1980年，521頁

4. 王明　《無能子校注》，北京，中華書局，1981年89頁

道教典籍（論文之部）

1. 馮達文　《《太平經》剖析》兼談《太平經》與東漢末年農民起義的若干思想聯系》，《中山大學學報》，1980年 3 期（總76期）， 1 -12頁。

2. 王利器　《道藏本《道德眞經撥歸》提要》，《中國哲學》　第 4 輯（1980年10月），337-360頁

3. 龍晦　《讀《中國科學技術史》第 5 卷第 2 ， 3 分册-兼評其有關煉金術和道家部分》，《世界宗教研究》，第 2 集（1980年 8 月），99-105頁

4. 楊曾文　《道教的創立和《太平經》》，《世界宗教研究》，第 2 集（1980年 8 月），115-122頁

5. 黃烈　《略論吐魯番出土的"道教符號"》《文物》，1981年 1 期（總296期）（1981期 1 月）51-55頁

6. 陳國符《《石藥爾雅》補與注》，《世界宗教研究》，1981年3期，（總 5 期）（1981年 9 月），14-35頁及87頁

7. 王明　《論太平經的成書時代和作者》，《世界宗教研究》，1982年 1 期，（總 7 期）（1982年 2 月），17-26頁

8. 朱伯昆　《張角與太平經》，《中國哲學》　第9輯（1983年 2

月），169-190頁

9. 朱越利　《試論《無能子》》，《世界宗教研究》　1983年1期
　　（總11期）（1983年2月）107-122頁

10. 朱越利　《《太上感應篇》與北宋末南宋初的道教改革》，《世
　　界宗教研究》，1983年4期（總14期）（1983年11月），81-94
　　頁

11. 世界宗教研究所道教研究室　《《道藏提要》選刊》
　　（上）《世界宗教研究》，1984年2期（總16期）（1984年5月）
　　1-29頁
　　（下）《世界宗教研究》，1984年3期（總17期）（1984年9月）
　　84-101頁

12. 陳國符　《有關中國對丹黃白術之著作之撰述經歷若干時期之特
　　例《蓬萊山西灶還丹歌》諸撰述時期考稿》，《世界宗教研究》，
　　1984年4期，（總18期）（1984年11月），45-64頁

13. 孟乃昌　《中國煉丹術原著評介》，《世界宗教研究》，1984年
　　4期（總18期）（1984年11月）72-84頁

14. 左景權　《《洞淵神咒經》源流考-兼論唐代政治與道教之關係》，
　　《文史》　第23輯（1984年11月），279-285頁

15. 卿希泰　李剛　《試論道教勸善書》，《世界宗教研究》，1985
　　年4期（總22期）（1985年12月），50-57頁

16. 朱越利　《《養性延命錄》考》，《世界宗教研究》，1986年1
　　期（總23期），（1986年3月），101-115頁

伊斯蘭教

伊斯蘭教（專書之部）

1. 馬堅譯 《古蘭經》，北京，中國社會科學出版社，1981年，493頁
2. 楊永昌 《漫談清眞寺》，銀川，寧夏人民出版社，1981年，81頁
3. 甘肅省民族研究所 《伊斯蘭教在中國-西北五省（區）伊斯蘭教學術討論會（蘭州會議）論文選編》，銀川，寧夏人民出版社，1982年，492頁
4. 白壽彝 《中國伊斯蘭史存稿》，銀川，寧夏人民出版社，1983年，421頁
5. 福建省泉州海外交通史博物館，泉州市泉州歷史研究會編 《泉州伊斯蘭教研究論文選》，福州，福建人民出版社，1983年，270頁
6. 福建省泉州海外交通史博物館，陳達生編，《泉州伊斯蘭教石刻》（中英文對照本），銀川，寧夏人民出版，福州，福建人民出版社聯合出版，1984年，324頁
7. 泉州伊斯蘭教史迹保護委員會中國文化史迹研究中心編 《泉州伊斯蘭史迹》（中英文本），福州，福建人民出版社，1985年，62頁
8. 李興華，馮今源編 《中國伊斯蘭教史參考資料選編》，銀川，寧夏人民出版社，1985年上下册，1823頁

伊斯蘭教（論文之部）

1. 金宜久 《《古蘭經》在中國》，《世界宗教研究》，第 3 集（1981年 2 月），128-132頁
2. 李興華 《《天方性理》哲學思想初探》，《世界宗教研究》，1981年 2 期（總 4 期）（1981年 6 月），109-118頁
3. 莊爲璣，陳達生 《泉州清眞寺史迹新考》《世界宗教研究》，1981年第 3 期（總 5 期）（1981年 9 月），99-106頁
4. 金宜久 《試論《古蘭經》的一神思想》，《世界宗教研究》，1981年第 4 期（總 6 期）（1981年12月），67-78頁
5. 勉維霖《寧夏回族伊斯蘭教教派分化淺談》，《世界宗教研究》，1981年第 4 期（總 6 期）（1981年12月），101-113頁
6. 王森譯《正理滴論》，《世界宗教研究》，1982年第 1 期（總 7 期）（1982年 2 月），1 - 7 頁
7. 楊化羣《正理滴論》，《世界宗教研究》，1982年第 1 期（總 7 期），（1982年 2 月），8 -16頁
8. 戴康生，秦惠彬 《試論伊斯蘭教在我國回族中傳播的特點》《世界宗教研究》，1982年 1 期（總 7 期）（1982年 2 月），103 - 117 頁

9. 馮增烈 《漢型制伊斯蘭文物述略》，《世界宗教研究》，1982
年 1 期（總 7 期）（1982年 2 月），118-124

10. 楊兆鈞《關於中國伊斯蘭教史的分期問題》，《思想戰綫》，
1982年 2 期，（總44期）（1982年 4 月），54-59頁及34頁

11. 沙秋眞《中國伊斯蘭教經堂教育》，《世界宗教資料》，1982年
2 期，（1982年 5 月），56-57頁及29頁

12. 金宜久 《伊斯蘭教復古傳統初探》，《世界宗教研究》，1982
年 2 期，（總 8 期）（1982年 5 月），56-66頁

13. 關建吉《西道堂歷史概述》，《世界宗教研究》，1982年 3 期，
（總 9 期）（1982年 8 月），143-151頁

14. 陳達生《泉州靈山聖墓年代初探》，《世界宗教研究》，1982年
4 期，（總10期）（1982年11月），116-121頁

15. 戴康生《伊斯蘭教社會運動淺析》，《世界宗教研究》，1983年
1 期，（總11期）（1983年2月），38-50頁

16. 秦惠彬《論王岱輿的宇宙觀和認主學》，《世界宗教研究》，
1983年 1 期，（總11期）（1983年 2 月），51-65頁

17. 李興華《明清之際我國回族等族伊斯蘭教新特點的形成》，《世
界宗教研究》，1983年 1 期，（總11期）（1983年 2 月），
66-77頁

18. 李耕硯，徐立奎《青海地區的托茂人及其與伊斯蘭教的關係》，
《世界宗教研究》，1983年 1 期，（總11期）（1983年 2 月）
132-138頁

19. 馬啓成《略述伊斯蘭教在中國的早期傳播》《中國社會科學》，
1983年 2 期（總20期），（1983年 3 月），205-216頁

20. 金宜久《蘇非派與漢文伊斯蘭教著述》，《世界宗教研究》，
1983年 2 期，（總12期），（1983年 5 月），100-109頁

21. 趙立人 《廣州伊斯教"懷聖寺"及"懷聖塔"之建築年代》，
《中國社會科學》，1983年 5 期，（總23期，（1983年 9 月）第
221頁

22. 馮今源《中國伊斯蘭教教坊制度初探》，《世界宗教研究》，
1984年 1 期，（總15期）（1984年 2 月），81-91頁

23. 羅萬壽 《關於中國伊斯蘭教歷史分期的管見》，《世界宗教研
究》，1984年 2 期，（總16期），（1984年 5 月），120-133頁

24. 王懷德《蘇非派的演變與門宦制度形成的特點》，《世界宗教研
究》，1984年 2 期，（總16期）（1984年 5 月），134-143頁

25. 馮今源《《來復銘》析》，《世界宗教研究》，1984年 4 期，
（總18期），（1984年11月），105-115頁

26. 馮今源 《試論儒家思想對中國伊斯蘭教的影響和滲透》，《中
國哲學史研究》，1985年 3 期（總20期）（1985年 7 月），43-51
頁

27. 魏長洪 《楊增新與新疆伊斯蘭教》，《世界宗教研究》，1985年2期，（總20期），（1985年6月），112-121頁

28. 努爾 《論中國東南沿海伊斯蘭碑刻之研究》，《世界宗教研究》，1985年2期，（總20期）（1985年6月），122-131頁

29. 陳國光 《《回回二十五世到中原》考—關於新疆伊斯蘭教神秘主義在內地傳佈問題》，《世界宗教研究》，1985年3期，（總21期）（1985年9月），80-90頁

30. 金宜久 《論漢文伊斯蘭教早期著述中的神光思想》，《世界宗教研究》，1985年4期，（總22期）（1985年12月）1-11頁

31. 高鴻君 《伊斯蘭教法的主要特點及伊斯蘭法系的現狀與前景》，《世界宗教研究》，1985年4期，（總22期）（1985年12月）12-20頁

32. 蘇北海 《伊斯蘭教在哈薩克族中的發展》，《世界宗教研究》，1986年1期，（總23期），（1986年3月），116-124頁

33. 買買提、賽來，馬品彥 《新疆的麻札和麻札朝拜-新疆伊斯蘭教地域特色初探》，《世界宗教研究》，1986年4期，（總26期）（1986年12月）93-102頁

基督教、天主教、猶太教及東正教

基督教、天主教、猶太教及東正教（專書之部）

1. 楊眞　《基督教史綱》，上冊，北京　三聯書店，1979年，540頁

2. 顧長聲　《傳敎士與近代中國》，上海　上海人民出版社，1981年，460頁

3. 江文漢　《中國古代基督敎及開封猶太人（景敎，元朝的也里可溫、中國的猶太人）》，上海，知識出版社，1982年，230頁

4. 潘光旦　《中國境內猶太人的若干歷史問題-開封的中國猶太人》，北京，北京大學出版社，1983年，307頁

5. 顧長聲　《從馬禮遜到司徒雷登-來華新傳敎士評傳》，上海，上海人民出版社，1985年，491頁

6. 張綏　《東正敎和東正敎在中國》，上海　學林出版社，1986年，361頁

基督教，天主教及猶太教（論文之部）

1. 陳申如，朱正誼　《試論明末清初耶穌會士的歷史作用》，《中國史研究》，1980年2期，（1980年6月），135-144頁

2. 林金水　《利馬竇在中國的活動與影响》，《歷史研究》，1983年1期，（總161期）（1983年2月），25-36頁

3. 楊欽章　何高濟　《對泉州天主敎方濟各會史迹的兩點淺考》，《世界宗敎研究》，1983年3期。（總13期）（1983年8月），148-151頁

4. 林金水　《利馬竇在廣東》，《文史》　第20輯（1983年9月）147-161頁

5. 楊欽章　《泉州景敎石刻初探》《世界宗敎研究》　1984年4期（總18期）（1984年11月），100-104頁

6. 劉建　《十六世紀天主敎對華傳敎政策的演變》《世界宗敎研究》1986年1期，（總23期）（1986年3月），90-100頁

少數民族宗教

少數民族宗教（專書之部）

1. 趙櫓 《論白族神話與密教》，雲南省， 中國民間文藝出版社，1983年，161頁
2. 宋恩常 《中國少數民族宗教初編》，昆明，雲南人民出版社，1985年， 532頁
3. 秋浦 《薩滿教研究》，上海，上海人民出版社，1985年，180頁

少數民族宗教（論文之部）

1. 耿世民 《回鶻文摩尼教寺院文書初釋》，《考古學報》，1978年第4期，（1978年10月），497-516頁
2. 耿世民 《古代新疆和突厥，回鶻人中的佛教》，《世界宗教研究》，第2集（1980年8月），73-81頁
3. 宋恩常《彝族的原始宗教》，《世界宗教研究》，第3集（1981年2月），150-156頁
4. 杜國林《景頗族的有神論觀念》，《世界宗教研究》，第3集（1981年2月），157-161頁及140頁
5. 滿都爾圖《中國北方民族的薩滿教》，《社會科學輯刊》，1981年第2期（總13期）（1981年3月），89-93頁
6. 劉建國《關於薩滿教的幾個問題》，《世界宗教研究》，1981年2期，（總4期）（1981年6月），119-124頁
7. 曾文瓊，陳泛舟《羌族原始宗教考略》，《世界宗教研究》，1981年2期（總4期）（1981年6月）123-135頁
8. 田繼周，羅之基《西盟佤族的自然宗教》，《世界宗教研究》，1981年第4期（總6期）（1981年6月），114-123頁
9. 林悟殊 《摩尼的第二宗三際論及其起源初探》，《世界宗教研究》，1982年第3期，（總9期）（1982年8月），45-56頁
10. 李經緯 《古代維吾爾文獻《摩尼教徒懺悔詞》譯釋》，《世界宗教研究》，1982年第3期（總9期），（1982年8月），57-78頁
11. 陳世良 《魏晉時代的鄯善佛教》，《世界宗教研究》，1982年第3期（總9期）（1982年8月），79-89頁
12. 李經緯《佛教"二十七賢聖"回鶻文譯名考釋》，《世界宗教研究》，1982年第4期（總10期）（1982年11月），28-46頁
13. 和志武 《略論納西族的東巴教和東巴文化》，《世界宗教研究》，1983年第1期（總11期）（1983年2月），1-15頁
14. 馬學良 《明代彝文金石文獻中所見的彝族宗教信仰》，《世界宗教研究》，1983年第2期（總12期）（1983年5月），60-72頁
15. 于錦綉 《彝族的"近祖崇拜"》，《世界宗教研究》，1983年

第 2 期（總12期）（1983年 5 月），89-99頁

16. 李經緯　《回鶻文景教文獻殘卷《巫師的崇拜》譯釋》，《世界宗教研究》，1983 年第 2 期，（總12期）（1983 年 5 月）143-151頁

17. 林悟殊　《摩尼教入華年代質疑》，《文史》，第18輯（1983年7 月），69-81頁

18. 樊圃《六至八世紀突厥人的宗教信仰》，《文史》，第19輯（1983年 8 月），191-209頁

19. 林悟殊《敦煌本《摩尼光佛教法儀略》的產生》，《世界宗教研究》，1983年第 3 期（總13期），（1983年 8 月），71-76頁

20. 莊爲璣　《泉州摩尼教初探》，《世界宗教研究》，1983年第 3 期（總13期）（1983年 8 月），77-82頁

21. 林悟殊　《《摩尼教殘經-》原名之我見》，《文史》，第21輯（1983年10月），89-99頁

22. 伍維武《關於古代傣族的原始宗教》，《哲學研究》，1983年11 期，（1983年11月），70-74頁

23. 林悟殊《摩尼教在回鶻復興的社會歷史根源》，《世界宗教研究》，1984年第 1 期（總15期）（1984年 2 月），136-143頁

24. 王叔凱　《古代北方草原諸游牧民族與薩滿教》，《世界宗教研究》，1984年 2 期，（總16期），（1984年 5 月），80-88頁

25. 滿都爾圖，夏之乾《察布查爾錫伯族的薩滿教》，《世界宗教研究》，1984年 2 期，（總16期）（1984年 5 月），97-104頁

26. 彭耀　《永寧半月談-摩梭人古代宗教考察漫筆》，《世界宗教研究》，1984年 2 期，（總16期）（1984年 5 月），105-119頁

27. 林悟殊　《摩尼教《下部贊》漢譯年代之我見》，《文史》第22輯（1984年 6 月），91-96頁

28. 林悟殊　《唐代摩尼教與中亞摩尼教團》，《文史》，第23輯（1984年11月），85-93頁

29. 陳世良　《龜茲白姓和佛教東傳》，《世界宗教研究》，1984年4 期，（總18期），（1984年11月），36-44頁

30. 林悟殊　《《老子化胡經》與摩尼教》，《世界宗教研究》，1984年 4 期（總18期）（1984年11月），116-122頁

31. 劉銘恕　《火祆教雜考》，《世界宗教研究》，1984年 4 期（總18期）（1984年11月），123-126頁

32. 烏蘭察夫　《蒙古族早期宗教和無神論思想萌芽》，《哲學研究》，1984年11期，（1984年11月），45-51頁

33. 林悟殊　《宋元時代中國東南沿海的寺院式摩尼教》，《世界宗教研究》，1985年 3 期，（總21期）（1985年 9 月）103-111頁

34. 史金波　《西夏佛教的流傳》，《世界宗教研究》，1986年 1 期，（總23期），（1986年 3 月），27-42頁

35. 李玉昆　《福建晉江草庵摩尼敎遺迹探索》，《世界宗敎研究》，1986年 2 期，（總24期），（1986年 6 月），134-139頁

36. 楊福泉　《納西族東巴經中的"黑""白"觀念探討》，《世界宗敎研究》，1986年 2 期（總24期）（1986年 6 月），140-147頁

37. 耿世民《甘肅博物館藏回鶻文《八十華嚴》殘經研究㈠》，《世界宗敎研究》，1986年 3 期，（總25期）（1986年 9 月），68-77頁

38. 張廣達，榮新江　《于闐佛寺志》，《世界宗敎研究》，1986年 3 期，（總25期）（1986年 9 月），140-149頁

39. 李玉昆，楊欽章，陳達生　《泉州外來宗敎文化之研究》，《世界宗敎研究》，1986年 4 期，（總26期），（1986年12月）111-120頁

40. 鄧光華　《貴州土家族儺壇考略》，《世界宗敎研究》，1986年 4 期（總26期），（1986年12月）121-127頁

民間宗教及其他

民間宗教及其他（專書之部）

宗力　劉羣　《中國民間諸神》，石家莊　河北人民出版社，1986年
944頁

民間宗教及其他（論文之部）

1. 喻松青　《羅教初探》，《中國哲學》　第2輯（1980年3月），
   225-240頁
2. 喻松青　《明清時代民間的宗教信仰和秘密結社》，《清史研究
   集》　第1集（1980年11月），北京，中國人民大學出版社出
   版，113-153頁
3. 周慶基　《"且"崇拜和祖先崇拜》，《世界宗教研究》，1982
   年1期（總7期）（1982年2月），125-129頁
4. 雷中慶　《史前葬俗的特徵與靈魂信仰的演變》，《世界宗教研
   究》，1982年3期（總9期）（1982年8月），133-142頁
5. 程嘯　《民間宗教與義和團揭帖》，《歷史研究》，1983年2期，
   （總162期）（1983年4月），147-163頁
6. 馬白沙　《略論明清時代民間宗教的兩種發展趨勢》，《世界宗
   教研究》，1984年1期（總15期）（1984年2月），22-33頁
7. 馬白沙《黃天教源流考》，《世界宗教研究》，1985年2期（總
   20期）（1985年6月），1-18頁
8. 陳志東　《殷代自然災害與殷人的山川崇拜》，《世界宗教研究》，
   1985年2期（總20期）（1985年6月），19-32頁
9. 楊曾文　《觀世音信仰的傳入和流傳》，《世界宗教研究》，
   1985年3期（總21期）（1985年9月），21-33頁
10. 韓秉方　《紅陽敎考》，《世界宗教研究》　1985年4期（總22期）
    （1985年12月），21-34頁
11. 馬白沙　《最早一部寶卷的研究》，《世界宗教研究》，1986年
    1期（總23期）(1986年3月)，56-72頁，編者按：指《佛說楊
    氏鬼綉紅羅化仙哥寶卷》
12. 王笠荃　《淺談龍神》，《世界宗教研究》，1986年3期（總25
    期）（1986年9月），150-153頁
13. 黃心川　《瑜伽哲學述評-兼論瑜伽與我國佛教，道教，武術，
    民間氣功等關係》，《世界宗教研究》，1986年4期，（總26期）
    （1986年12月），1-15頁
14. 韓秉方　《羅教"五部六册"寶卷的思想研究》，《世界宗教研
    究》，1986年4期（總26期）（1986年12月），34-38頁

Index

Protestant Church
Chrysanthemum flowers, symbolic meaning in folk jewelry, 114
Cicada, 109-10, symbolic meaning in folk jewelry, 109
Confucius, 210; deification, 212
Confucian Temple, Qufu, 210-1; ceremonies to celebrate Confucius' birthday, 212-4, 217-21 (illus.)
Confucianism, 210; government's attitude towards, 13-4
Constitution of People's Republic of China, religious freedom outlined, 6-7

DASHI, *see* King of the Ghosts
De Groot, J.J.M., 51
Dipper, prayer to in *Jiao*, 293
Dongyue miao, *see* Temple of the Eastern Peak.
Donkeys, symbolic meaning in folk jewelry unknown, 112
Door gods, 176-80, 187-203 (illus.); departures from traditional themes, 179-80
Dragons, symbolic meaning in folk jewelry, 111

EIGHT TRIGRAMS (*bagua*), symbolic meaning in folk jewelry, 111
Engels, Friedrich, views on religion, 46-7

FAN, BISHOP JOSEPH XUEYAN, 223
Fang Litian, 28
Fayuan monastery, Beijing (B), 308; Buddhist seminary in, 165
Feasts and festivals, renewal of observances in Fujian, 71-2
Feng Qi, 29
Feng Youlan, 34-5
Festivals, *see* feasts and festivals
Fish, symbolic meaning in folk jewelry, 110
Folk jewelry (*see also* attachments for clothing, hairpins, pendants): constituents of, 103; motifs of, 102, 104, 106-116 *passim*; symbolic meaning, 104; use of, 104
Folk religion (*see also* shrines and temples) in Hong Kong, 259-270; meaning of, 260; revival of, 175-84
Freeing the living creatures, at *Jiao*, 288
Frogs and toads, 110, 113; symbolic meaning in folk jewelry
Fu Tieshan, Bishop, 227, 229, 232, 237; spokesman for the CCPA, 225-6

Fujian, celebration of *Jiao*, 51-77

GAO CHONG as door god, 177
Gods, invitation to *Jiao*, 280; procession of, 66, 70, 288; sending off, 288-9
Gong Pinmei, Bishop, 224, 225, 228
Guan Gong, as door god, 178
Guangji Monastery, Beijing (B), 165 evcellent library 165
Guanyin, (Cihang Daoren), 160
Guiyuan Monastery, Wuhan (B), 174

HAIR, HUMAN, as charm, 109
Hairpins, 115-6, 156-7 (illus.)
Heavenly Envoy, role in *Jiao*, 287
Heretic religious sects (*see also* White Lotus Sect, Yellow Turban Rebellion), role in peasant rebellions, 45, 46
Hong Kong (*see also* Changzhou, Longyuetou): Buddhism, 299-311 celebration of *Jiao*, 271-90; effect of urbanization on religious practices, 259; folk religion in, 259-70; investment in Guangdong's restoration of Taoist monasteries, 82.
Hong Kong Buddhist Association: ventur-into education & social work, 304, 308.
"House Churches" vs. "Three-Self Churches", 250-8
Hu, Yaobang, 223, 236
Huagu, drum troupes, 69
Huishou, see leaders of worship.
Hungry ghosts, 65, 285, 288
Huo Binzhang, Father, 223

IMPERIAL GOVERNMENT, relations with religion, 3-4
Institute for Research on World Religions, academic study of religion, 17
Islam, government's attitude towards, 14
Islamic studies, literature review, 32-3

JADE BUDDHA TEMPLE, Shanghai (B), 161
Jade Emperor, 63, 66, 67, 69, 159, 285-6, 293
Jan Yünhua, 92, 162 .
Jewelry, *see* folk jewelry
Ji Xianlin, 29
Jiang Wenxuan, on existence of religions in socialist societies,
Jiao: comparison between Fukinese and Hakka-Cantonese Tradition, 272-90;

religion, 116
Priests, Buddhist: ordination of, 92-4: training of, 93-4, 165, 166, 172, 308-9
Priests, Taoist: 51-2, 57, 261; definition, 86; number of, 95-7; ordination, 90-1; recall to temples, 82; training of, 83-4, 91, 162-3, 169-70, 173-4, 175; in Zhejiang, 85-7
Procession for Offering Incense, ritual of at *Jiao*, 61-2, 280, 291-2
Protestant Church, 243-58, number of Christians and churches, 246

QILINS, 107-8, 128-32 (illus.); symbolic meaning as amulets, 107
Qin Qiong, as door god, 176-177
Qing Xitai, 29-30
Qingchengshan, Sichuan (T) 168-70; Taoist seminaries in, 169
Qingyang Gong, Chengdu (T) 167-8

RELIGION (*see also* Buddhism, Christianity, . . . ; lower class religion, upperclass religion, academic study of, 17; antireligious bias in literature, 17-8; coordination with socialist society, 37; existence in socialist societies, 36, 38; as foreign policy tool, 15; lack of interest shown by people, 16; as opiate of people, 43, 45; new interpretation of, 48-9; place of, in socialist period, 37-8; prospects of, 19-22; rallying believers into realizing modernizations, 8, 21; relations with Imperial government, 3-4; roles in peasant rebellions, 43-50; tolerated but not encouraged, 13, 21; uncertainties facing, 22
Religious Affairs Bureau, 9, 12, 82, 91, 222
Religious freedom, 8, 13, 38; outlined by Constitution of the PRC, 6
Religious mysticism, serves as stimulant to peasant rebellions, 44
Religious organizations, 9-13 *passim;* foreign donations to, 12; functions, 9, 10
Religious personnel: foreign donations to, 12; number of, 8-9; training of, *see* Priests, Buddhist *and* Priests, Taoist
Religious policy: of Chinese Communist Party, 7-12; implementation, 12-15; impact on Hong Kong, 308-9
Religious practices, authorities' view of, 88-89; in Fujian, 51-77; government objections to, 72-5; nation-wide revival, 89
Religious studies (*see also* Buddhist studies,

Christianity, Islamic studies, Minorities, Taoist studies), literature review, 25-42; new direction in, 35-6
Ren Jiyu, 26, 27, 35; new direction in religious studies, 35-6
Ren Wuzhi 224
Repentance, ritual of in *Jiao*, 60-1, 282-3, 284, 286
Rituals, Buddhist, in Hong Kong, 305-7; *pudu*, 161-2
Rituals, Taoist: 52, 158-61; comparisons between Fujian and Taiwan, 57; functions and nature of, 75; meaning to villagers in Fujian, 75-6
Ruan Renze, 20-1
Ruyi scepters, 107-109; symbolic meaning in folk jewelry, 107

SCHIPPER, KRISTOFER, 20, 51
Secret societies, association with peasant rebellions, 44
Sending off paper boat at *Jiao*, 286-7, 289
Shamanism, study of, 33-4
Shanghai Academy of Social Sciences, 17
Shanghai Buddhist Institute, 308
Shao Xungzhen, views on peasant rebellions, 46-7
Shing Wong Temple, Shaukeiwan, 262
Shing Wong Temple, Western District, 263
Shrines and temples: in folk religion, 261-2; as middlemen between Hong Kong Government and Chinese communities before World War II, 264
Silvestrini, Archbishop Achille, 234
Strickmann, Michel, 51
Sun Changwu, 28
Sun Zuomin, views on peasant rebellions, 43-4, 45

TA-TSIU, see Jiao
Taam Kung Temple, Shaukeiwan, 262
Tan, see altars.
Tang Ludao, 323, 234
Tang Yongtong, 27
Tanzhe Monastery, Beijing (B) 164
Taoism, government's attitude towards, 13-14
Taoist Association of China, 9, 13, 162-3
Taoist associations, 82-3; functions, 83
Taoist monasteries, *see* Monasteries, Taoist
Taoist priests, *see* Priests, Taoist
Taoist rituals, *see* Rituals, Taoist
Taoist studies, literature review, 30-1